×

Medieval History and Archaeology

General Editors

JOHN BLAIR   HELENA HAMEROW

Food in Medieval England

FOOD IN MEDIEVAL ENGLAND

Diet and Nutrition

Edited by

C. M. WOOLGAR, D. SERJEANTSON,
AND T. WALDRON

UNIVERSITY PRESS

OXFORD
UNIVERSITY PRESS

Great Clarendon Street, Oxford OX2 6DP

Oxford University Press is a department of the University of Oxford.
It furthers the University's objective of excellence in research, scholarship,
and education by publishing worldwide in

Oxford New York

Auckland Cape Town Dar es Salaam Hong Kong Karachi
Kuala Lumpur Madrid Melbourne Mexico City Nairobi
New Delhi Shanghai Taipei Toronto
With offices in
Argentina Austria Brazil Chile Czech Republic France Greece
Guatemala Hungary Italy Japan South Korea Poland Portugal
Singapore Switzerland Thailand Turkey Ukraine Vietnam

Oxford is a registered trade mark of Oxford University Press
in the UK and in certain other countries

Published in the United States
by Oxford University Press Inc., New York

Reprinted 2011

ISBN 978-0-19-956335-7

Printed in the United Kingdom by
Lightning Source UK Ltd., Milton Keynes

Acknowledgements

Many of the papers in this volume had their genesis in meetings of the Diet Group, at Somerville College, Oxford. The Group would especially like to thank Barbara Harvey for hosting its meetings there over the last decade and Helena Hamerow for suggesting that this volume might form a part of Oxford University Press's 'Medieval History and Archaeology' series. For permission to use illustrative material, the contributors wish to thank Oxford Archaeology (Plate 1.1), the Dean and Chapter of Westminster Abbey (Plate 2.1), the Controller of Her Majesty's Stationery Office (Plate 3.1), the British Library (Plates 6.1, 9.1, 10.2), the Bodleian Library, University of Oxford (Plate 7.1), Marshall Laird and Winchester College (Plate 9.2), the Fitzwilliam Museum, University of Cambridge (Plate 11.1), and William Muirhead and the National Trust (Plate 14.1). The editors are grateful to the Hartley Institute, University of Southampton, for a grant towards the costs of preparing this volume for publication.

C. M. W.
D. S.
University of Southampton
T. W.
University College London
August 2005

Contents

PART II: STUDIES IN DIET AND NUTRITION

List of Illustrations

List of Figures

List of Tables

Abbreviations

AD	Archives départementales
AML	Ancient Monuments Laboratory
ANTS	Anglo-Norman Text Society
BAR	British Archaeological Reports
BAV	Biblioteca Apostolica Vaticana
BL	British Library
CBA	Council for British Archaeology
CCR	*Calendar of Close Rolls*
CLR	*Calendar of Liberate Rolls*
CUL	Cambridge University Library
EconHR	*Economic History Review*
EETS, os	Early English Text Society, Original Series
EETS, ss	Early English Text Society, Supplementary Series
JRULM	John Rylands University Library of Manchester
NA	National Archives, Kew
RO	Record Office
VCH	*Victoria History of the Counties of England*
WCM	Winchester College Muniments

Notes on Contributors

UMBERTO ALBARELLA is Research Officer in Zooarchaeology, Department of Archaeology, University of Sheffield.

JEAN BIRRELL is an Honorary Fellow of the Institute for Advanced Research in the Arts and Social Sciences at the University of Birmingham.

CHRISTOPHER DYER is Professor of English Local History at the University of Leicester.

BARBARA HARVEY is an Emeritus Fellow of Somerville College, Oxford.

LISA MOFFETT is an archaeobotanist and Research Fellow at the University of Birmingham.

GUNDULA MÜLDNER is Lecturer in Palaeodiet in the Department of Archaeology, University of Reading.

MICHAEL RICHARDS is Professor of Bioarchaeology, Department of Archaeological Sciences, University of Bradford, and Professor, Department of Human Evolution, Max Planck Institute for Evolutionary Anthropology, Leipzig.

PHILLIPP SCHOFIELD is Professor in the Department of History and Welsh History at the University of Wales Aberystwyth.

DALE SERJEANTSON is a Research Fellow in the Archaeology Division, School of Humanities, University of Southampton.

DAVID STONE teaches history at Dulwich College.

NAOMI SYKES is a Research Fellow at the Centre for Applied Archaeological Analysis, Archaeology Division, School of Humanities, University of Southampton.

TONY WALDRON is a Consultant Physician at St Mary's Hospital, London, and an Honorary Professor at the Institute of Archaeology, University College London.

CHRISTOPHER WOOLGAR is Reader and Head of Special Collections, University of Southampton Library.

Notes on Contributors

1

Introduction

C. M. WOOLGAR, D. SERJEANTSON, AND T. WALDRON

Food and diet are rightly popular areas of research, central to understanding daily life in the Middle Ages. In the past twenty years their study has changed a great deal and a multi-disciplinary approach has become essential to encompass the historical, archaeological, and scientific record. During this time, historians have opened up sources in new ways; zooarchaeologists and archaeobotanists have processed and assimilated archaeological material from a wide range of sites; and scientific techniques, applied to the medieval period, have begun to allow an assessment of the cumulative impact of diet on the human skeleton. Nonetheless, the wide variety of information about diet and nutrition has rarely been drawn together. Even for a single country this project is a daunting task, yet it is one that is crucial to establishing how much more is now known about changes in patterns of eating, the general levels of nutrition and the consequences for health and life expectancy. This is the first ambition of this volume. At the same time, the interchange between the methodological approaches of historians and archaeologists has produced important developments and is equally central to this book. Looking at both strands together, we can reassess the state of our knowledge of this complex subject and see where deficiencies lie in our approaches, planning research accordingly.

To this end, the text brings together much original and unpublished research, marrying historical and archaeological approaches with analyses from a range of archaeological disciplines including archaeobotany, archaeozoology, osteo-archaeology, and isotopic studies. The volume covers the whole of the Middle Ages from the Early Saxon period up to *c.*1540. Inevitably, the greatest contribution to the period before 1066 has come from archaeology, while the emphasis for the historical essays lies in the period between 1066 and the Reformation. From the eleventh century onwards, the contributions of the historians and archaeologists complement each other. While the focus of the volume is England (Fig. 1.1), reference is made to wider European developments, although research in comparable depth is not available for many parts of the Continent.

Fig. 1.1 Map of England showing selected places referred to in the text

The contributors to this volume are members of an informal group of historians, archaeologists, and archaeological scientists, the Diet Group, which has met in Oxford over the last decade to pursue the study of food, diet, consumption, and health in the past. The different disciplines not only bring different kinds of data, but also different approaches and styles of scholarship and presentation. A combination of these is now a virtue essential to the achievement of a holistic view of this subject.

The study of medieval food and diet

Food has been perennially of interest in the study of the Middle Ages, but the context of that research has changed. Its presence in the collections of recipes and descriptions of banquets, prominent in the works of eighteenth- and nineteenth-century antiquaries, is markedly different in emphasis from its place in the discussion of living standards, prices, and wages of twentieth-century social and economic history and in the great regional studies of the 1950s and 1960s that considered the nutrition, calorific intake, and the vulnerability of populations to starvation. In England, the survival of large numbers of documents for the administration of landed estates (or manors) from the thirteenth century onwards has led to a concentration on production in medieval agriculture, rather than the consumption of food; but in the last two decades there has been a significant change in perspective. Outlines of diet in late medieval England were succinctly mapped by Christopher Dyer;[1] detailed studies, such as Barbara Harvey's work on the monks of Westminster, have shown the rich potential of monastic sources;[2] work on large numbers of manorial accounts has produced new conclusions about regional patterns of production, marketing, and consumption;[3] and new examinations of sources, such as household accounts, have broadened the material available for the study of diet.[4] Alongside conclusions from much other work, it is now possible, from historical sources, to make a wide range of statements about consumption.

The analysis of archaeological plant and animal remains with a view to demonstrating food production and consumption was almost unknown before the 1960s. It was then a decade before the techniques were applied to medieval sites. There were very few excavations in medieval towns before the 1970s, and it was not until that time that the importance of consistent recovery of plant and animal remains was recognized. The earliest reports on medieval bone assemblages were published in the 1970s;[5] and it was only in the 1980s that attempts were made to address general issues associated with the medieval food economy and the theoretical problems of intepreting finds from complex sites.[6] No survey has been attempted before on the scale of the chapters here.[7]

The systematic analysis of human bones from medieval sites for direct evidence of the consequences of diet is also a very recent development. Some results are brought into the discussion here, but further work is necessary to underpin comparative work, such as the criteria to be used for diagnosing diseases associated with malnutrition, for example, scurvy and rickets; to unify the approach to

[1] Dyer (1983). [2] Harvey (1993).

[3] Campbell, Galloway, Keene, and Murphy (1993); Campbell (2000).

[4] Woolgar (1992–3; 1995). [5] e.g. Kings Lynn: Noddle (1977); Exeter: Maltby (1979).

[6] Grant (1988); Bourdillon (1988); Serjeantson and Waldron (1989); O'Connor (1992).

[7] Since the mid-1990s, English Heritage has commissioned surveys of environmental archaeology of the prehistoric and historic period in England, a vast undertaking as far as animal remains are concerned. These surveys form the basis for some of the chapters in this book.

determining final achieved height; and to develop reliable methods for estimating body weight and changes in weight over the adult lifetime. The application of archaeological science to the medieval period has much to offer, but work is at an early stage. The analysis of stable isotopes as a means of identifying the major foodstuffs that contribute to the human skeleton has been developed only since the 1990s. The discussion in this volume is one of no more than a handful of cases, some in France and some in England, where the technique has been applied to the medieval period.

The evidence and its limitations

To unite and interpret this wide range of evidence is far from straightforward. In terms of historical sources, we know most about agricultural production and the seigneurial economy. We need to look beyond this, however, to discover information about consumption, particularly for the peasantry and urban populations. Even the evidence for the great households or monasteries is not uniformly spread: it is much stronger for churchmen, especially from the late fourteenth and early fifteenth centuries, and for widows, than for secular lords; and some major monasteries have left little historical record. There are also clusters in the documentation: little is available for the period before *c*.1200; there is much less again after *c*.1430 and it is less systematic in both form and content. There is also an uneven geographical distribution: the information in some categories of document, such as manorial accounts, is at its best for estates where land was managed directly, rather than by leasing out farms, a practice that varied both spatially and temporally. The net has therefore to be cast wide for information on diet, from accounts to wills, records of markets, chronicles, and collections of miracles. None of this evidence may be typical in itself—single accounts or isolated references may be difficult to interpret in a wider context—but cumulatively it both provides a wide range of information and demonstrates general patterns.

The chronological perspectives of historians and archaeologists exhibit important contrasts. Historians may discuss some aspects of consumption at a daily level, with a view that might encompass monthly, seasonal, and annual arrangements, as well as the longer term; archaeologists deal in tens to hundreds of years. In the surveys in this book, only those archaeological assemblages which could be dated to within 200 years have been considered. Since the dating of most deposits is based on pottery, the styles of which changed slowly, with old pots remaining in use, it is likely that archaeological data will continue to be analysed within this pattern. This has the advantage of showing trends over the *longue durée*, but misses short-term fluctuations which may have been of considerable importance to human health, such as the consequences of famine. Sometimes, in particularly fortunate contexts, deposits can be dated more closely,[8] but it is exceptionally rare for any deposit to be identified as a single episode, while the historian may not uncommonly have evidence for an individual feast.[9]

Archaeologists base their interpretations of human behaviour on physical remains. In this volume, these are mostly those of plants and animals, along with human skeletons from excavations of cemeteries. Other materials can be informative in relation to social and cultural aspects of food: the size and shape of pottery vessels, for example, can suggest how consumption changed from food shared on communal dishes to presentation on individual plates; different patterns of food preparation can be determined from the use of new styles of cooking vessels, such as frying pans; arrangements for food presentation are implicit in, to take one example, the use of chafing dishes, and the quality of vessels for serving food tells us much about the context in which food might form an element in conspicuous consumption.[10] The emphasis of this volume, however, is on the use of archaeological material in a quantitative and comparative framework to indicate overall patterns of diet and nutrition.

Physical remains have to be interpreted in the light of patterns of disposal, preservation, and recovery. Historical records and biological and ethnographic models can help with the first, as they illuminate cultural processes. To take the remains of bones as an example: in towns and in other households distant from the processes of food production, people often put rubbish from food preparation and meals into pits which were rapidly covered. If bones were discarded elsewhere, into general refuse layers, they might suffer the attention of dogs and other scavengers before they became buried, and for this reason fish and bird remains are found in greater quantities in pits than in general layers of rubbish. Where kitchen floors have survived (Plate 1.1), they often exhibit an accumulation of rubbish. Features such as latrines and cesspits, which may preserve a range of bones and environmental material well, needed an investment to create them. In towns in the later Middle Ages, rubbish was carted away from the households where it was generated, and dumped outside the town, sometimes into rivers, where deposition will have occurred at points where there was silting. Deposits that have been transported in this way usually lack the bones of small animals. Few bone remains are found in villages. For example, at Dean Farm, Cumnor, a fourteenth-century cottage was excavated, with archaeological material mostly found in ditches. A large quantity of pottery was recovered, but very little animal bone.[11] This could suggest that the peasants ate very little meat, but it may reflect the fact that bones were discarded onto the dungheap, the contents of which were later spread on the garden or carted to the fields. Material from some archaeological contexts is therefore more likely to document consumption by some social groups than by others.

[8] See Chapters 8 and 9. [9] See Chapter 10.
[10] Brown (2002); Hinton (2005: 185, 234–6, 255). [11] Jones (1994*a*).

Plate 1.1 A succession of kitchen floors at Eynsham Abbey, thirteenth to fifteenth centuries. Each time the floor was relaid, a dense layer of debris, including small bones and eggshell, was preserved. Photograph: © Oxford Archaeology.

The way material is preserved is equally complex. Acid soils normally destroy all bones, while in neutral, chalky, or alkaline soils larger bones and those of mature animals survive better than smaller bones and those of immature animals. In suitable sediments, pits dug for rubbish and rapidly filled tend to preserve bones well. The ratio of birds and fish to the larger animals is therefore dependent on the deposit. Soils which favour the survival of small fish bones, for example, are anaerobic sediments, such as those often found in cesspits where there has been no disturbance and little water percolation.

These biases are most relevant at the level of an individual assemblage: they have been mitigated in this book by the selection of the groups of bones which can be interpreted most reliably. For this reason the surveys of the larger mammals (cattle, sheep, pigs, and deer) draw on a wider range of sites than those of birds and fish. The number of published bone assemblages is now very great, with the surveys of the larger food animals based on hundreds of samples—which may, nevertheless, represent relatively few animals—and we can now have confidence in the general trends they display.

At a further level, interpretation needs to consider questions of recovery. One has to assess the sites that have been excavated and why that work has been carried out: have the requirements of rescue archaeology, for example, privileged or disadvantaged some classes of site ? Is there now a suitable range of sites for assessment ? The process of excavation itself also requires scrutiny. In excavations of medieval sites in England it is usual to retrieve finds by hand and to take limited samples of the sediments for sieving. These last are not a constant proportion of the whole deposit excavated and can be very small. While this has been seen as the best compromise, faced with decisions about the overall area which it is desirable to excavate or the levels of post-excavation processing that may be possible, inevitably some information will be lost as finds may be sacrificed. The implications for the study of diet are important: it can, for example, have an impact on the relative numbers of large and small mammals retrieved—cows as opposed to sheep, red deer as opposed to roe deer; but the loss when deposits are not sieved is greatest for birds, fish, and other environmental evidence, particularly archaeobotanical material.

Archaeologists working with food remains have always attempted to take into account the historical evidence for food production and consumption. Inevitably they have relied on secondary sources, but assessing the value of these is difficult without experience in the interpretation of historical documents. Some zooarchaeologists have turned to primary historical material to answer specific problems,[12] but this is usually impractical. Few reports explicitly tackle the question of why the numbers of livestock referred to in accounts do not correspond to the archaeological evidence;[13] many do no more than acknowledge anecdotal scraps of historical data as a context for discussion. The tensions between archaeological and historical evidence merit careful consideration.[14] This volume makes apparent a number of cases where these differences arise and where resolution may remain a matter of debate.

Equally, historians recognize that it is difficult for them, trained in different methods of analysis, to interpret archaeological data, especially reports on animal, human, and plant remains. To what extent can any sample be taken as representative? Are filters at work, such as differential survival and preservation, which will bias the results in ways not made explicit? Individual historians, including the contributors to the present book, have made greater or lesser use of archaeological data in the past, according to taste and experience. This recognition was part of the initial impetus for the formation of the Diet Group, together with the acknowledgement by archaeologists that their work required the context of historical knowledge.

This book therefore comes at a point of reappraisal. To obtain a much greater understanding of the evidence, we need to consider together the literature and sources of all disciplines involved. In order to cover this breadth, a group of contributors has been required, and the book has had to focus closely on diet and nutrition. The volume is divided into two parts. The first surveys foodstuffs, combining both historical and archaeological evidence, to give an up-to-date synthesis across a wide range of materials. The second section contains a series of short studies examining the evidence for the effects of diet, the cumulative

[12] Biddick (1989). [13] e.g. Jones (2002). [14] Albarella (1999); Coy (1996).

impact of foodstuffs, group diets, the consequences of social distinction, virtue, and religion, as well as seasonal patterns of consumption and the effects of diet on health, mortality, and the skeleton. It looks as well at two further categories of evidence: the direct evidence from human bones, and the assessments that can be made at a macro-level from the point of view of diet and demography. There is much that might be written about medieval foodways and anthropology, about diet, social competition, and display. Some aspects are addressed in the thematic studies in the second part of the volume, but the book does not aim to be comprehensive in its coverage of these topics. It has focused largely on historical material in these discussions, with the intention of stimulating further work; and it also outlines some of the contributions that archaeological science and the study of human bones can make to the debate. Indeed, it is our hope that, beyond reappraisal, the volume will lead to new directions in the research and study of diet.

PART I

SURVEY OF FOODSTUFFS

2

The Consumption of Field Crops in Late Medieval England

D. J. STONE

It is hard to avoid platitudes when describing the place of grain in medieval diet, for in both absolute and relative terms it towered over any other foodstuff. This may not have been the case in every part of medieval Britain, as Gerald of Wales informs us in his *Description of Wales* of *c.*1200,[1] but for the vast majority of people in England grain provided the bulk of their calorific intake. It has been estimated that at the start of the fourteenth century grain accounted for up to 80 per cent of a harvest worker's calories and 78 per cent of a soldier's; even among the lay nobility of medieval England, grain provided 65–70 per cent of their energy intake.[2]

Medieval people consumed grain in three main ways: as bread, as ale, and—among the poorer sections of society—in pottage, a thick soup. On balance, bread was the most important—the monks of Westminster Abbey, for instance, gained 35–46 per cent of their calories in this way at the turn of the sixteenth century[3]—but, for many, ale was not far behind. The basic allowance for these monks was a gallon per monk per day, and great households consumed ale in vast quantities: Henry de Lacy bought an average of 85 gallons of ale a day for his household in 1299, while at Framlingham Castle 78 gallons were consumed per day in 1385–6.[4] A sharper picture of consumption per person emerges from the allowances of food and drink given to lay folk who retired to monastic houses. At Selby Abbey in 1272, for example, Adam of Fleyburgh and his wife Emma received two white loaves, one brown loaf, and two gallons of ale every day.[5]

We have a great deal of written information about the production of crops from the thirteenth century onwards and this has understandably been the focus of much historical work, but we know less about the consumption of these

[1] Quoted in Hallam (1988*d*: 841). [2] Murphy (1998: 120).
[3] Harvey (1993: 57).
[4] Harvey (1993: 58); Woolgar (1992–3: i. 164–7); Ridgard (1985: 109).
[5] Hallam (1988*d*: 826).

foodstuffs at that time. The main aim of the present chapter is to explore the documentary evidence that survives for the consumption of grain, looking in particular at the types of crops that were consumed, the process of turning grain into bread and ale, consumption at different levels of society, and how the consumption of bread and ale varied over time; it is in these areas that historical evidence has much to offer that cannot be gleaned from other sources. Archaeological evidence, especially archaeobotanical evidence, is crucial to our understanding of field crops and plants for the centuries preceding this and broadens our interpretation of plant foods in the later medieval period. While this chapter therefore focuses on the historical evidence for field crops for the period 1250 to 1540 and the next chapter looks at the evidence for horticulture in broadly the same period, Chapter 4 reviews the archaeobotanical evidence for both over a longer timescale.

From grain to bread and ale

Table 2.1 shows the main field crops cultivated in medieval England and their major uses. Medieval documents are usually precise when it comes to distinguishing one type of crop from another, but unfortunately provide only rare glimpses of the botanical diversity that doubtless existed within each of these categories. For example, although in the early sixteenth century Fitzherbert mentioned seven different types of wheat,[6] manorial records uniformly refer to wheat by only one name: *frumentum*. For other crops, there is occasionally more information. Spring-sown barley, referred to as *ordeum*, was grown to a much greater extent than its winter-sown variety, *bere*, but the latter was not unknown and at Wisbech was divided into two types, *hastibere* and *rackbere*.[7] Similarly, oats (*avena*) were sometimes divided in manorial accounts into large and small oats, the latter probably being synonymous with 'naked oats' or 'pillcorn'.[8] As we see in Table 4.3, which shows the species found on excavated medieval sites, the types of cereals identified by estate managers and farmers did not correspond exactly with the botanical species. The different types of wheat, for instance, may refer to mixtures of bread and rivet wheat, but may also refer to different landraces of wheat. Similarly, the two types of oats recognized may correspond to the two botanical species, common and bristle oats, but may refer to landraces of the common oat, which was much more prevalent. Peas (*pisa*), beans (*faba*), and vetches (*vicia*) were all cultivated in this period as well, the last mainly for fodder and probably not for human consumption. Peas were occasionally distinguished by their colour, presumably when dry: white, black, green, and grey.[9] Medieval farmers commonly planted mixtures of these different crops, particularly

[6] Skeat (1882: 40–1); see also Chapter 4.
[7] *Polbere* is also mentioned and may be synonymous with *rackbere*: CUL, EDR D8/1/5-6.
[8] Finberg (1951: 95–7). [9] Campbell (2000: 228).

Table 2.1. The composition of the medieval crops referred to in documents and their uses

Sowing season	Nature	Crop	Composition	Main uses
Winter	Pure	Wheat		Bread, ale
		Rye		Bread, thatch, fodder
		Winter barley		Ale, bread, fodder
	Mixed	Maslin/mancorn	Wheat/rye	Bread
		Mixtil	Wheat/winter barley	Bread, ale
Spring	Pure	Oats		Pottage, fodder, ale, thatch, bread
		Spring barley		Ale, bread, pottage, fodder
		Legumes	(Beans, peas, vetches)	Fodder, nitrogen-fixing, pottage, bread, vegetables
	Mixed	Dredge	Spring barley/oats	Ale
		Bulmong/harascum	Oats/beans/peas	Fodder, nitrogen-fixing, pottage
		Mengrell/ pulmentum	Oats/legumes (inc. vetch)	Fodder, nitrogen-fixing, pottage

Note: See also Table 4.1.

winter-sown 'maslin', a mixture of rye (*siligo*) and wheat, and spring-sown 'dredge', a mixture of barley and oats.

The written record is more explicit about the uses to which the various crops were put. Much as today, wheat was considered the premier bread grain, producing the whitest and lightest loaf, though almost all the other crops were used for this purpose as well. Rye—which could successfully be grown in comparatively adverse environments—and maslin were used to produce loaves of a darker hue and inferior value, while barley and oats were milled and baked to produce coarse, cheap bread; even dried and ground-up peas and beans were used in the cheapest of loaves. In terms of ale production, barley was thought to produce the best malt and was used in quantity for this purpose, although other grains, especially oats—which were more tolerant of growing conditions than any other crop—and dredge, were used as well. Ale brewed from malted oats was particularly common in the north and south-west of England, although it appears to have been something of an acquired taste: in the sixteenth century Cornish ale was said to be 'lyke wash as pygges had wrestled dyrn'.[10] Wheat was occasionally malted for ale as well during the Middle Ages, producing a far superior brew. The main cereal ingredient of pottage was usually oats, small oats (*avene minute*) being easy to turn into oatmeal without need of a mill, though husked barley might also be consumed in this way. Peas and beans were often added to pottage, but at Cuxham in 1289–90, peas were simply provided 'as vegetables for the *famuli*', the permanent staff on the lord's demesne farm.[11]

[10] Fox (1991: 304). [11] P. D. A. Harvey (1965: 78).

Pottage was comparatively simple to make, while the production of bread and ale was more complex. The first stage in bread making is to mill the grain, producing coarse flour on the one hand and bran on the other. Wheat flour was sometimes then sieved or 'bolted' to make certain types of bread, every bushel of wheat producing an estimated 34.5 lb of fine flour.[12] Nevertheless, medieval methods of milling—and perhaps also the different botanical characteristics of the crop—produced a coarser, less absorbent flour than today, and the water content of medieval dough was consequently comparatively low.[13] Only the wealthiest households and demesne farms baked bread in their own purpose-built bakehouses; even Dame Katherine de Norwich paid to have her loaves baked in 1337 at 4*d.* per quarter.[14] In great households, baking was done in batches, six times a month in Alice de Bryene's household in 1412–13 (averaging 297 loaves each time), but more frequently in larger households: the Abbot of Peterborough's kitchen, for instance, baked more than eleven times a month in 1371 (averaging 410 loaves each time).[15] The quality of the bread was chiefly affected by the crop from which the flour was derived. Yet various qualities of bread could be produced from wheat alone, depending on the quality of the grain, the extent to which the flour had been sieved, and the amount of bran that was discarded (Plate 2.1). In Alice de Bryene's household, 89 per cent of wheaten loaves produced were white, the remainder black, probably having a much higher bran content.[16] The nature of the bread also depended on oven temperature. At Westminster Abbey, ten faggots were used as fuel to make 100 loaves of standard wheaten bread, but in making high-quality wastel bread, the biscuity texture of which required a much hotter oven, thirty faggots were required per 100 loaves.[17]

Medieval loaves not only differed in nature, but they varied considerably in size and weight too. For example, the loaves distributed to paupers by Katherine de Norwich in 1336–7 were comparatively small, each probably weighing 1.22 lb on average. Bread baked for customary workers on demesne farms was generally bigger and heavier: in the early fourteenth century, the loaves baked for harvest workers at Wisbech weighed 2.88 lb each, while those for plough boons at Hinderclay weighed 3.58 lb.[18] The weight of loaves also varied over time. This was chiefly a result of the assize of bread of 1256, under which the cost of a loaf to the buyer remained the same from year to year through a system which ensured that the weight of a loaf changed in inverse proportion to the price of grain. Thus, when the price of wheat stood at 4*s.* per quarter, the assize dictated that 284 wastel loaves should be baked from a quarter of wheat, each weighing

[12] Prestwich (1967: 537). [13] Campbell, Galloway, Keene and Murphy (1993: 191).
[14] Woolgar (1992–3: i. 203–25). [15] Dale and Redstone (1931: 1–102); Greatrex (1984: 56–83).
[16] Dale and Redstone (1931: 1–102, 128). [17] Harvey (1993: 59).
[18] Woolgar (1992–3: i. 179–227); CUL, EDR D8/1/5-19; Chicago University Library, Bacon 416, 435–44. These weights have been calculated on the assumption that a bushel of mixed grain produced 57.6 lb of coarse bread: Campbell, Galloway, Keene, and Murphy (1993: 191).

Plate 2.1 The Feeding of the Five Thousand, a detail from the Westminster Abbey Retable, *c*.1270–80. The round loaves were typical of the bread in aristocratic households and elsewhere. Photograph: © Dean and Chapter of Westminster Abbey.

on average 1.48 lb; but when the price of wheat stood at 6*s*. per quarter, 453 wastel loaves were to be baked from each quarter, each loaf weighing 0.93 lb.[19] When we have data about the number of loaves baked for private consumption rather than for sale, variation over time is also evident (Table 2.2). In general the weight of these aristocratic wheaten loaves lay between that of today's large and small loaves, though it seems to have become increasingly fashionable to serve loaves that were at the lower end of the scale.

To produce ale, grain was first soaked to allow it to germinate and release natural sugars, and then heated in a kiln to prevent further germination; at Cuxham, the demesne had a separate malting oven, which was slower burning and thus cooler than the bread oven.[20] Then in the brewhouse, which at Hanley Castle was equipped with large vats and a lead-lined cistern,[21] the malt was

[19] Davis (2004: 479). [20] P. D. A. Harvey (1965: 37–8). [21] Woolgar (1999: 144).

Table 2.2. Approximate weight of wheaten loaves baked for eleven great households

Year	Household	Loaves per quarter	Approximate weight per loaf (lb)
1240–2	Bishop of Lincoln	180	1.53
1299	Henry de Lacy	235	1.17
1336–7	Dame Katherine de Norwich	281	0.98
1337–8	Bishop of Bath and Wells	256–64	1.05–1.08
1370–1	Abbot of Peterborough	237	1.15
1378	Earl of March	275	1.00
1381–4	Bishop of Ely	272–96	0.93–1.01
1382–3	Sir William Waleys	253	1.09
1385–6	Countess of Norfolk	256	1.08
1412–13	Dame Alice de Bryene	297	0.93
1431–2	Earl of Oxford	312	0.88

Note: The weight per loaf has been calculated on the assumption that each bushel of wheat produced 34.5 lb of fine flour and that the weight of water added to make dough was cancelled out by the loss of weight during baking.

Sources: Dale and Redstone (1931: 1–102); Greatrex (1984: 56–83); Ridgard (1985: 106); Woolgar (1992–3: i. 165–7, 180–225, 256, 259–61; ii. 539); Woolgar (1999: 124).

crushed and mixed with hot water to allow the sugars to dissolve; finally the liquid was drained off, cooled, and allowed to ferment. Ale did not have good keeping qualities and was thus brewed regularly, although the frequency varied from one household to another, from on average 2.7 times a month in Katherine de Norwich's household in 1336–7 to 6.4 times a month at Bolton Priory in 1307–8.[22] It has been estimated that the brewhouse at Castle Acre Priory was capable of making 700 gallons of ale at each brewing, while those belonging to Katherine de Norwich and Alice de Bryene produced approximately 130–40 gallons a time.[23] Several strengths and qualities of ale were often produced—for instance, three at Dunstable Priory[24]—though in general much depended on the number of gallons brewed per quarter of malt. Between 50 and 75 gallons of ale per quarter was usual, although there is some suggestion that a taste for stronger ale developed in the later Middle Ages: in the 1330s, for instance, two households produced 60–75 gallons per quarter; in the 1380s, two other households were producing 53–6 gallons per quarter; and in 1500 the monks of Westminster Abbey produced 45–50 gallons per quarter.[25] Indeed, the strongest ale in the later Middle Ages may not have been dissimilar in strength to some modern beer; after all, in *Piers Plowman*, Glutton collapsed drunk with just over a gallon of ale inside him.[26] Nor is this the only recognizable feature of medieval drinking, for medieval ale was also cheaper the further north you were: on a journey from Hertfordshire to Scotland in 1378, the Earl of March was able to buy a gallon of

[22] Woolgar (1992–3: i. 180–226); Kershaw (1973*a*: 147).
[23] Wilcox (2002: 51); Woolgar (1992–3: i. 180–226); Dale and Redstone (1931: 1–102).
[24] Hallam (1988*d*: 827).
[25] Bennett (1996: 18); Woolgar (1992–3: i. 180–226, 259–61); Ridgard (1985: 108); Harvey (1993: 58).
[26] Schmidt (1992: 53–4).

ale for 2*d*. between Royston and Pontefract, for 1½*d*. between Boroughbridge and Newcastle, and for 1*d*. between Morpeth and Jedburgh.[27]

The nature of bread and ale before the Black Death

The bread consumed by the great lay and ecclesiastical lords of medieval England was made almost exclusively from wheat. Although they sometimes had to make do with maslin, wheat was also the bread grain of choice among lesser lords: Lionel de Bradenham's household received 18¾–25½ quarters of wheat a year from his only demesne farm of Langenhoe; and the lord of High Hall manor in Walsham-le-Willows had 'white bread' stolen from his bakehouse in 1344.[28] Lower down the social ladder the balance shifted markedly towards other grains, especially at the beginning of the fourteenth century, a time of great pressure on resources and immense social stress. Even in London, bakers of brown loaves outnumbered bakers of white in 1304, while a resident of Lynn seems to have consumed mainly rye bread.[29] For peasants in the countryside, white bread must have been a rare treat at this time. Harvest workers in some counties, such as Oxfordshire and Sussex, were given wheaten bread, but in many parts of the country harvest loaves were of a lower quality. Bread for harvest boons at Mildenhall was composed chiefly of maslin and rye and at Hinderclay mostly of rye and barley, but on other manors barley bread was the norm: barley made up 94 per cent of the harvest bread at Sedgeford in 1256, and was the only bread grain given to the harvest boon workers at Crawley and Bishopstone in 1302.[30] Similarly, in 1328, a maintenance agreement for a retired peasant from Oakington laid down that his annual grain allowance should consist of two bushels of wheat, two of rye, four of barley, and four of peas, all of which was probably consumed as bread or pottage.[31]

Even so, maintenance agreements and harvest bread are unlikely to be representative of the normal diet of most peasants; a more accurate sense of the nature of their bread intake can be gained from the provisions given to *famuli* on demesne farms. In 1346–7, *famuli* at Cuxham were given grain composed of 50 per cent *curallum*, the poorest part of threshed wheat, 29 per cent barley, and 21 per cent peas; in 1297–8, the *famuli* at Wellingborough received 45 per cent rye, 33 per cent barley, and 22 per cent *bulmong*, a mixture of oats, beans, and peas; in 1324–5, the bread consumed by *famuli* at Framlingham must have been even coarser still, their allowance made up of 70 per cent barley, 25 per cent beans and peas, and 5 per cent *curallum*.[32] Alms payments provide some insight into the crops consumed by the poorest members of medieval society. Katherine de Norwich provided wastel bread for the poor on Good Friday 1337, but most

[27] Woolgar (1992–3: i. 247–56).
[28] Woolgar (1999: 124); Britnell (1966: 380–1); Lock (1998: 274).
[29] Campbell, Galloway, Keene, and Murphy (1993: 26); Hanawalt (1976: 118).
[30] Dyer (1994*b*: 83, 88); Chicago University Library, Bacon 416, 436–44; Page (1996: 75, 89).
[31] Dyer (1998*b*: 55–6). [32] Harvey (1976: 423); Page (1936: 77); Ridgard (1985: 71).

alms were of a much more lowly form: a pottage made from peas was given in alms at Wellingborough in 1321–2; beans were given to the poor at St Leonard's Hospital, York, in 1324; while in 1346 the alms payments made by Norwich Cathedral Priory consisted of 46 per cent barley, 23 per cent peas, 23 per cent rye, and 8 per cent wheat.[33] Despite its standing today, bran was baked into bread either for horses or for the very poor.[34]

A similar diversity is apparent in the character of ale consumed during this period. Massive quantities of barley were clearly malted for brewing, for manorial accounts show barley being processed on demesne farms and either sent for the use of the lord's household or sold at market, and other accounts record barley malt arriving at the estate centre. The Norwich Cathedral Priory manors of Sedgeford, Martham, and Hemsby, for example, malted 33, 57, and 70 per cent of their available barley (after deduction of tithe and seed), and the Priory's granger annually accounted for up to 2,020 quarters of barley malt received from the estate in the late thirteenth century.[35] Even some wheat was malted for ale, for instance for the Dean and Chapter of St Paul's Cathedral in 1286,[36] but such an extravagant use of this grain was probably rare at this time. A considerable proportion of the ale brewed before the Black Death was in fact derived from inferior grains. In providing for their servants as well as for themselves and their guests, many great landlords malted a mixture of grains: in 1297–8, for example, the malt sent from Wellingborough to Crowland Abbey consisted of 40 quarters of dredge, 32 quarters of barley, and 30 quarters of oats; and in 1287 Glastonbury Abbey received 328 quarters of barley, 364 quarters of wheat, and 825 quarters of oats from its estate for making ale.[37] Though lords were invariably keen to maintain the high quality of the bread that they ate, some even growing wheat in environments ill suited to its cultivation, more were prepared to compromise in terms of the quality of ale. The canons of Bolton Priory, for example, grew wheat for their bread, but made their ale almost entirely from oats, which—though inferior to barley as a brewing grain—could be grown in the most testing conditions.[38]

Compromise in this respect was even more of a feature lower down the social scale. While good-quality ale was clearly consumed by some country folk, we should not assume that this was generally the case. It is unsurprising to find oaten ale on the manor of Cockerham in the 1320s, but even in Norfolk the rent paid by a twelfth-century tenant of the abbey of St Benet of Holme included six times as much malted oats as malted barley.[39] In fact, by the beginning of the fourteenth century many rural poor may not have drunk ale on a regular basis at all. Quarter for quarter, ale provides considerably fewer calories than bread or pottage, and many peasants may have been forced by their circumstances to consume grain in

[33] Woolgar (1992–3: i. 223); Page (1936: 130); Ashley (1928: 104, 106).
[34] Richardson and Sayles (1955–83: i. 258).
[35] Campbell (2000: 200, 223). [36] Campbell, Galloway, Keene, and Murphy (1993: 203–4).
[37] Page (1936: 76–7); Campbell, Galloway, Keene, and Murphy (1993: 203–4); Hallam (1988*c*: 368).
[38] Kershaw (1973*a*: 146). [39] Bailey (2002: 66); Hallam (1988*b*: 294).

as efficient a form as possible. Indeed, the Oakington maintenance agreement of 1328 would not have provided sufficient calories if all the barley had been consumed as ale.[40]

Patterns of consumption naturally have important implications for crop choice and vice versa. Wheat, for instance, would probably have been found to a much greater extent on demesne farms than on peasant land. Indeed, in the 1283 tax returns for the village of Ingham, wheat comprised 12.8 per cent of the lord's crops, but only 0.4 per cent of the peasants'.[41] Generally, peasants focused their attentions on inferior bread grains. On the Bishop of Winchester's manor of Burghclere, for example, payments made by peasants for grinding their corn at the lord's mill in 1301–2 included 158 bushels of maslin but only 2 bushels of wheat.[42] In many areas, peasants must have made their bread and pottage from barley. On a Hampshire manor of Winchester Cathedral Priory in 1338, wheat, barley, and oats were all important crops on the demesne, but the issue of the parsonage, presumably consisting largely of tithe corn collected from villagers' lands, contained twice as much barley as either wheat or oats.[43] Peasant payments for grinding corn sometimes provide a clear indication of how this barley was consumed; in Hampshire, for instance, some malt was ground in preparation for brewing, but a much larger amount of unmalted barley was often milled into flour.[44] Equally illuminating are the cropping data for the 1,238 households in the Suffolk Hundred of Blackbourne assessed for the 1283 tax (Fig. 2.1). Barley was hugely prominent in both Breckland and non-Breckland households; some may have been sold or given to the lord as rent in kind, but much was probably consumed as bread, for it is notable that the wealthier the household the lower the proportion of barley and the higher the proportion of wheat or rye. Barley may have made comparatively coarse bread, but its flour extraction rate was virtually identical to other grains and its yields were often considerably higher than those of other crops: on the demesne of Hinderclay (also in Blackbourne Hundred) net barley yields before the Black Death were 31 per cent higher than wheat yields.[45]

The significance of these points extends beyond our understanding of diet and farming. Most historians agree that the population of medieval England peaked at between five and six million in 1300, but—based on the amount of grain, and thus calories, that the country could produce—Bruce Campbell has challenged this, arguing that the population at that time cannot have been higher than 4–4.25 million.[46] However, his calculations are based on demesne yields and cropping proportions, and on the assumption that all barley and dredge was brewed for ale (ale has a kilocalorie extraction rate of 30 per cent, rather than 78 per cent for barley flour). It now seems probable that peasant yields were significantly higher than those from demesnes,[47] and peasants produced and consumed crops in

[40] Dyer (1998*b*: 56). [41] Bailey (1989: 141). [42] Page (1996: 117).
[43] Hallam (1988*c*: 357). [44] Page (1996: 241, 247, 274–5, 316, 326).
[45] Campbell (2000: 215); Chicago University Library, Bacon 416, 423–65.
[46] Campbell (2000: 386–410). [47] Stone (2005: 262–72).

Fig. 2.1 Proportions of field crops in peasant households in Blackbourne Hundred in Suffolk, 1283. Top: 280 Breckland households; bottom: 958 non-Breckland households

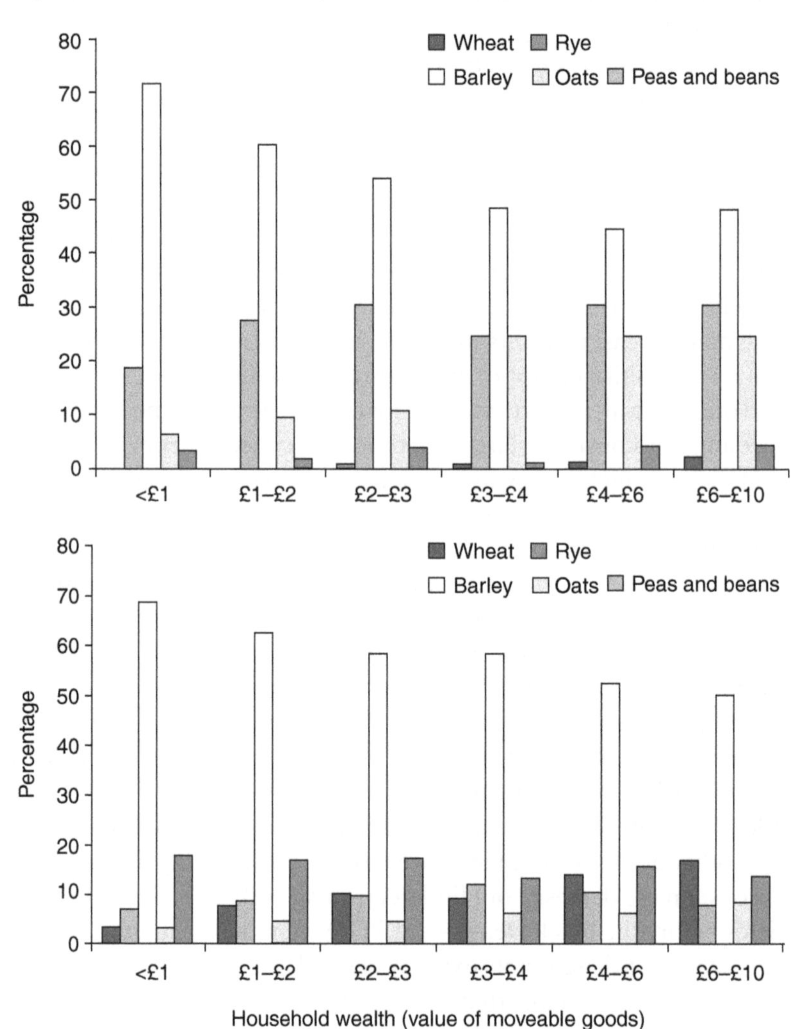

Source: Miyoshi (1981: 46, 51, 55).

different proportions from lords. By using assumptions that take account of these differences, a new population estimate of nearly 5.5 million is reached, which fits very well with orthodox demographic estimates (Table 2.3).

Change over time

The consumption of bread and ale changed considerably over time, even in the short term. In great households, bread consumption could vary significantly from meal to meal and day to day. In 1412–13, for example, Alice de Bryene's household consumed more bread at meals on fish days: an average of 1.14 lb of bread was consumed per person per meal on Sundays, Mondays, Tuesdays, and

Table 2.3. A re-estimate of national grain output and the population it was capable of feeding, *c.*1300

	Wheat	Rye and rye mixtures	Barley and dredge	Oats	Total
Percentage national grain area[a]	16.7	19.1	48.0	16.2	100.0
Total national grain area (million acres)	1.04	1.19	2.99	1.01	6.23
Net yield per acre (qtrs)[b]	1.32	1.14	1.81	1.09	
Total net grain output (million qtrs)	1.37	1.36	5.41	1.10	9.24
Kilocalories per quarter	644,480	620,160	534,336	482,688	
Total net kilocalorie grain output (million)	885,091	841,260	2,892,160	531,002	5,149,512
Uses[c]	100% bread	100% bread	50% ale 40% bread 10% pottage	57% pottage 10% ale 33% fodder	
Food extraction rate	0.80	0.80	0.56	0.60	
Total net food output (million kcal)	708,073	673,008	1,619,610	318,601	3,261,448
Less 10% wastage	637,265	605,707	1,457,649	286,741	2,935,303
Total daily supply of kilocalories (million)	1,746	1,659	3,994	786	8,042
Total population in millions capable of being fed at 1,500 kcal per person per day					5.46

Notes: This table uses the framework in Campbell (2000) for estimating population, but adjusts his assumptions in the following ways:

[a] National grain area takes account of peasant crop preferences, using the 1283 tax returns for Blackbourne Hundred (Suffolk), the pre-plague yield figures for the East Anglian Breckland and for Hinderclay (Suffolk), with a weighting of 80% peasant land and 20% demesne land.

[b] Net yields per acre are inflated by 11% to reflect higher peasant yields, on the assumption that yields on half yardlands would have been 10% higher and those on smaller holdings 25% higher than those on demesnes; the weighting of size of tenant holdings is taken from the Hundred Rolls of 1279–80.

[c] The use of grain reflects the likelihood that peasants consumed much of their barley as bread and pottage, and that some oats were brewed into ale.

Sources: Campbell (2000: 222–4, 392–3); Miyoshi (1981: 53); Bailey (1989: 103–5); Chicago University Library, Bacon 416, 423–65; Dyer (1998*a*: 119).

Thursdays, but this increased to 1.36 lb on Fridays. However, as many members of the household may have had only one meal on Fridays, the amount of bread they consumed per day was probably higher when meat was eaten. Likewise, the consumption of bread at each meal increased steadily during Lent, when the household abstained from meat (Fig. 2.2), although for the same reason consumption per day may often have been reduced at this time. Nor did the nature or consumption of ale remain constant over the course of a year. In this household, ale was made half from barley and half from dredge between 3 October 1412 and 11 January 1413, but just from barley between 12 January and 1 March. Then a stock of 'new' barley and dredge was begun, and the half and half mixture

Fig. 2.2 The consumption of bread in the household of Dame Alice de Bryene, February to May 1413

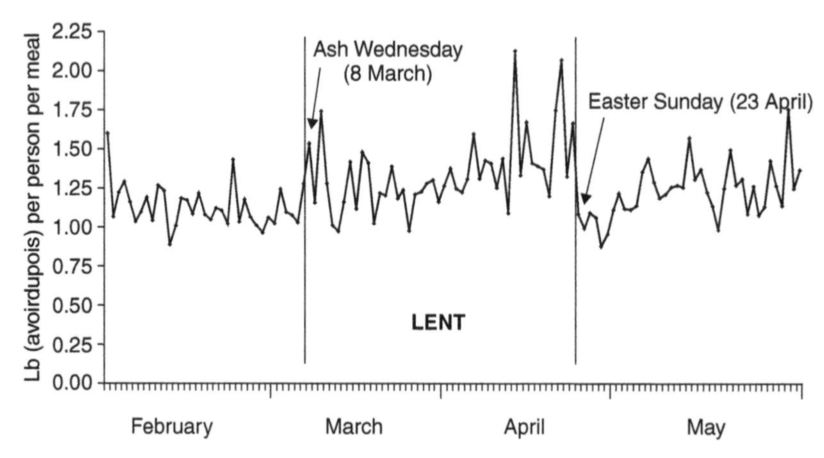

Source: Dale and Redstone (1931: 36–68).

was resumed.[48] The quantity of ale consumed by a household also fluctuated during the course of a year, rising considerably during the Christmas period. For example, in the Bishop of Salisbury's household, 42 gallons of ale were consumed daily between 1 October and 24 December 1406, but from Christmas Day to Epiphany this rose to 100 gallons.[49]

Harvest failure, of course, prompted sudden changes in patterns of consumption, notably during the Great Famine of 1315–17. At Bolton Priory, the amount of grain provided for making bread and ale plummeted at this time and its composition was adjusted: bread for these monks was normally made out of wheat, but in 1315–16 13 per cent of their bread was made from mixed grains and in the following year 21 per cent.[50] Lower down the social scale the problems were magnified and the response more dramatic. The grain allowance for *famuli* at West Wratting changed from 65.9 per cent rye, 25.6 per cent wheat, and 8.5 per cent barley in 1313 to 45.5 per cent rye, 43.4 per cent barley, 7 per cent beans, and 4.2 per cent wheat three years later, while harvest workers at Wisbech were given only bread made from winter barley in the years 1314–20.[51] Significantly, crimes of desperation were common in these years. In a case from 15 March 1316, a Norfolk plasterer was accused of breaking into the house of a fisherman to steal just a pennyworth of bread.[52] Later on that year, at Wakefield, a father and son attacked and drew blood from Thomas son of Peter to steal just three sheaves of barley.[53]

The Black Death of 1348–9 brought rising standards of living for many of the survivors and ushered in an era of significant changes in consumption. Qualitative

[48] Dale and Redstone (1931: 1–102). [49] Woolgar (1992–3: i. 264–320).
[50] Kershaw (1973*a*: 144–7). [51] Palmer (1927: 66); CUL, EDR D8/1/1-4.
[52] Hanawalt (1976: 99). [53] Bailey (2002: 231).

change is evident at the highest levels of society: in the 1380s, the Bishop of Ely had fresh bread baked for him every day; in 1416–17, the household of Robert Waterton of Methley baked considerable quantities of pain-demaine, the loaf of medieval kings; and by the end of the Middle Ages, the monks of Westminster Abbey not only consumed wheaten bread, but on special occasions wastel bread and sometimes enriched buns as well, and by then their ale was made almost exclusively from barley malt.[54] But the transformation was more emphatic for workers. In 1394, one Lincolnshire ploughman was given fifteen loaves of bread a week, seven of them made from wheat.[55] Harvest workers at Sedgeford received more ale and ate much higher-quality bread: in 1256 they received 2.8 pints of ale per person-day and their bread was composed mainly of barley; by 1424 they were each getting 6.4 pints of ale a day and their bread was entirely wheaten.[56] The diet of the *famuli* also improved: at Cuxham, for example, the use of peas and *curallum* ceased at the Black Death and the provision of pure wheat increased.[57] In village markets, too, the quality of wares improved. In 1374, for instance, 'cokett', 'treat', and 'wastall' loaves were all being sold in Pershore.[58] According to Langland, even beggars now turned up their noses at bread made from beans, holding out instead for the finest breads and best ales.[59]

People's expectations were clearly increasing, a phenomenon which is most readily appreciable in terms of ale consumption. At Appledram, for example, more ale had to be bought in 1354 'because the reap-reeve would not drink anything but ale in the whole of the harvest-time'.[60] The general quality of the drink itself also improved. Barley consolidated its position as the main malting grain, although some high-quality wheaten ale was also produced: in the last quarter of the fourteenth century, 15 per cent of manors in the ten counties around London malted wheat, while on the estate of Tavistock Abbey, wheat malt was even produced for farm labourers at Christmas and Easter.[61] Hopped beer also began to appear in the later Middle Ages. While it never threatened the dominance of ale in this period, it is indicative of changing consumption that two barrels 'de Holond beer' were bought for the daughters of the Duchess of Clarence in 1419–21, and that the Duke of Norfolk purchased 562 lb of hops in 1481 to make his own beer.[62] In fact, brewing became increasingly professional at this time, and alehouses a more permanent feature both of the landscape and of people's lives.[63] In 1365, even the statutes governing a chantry in Chesterfield had to be amended so that 'Where the ordinances say that the chaplain shall totally abstain from visiting taverns, this is to be understood as meaning that he shall not visit them habitually.'[64] Increased consumption meant increased production as well. At Castle Acre Priory,

[54] Woolgar (1999: 124–5); Harvey (1993: 58–9).
[55] Penn and Dyer (1994: 185). [56] Dyer (1994*b*: 83).
[57] Harvey (1976: 423, 440, 456, 466, 475, 489, 538, 584).
[58] Dyer (1998*b*: 68). [59] Schmidt (1992: 73). [60] Dyer (1994*b*: 96).
[61] Campbell (2000: 218); Finberg (1951: 100).
[62] Woolgar (1992–3: ii. 672); Woolgar (1999: 128). [63] Clark (1983: 20–38).
[64] Horrox (1994: 306).

Plate 2.2 The kilnhouse in the grain-processing complex at Castle Acre Priory, *c.*1360–1400. Adjacent were a granary, malthouse, and brewhouse. Photograph: C. M. Woolgar.

the grain-processing complex, including a malthouse and kilnhouse (Plate 2.2), was expanded in *c.*1360, presumably in part as a commercial enterprise, while sales of malt from Bromholm Priory brought in £54 4*s.* 8*d.* in 1416–17.[65]

Changes in the consumption of both bread and ale were also reflected in agriculture. Nationally, the proportion of demesne land under rye and maslin shrank from 17 per cent at the start of the fourteenth century to 7 per cent a century later, while the proportion of land occupied by brewing grains rose from 18 per cent to 27 per cent. Indeed, in 1391–2 all of Merton College's local demesne at Holywell was under barley, presumably to make ale for the fellows and undergraduates.[66] Similar changes in the cultivation of bread grains occurred on peasant land. In contrast to the low proportion of wheat and high proportion of inferior bread grains found around 1300, tithe corn at Oakham in the early 1350s contained 22.5 per cent wheat and 2.9 per cent rye.[67] In 1380, 40 per cent of one 12.5-acre holding at Hesleden was devoted to wheat.[68] Probably the clearest indication of change in peasant consumption and production comes from the proportions of corn ground at the lord's mill. On the Bishop of Winchester's estate, the mills on the manor of Taunton had produced 15 per cent wheat, 31 per cent maslin, and 53 per cent malt in 1301–2, but in 1409–10 this had changed to 24 per cent wheat, 15 per cent maslin, and 61 per cent malt.

[65] Wilcox (2002: 47); Redstone (1944: 59–61). [66] Campbell (2000: 240, 291).
[67] King (1991: 217–18). [68] Tuck (1991: 178).

Likewise, the Bishop's mill at Downton produced 8 per cent wheat, 40 per cent malt, and 50 per cent barley at the start of the fourteenth century but 15 per cent wheat, 58 per cent malt, and 25 per cent barley a century later.[69] Similarly, the accounts for the manor of St Columb show that by the mid-fifteenth century 'wheaten bread predominated in the diet and barley had partially replaced oats in brewing'.[70]

The later Middle Ages saw many shifts in the consumption and production of field crops, but change was not always wholesale or swift. In parts of the south-west, for example, the malting of oats for ale and the baking of rye for bread persisted, seemingly out of preference rather than as a result of environmental constraints.[71] Similarly, both brown and white bread were made in the Abbot of Peterborough's kitchens in 1370–1 (although a large number of the brown loaves were doubtless consumed by the Abbot's forty-nine mastiffs), and was sold by the bakers of Tamworth and Leicester in the fifteenth century. Even in the early sixteenth century, the monks at Thetford Priory consumed bread made from 55 per cent wheat, 43 per cent rye, and 2 per cent barley.[72] In this context, we should not forget the subtlety of Chaucer's characterization of grain consumption, for while the Cambridge scholars in the Reeve's Tale took wheat and malt to be milled, the friar in the Summoner's Tale begged for 'a bushel whete, or malt, or rye', and the poor widow of the Nun's Priest's Tale still made do with 'milk and broun bread'.[73]

Conclusion

While documentary evidence allows us to reconstruct agricultural production in late medieval England in great detail, manorial records, household accounts, and other sources, including surviving grains themselves, cast considerable light on the consumption of field crops. It is well known that grain, whether consumed in the form of bread, ale, or pottage, contributed more to the calorific intake of medieval people than any other foodstuff, but it was also the case that the nature and scale of consumption varied significantly from person to person and over time. Indeed, for much of the Middle Ages, wheaten bread and ale brewed from barley were chiefly the preserve of relatively high social groups. When pressure on agricultural resources was greatest, at the turn of the fourteenth century, most of the population would have eaten much coarser bread, made from barley, rye, and legumes, consumed little ale, and gained a considerable proportion of their calories from pottage. Even some lords were forced to compromise on the quality of their ale at this time, though it was only economic

[69] Page (1996: 13–14, 69); Page (1999: 11–12, 66).
[70] Fox (1991: 308). [71] Fox (1991: 303).
[72] Greatrex (1984: 56–83); Davis (2004: 487); Dymond (1995–6).
[73] Quoted in Ashley (1928: 96–7).

disasters such as the run of poor harvests in the 1310s that compelled them to reduce the quality of their bread as well. Documentary sources allow us to glimpse daily and weekly variations in the consumption of bread and ale, too, but by far the most significant temporal shift was the longer-term change in the aftermath of the Black Death. The standard of living of many people had improved by the late fourteenth and fifteenth centuries and this is reflected not just in the greater quantity of bread and ale that they consumed but also in its superior quality. In the higher echelons of society there is even evidence that fresh bread was consumed on a more regular basis and that the strength of ale increased as production could afford to employ more grain. These variations in patterns of consumption naturally affected agricultural production. Because of the nature of medieval documents it is frequently the case that inferences about consumption are drawn from patterns in production. Yet this brief survey of the historical evidence for the consumption of field crops suggests that this should be a two-way process. Most importantly, differing patterns of consumption imply that the agricultural profile of lords and peasants must have been very different, a conclusion that has significant implications for our understanding of the medieval economy at the broadest of levels.

3

Gardens and Garden Produce in the Later Middle Ages

C. C. DYER

The lack of much modern writing about medieval food production in gardens and orchards, or the consumption of vegetables and fruit, is easily explained. First, these matters have been dismissed by historians as marginal and trivial; secondly, full and detailed written evidence can be rather scarce.[1]

In fact gardens and their produce, far from being small matters best left to antiquarians, are essential to any assessment of the quantity and quality of medieval diets. In considering quantity, in a period of food shortages and potential malnutrition, we must enquire about the contribution that horticulture made to the total volume of food production. Quality can be measured partly in the sense of nutritional value, given the current understanding that fresh fruit and vegetables are an essential component of a healthy diet. The quality of a diet can also be judged in terms of medieval ideas about balanced eating as defined in the theory of humours, and the pleasure and satisfaction that were derived from consuming garden produce. The contribution that vegetables and fruit made to diet cannot be separated from the cultural importance of gardens, which figure prominently in medieval literature, and for which there is archaeological evidence.

Documents informing us fully about horticulture, the trade in garden produce, or the consumption of vegetables and fruit are indeed scarce. Yet almost every source commonly used by historians of the late medieval period—deeds and charters, surveys of manors, manorial and household accounts, the records of royal, seigneurial, and borough courts, wills, and narrative sources, such as chronicles and saints' lives—contains at least brief references to gardens and their produce. Using hundreds of such fragments of data, the subject will be discussed under four headings: the scale of gardening and access to gardens; the distribution of garden produce; the contribution of vegetables and fruit to diet; and the overall significance of gardens and the crops grown in them.

[1] Exceptions to this dismissive attitude include Harvey (1981); McClean (1981); Harvey (1984); Macdougall (1986); Brown (1991); Landsberg (1996); Higham (2002).

The scale of late medieval gardening

As we would expect, the largest gardens, which were most likely to be managed by full-time specialist gardeners and to produce large surpluses for sale, were those attached to royal palaces, to the principal residences of great aristocrats, and to religious institutions, such as monasteries and colleges. An especially well-documented lord's garden was that belonging to the London residence of the Earl of Lincoln in Holborn, for which a financial account survives for the year 1295–6.[2] The main function of the house was to provide the Earl with a residence on his visits to London to attend Parliament and the royal court. It also gave him and his officials a convenient base for buying imported goods, such as cloth and furs, from London merchants. The lord and his household's stay in the capital was made more pleasurable by the presence of the garden, both for recreation and as a source of fruit and vegetables for the table. In his absence, which was for most of the time, the produce could be sold and the proceeds added to the income of the estate.

The garden was in the charge of Robert Gardener, whose wage of 52*s.* 2*d.* can be judged to be similar to that of a carpenter or other skilled artisan who earned between 2½*d.* and 3*d.* per day, assuming that they worked for about 240 days in a year.[3] Unlike most artisans, however, the gardener's income was guaranteed for the whole year, and he would have had opportunities for additional earnings from the sale of produce, seeds, and plants. In addition to this manager, the Earl paid 'various workers' a sum of 41*s.* 6*d.*, probably for a total of 500 person-days at about 1*d.* per day, on such tasks as manuring and weeding the vegetable and leek beds, and pressing grapes to make verjuice, an unfermented grape juice commonly used in cooking for its acidic qualities. A sum of 8*s.* 9½*d.* was spent on buying seed and plants for beans, hemp, onions, garlic, and fruit trees. Mending the garden fence cost 2*s.* 6*d.* Sales of pears, apples, walnuts, cherries, beans, onions, garlic, 'vegetables', verjuice, roses, hemp, and vine stocks amounted to about £11, which represents a great volume of produce when good-quality apples fetched 12*d.* per hundred and onions were usually sold for 4*s.* per quarter.[4] We can confidently state that this garden contained hundreds of fruit trees and at least an acre devoted to vegetables and vines.

More precise indications of the size of the garden and the scale of cultivation are given for a country garden belonging to a wealthy churchman, that of the Bishop of Winchester at Rimpton, which was enlarged at some cost in 1264–6.[5] The perimeter hedge, planted on a substantial bank of earth, was 113 perches in length, suggesting an area of 4 acres, and 129 pear and apple trees were planted, as well as flax and vegetables.

These two gardens are characteristic of those attached to castles, manor houses, and monasteries in that they combined the functions of supplying the household and the market. The balance between these activities would vary from year to year.

[2] NA DL 29/1/1. [3] Farmer (1988: 768).
[4] Rogers (1866–1902: i. 223, 418–19, 445–50; ii. 175–7, 379–82). [5] Hunt and Keil (1959–60).

A household moving round the country might stay in a particular residence for a month and eat every apple and onion available. In another year the lord would not visit the manor house or castle at all, and all the produce would be sold. Even in the case of a monastic community, the place of residence of which was permanently fixed, the gardeners would find themselves with gluts of produce in the appropriate season, so they would sell surplus fruit and vegetables.[6] All gardens were evidently designed partly to raise money, as among the crops are found flax, hemp, nettles, madder, teasels, and other plants for industrial use. Beehives were often kept in gardens, and the honey and wax might be produced for sale.

The most numerous gardens were much smaller than these, and were kept by almost all householders in the countryside and many town-dwellers. A description of a peasant holding in a manorial survey, rental, or court roll will commonly call it a messuage or a cottage, which itself implies the existence of a plot of ground as well as buildings. Often the phrase 'a messuage and a curtilage' or 'a cottage and a close' is found, and that again shows that land capable of being used for growing garden produce formed part of the holding. The boundaries of these enclosures can be seen fossilized as earthworks in the plans of abandoned settlements, and they are known to archaeologists as tofts and crofts (Plate 3.1). In towns a burgage or fraction of a burgage would have land at the back of the house, part of which would commonly be planted with vegetables and fruit trees. In addition, in both towns and villages, separate parcels of garden ground could be rented (Fig. 3.1). As there were approximately a million households in England in 1300, there must have been a similar number of gardens.

These many gardens did not constitute the mainstay of food production, but they provided the population with a proportion of their diet. In the case of the gardens belonging to aristocratic houses the majority were smaller and less carefully tended than the Holborn garden of the Earls of Lincoln or the garden at Rimpton. When most manorial gardens were valued as part of the lords' demesnes in inquisitions in the late thirteenth and early fourteenth centuries the sum was usually between *6d.* and *3s.* per annum, which accounted for no more than 2 per cent of the total value of the manor. Agricultural production was largely based on the substantial income to be expected from corn and livestock, which fetched much higher prices in the market.[7] The garden aided the major agricultural operations in that its vegetables were served to the farm servants in a pottage containing oatmeal or other cereals (Table 2.1) and leeks or onions.[8] The manorial garden also contributed to the meals of the lord or lady in their periods of residence and generated some income with industrial crops, which on occasion could exceed a pound or two. The labour expended on the garden on many manors was often hidden from view, as few manors employed a gardener

[6] e.g. Noble, Moreton, and Rutledge (1997: 31, 33, 35–6, 37, 39, etc.); Kirk (1892: 51–8, 73–7).

[7] The generalizations about the value of gardens in inquisitions post mortem come from those for Gloucestershire and Staffordshire: Madge (1903); Wedgwood (1911).

[8] e.g. Northamptonshire RO, Finch Hatton MS 519 (reeve's account at Maidwell) refers to purchases of vegetables, onions, and oatmeal for the pottage of the *famuli* in 1290–1.

Plate 3.1 Aerial photograph of the deserted village of Holworth in Dorset, showing earthworks defining a row of square tofts facing the village street, and larger rectangular crofts, each at the rear of a toft. There was room for a small garden in the toft, and sometimes the croft was used for horticulture, though it could also serve as an animal pen. Photograph: © Crown Copyright/MOD.

alongside the ploughmen, carters, shepherds, and dairymaid. Instead work on digging plots, planting leeks, or sowing onion seed was a part-time activity for one or more of the servants. A few labour services were occasionally expended on garden work and this task might be specified formally as an obligation of tenants.[9]

Horticulture appears fleetingly in the treatises on estate management written in the thirteenth century: they tell us that 9 or 10 quarters of apples and pears could be expected to yield a tun (*c.*240 gallons) of cider.[10] This would help an auditor check on the accuracy of a reeve's account if it gave such details. Gardens and orchards required little expensive specialist equipment; a cider mill and

[9] For example, on Battle Abbey manors: Scargill-Bird (1887: 6, 8, 10).

[10] Oschinsky (1971: 428–9, 474).

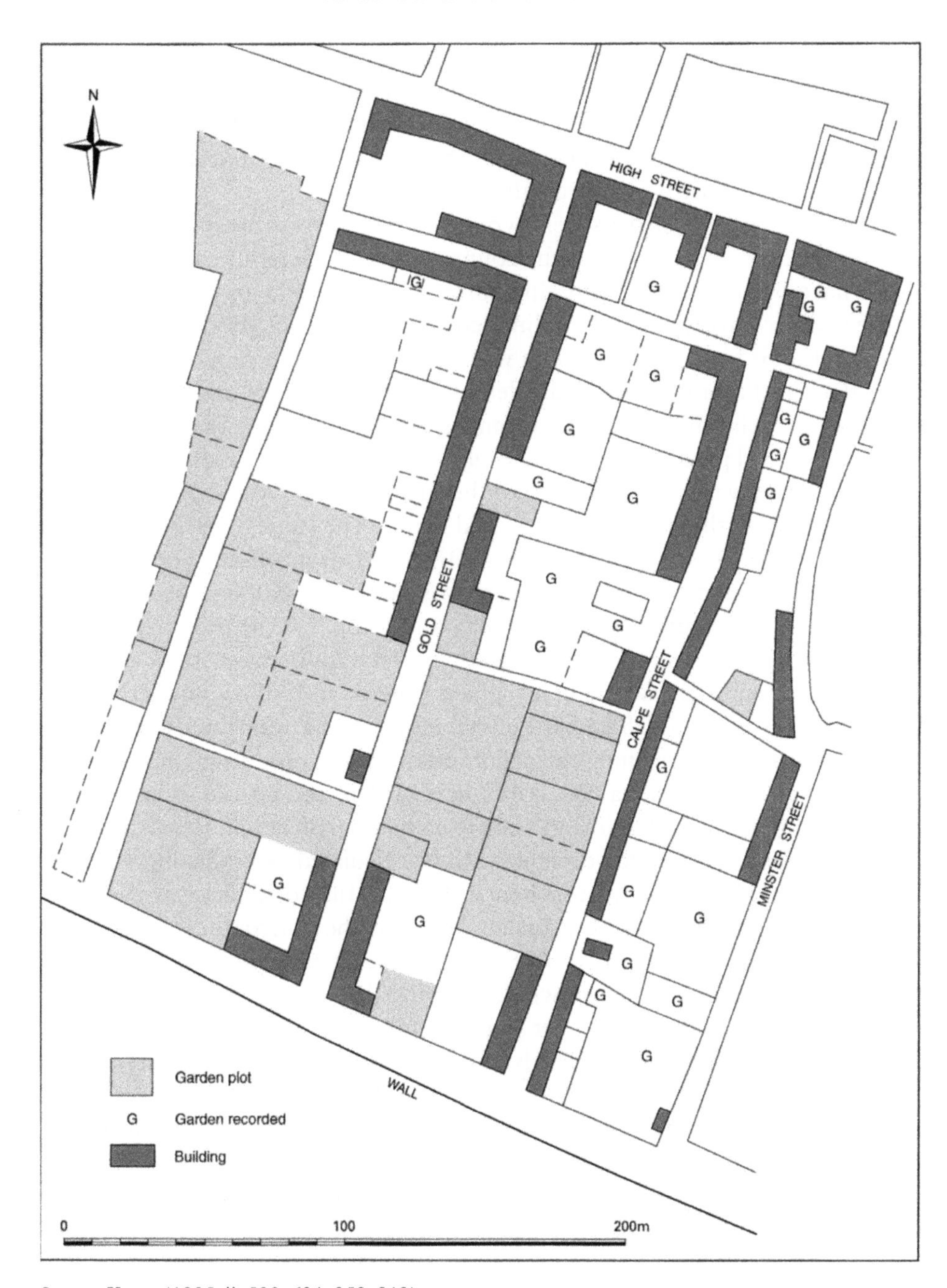

Source: Keene (1985: ii. 583–634, 858–912).

Fig. 3.1 A plan reconstructed from documents of part of the city of Winchester, based on Gold and Calpe Streets (now Southgate Street and St Thomas Street), *c.*1400. It shows streets, buildings, parcels dedicated entirely to horticultural use, here called garden plots, and gardens which mostly lay at the near of houses.

press, the most elaborate machinery used, was built at Clare in 1330–1 for 19*s.*[11] Specialist labour was not usually needed in the garden, and presumably everybody was aware of the basic techniques of growing fruit and vegetables from experience of their own small gardens. Grafting fruit trees needed more skill, and instructional literature was written about this.[12]

Peasant and artisan gardens were typically small, with as much as a half-acre in some places; but over most of the country a quarter-acre for each household would have been the maximum, and some were tiny plots of only a hundred square yards.[13] Larger peasant holdings consisted mainly of grain-growing land in the fields, but for them the garden was a more significant proportion of their assets than in the case of the lords' demesnes—for cottagers and townspeople this could be their only land. Tithe revenues provide a guide to the contribution that horticultural produce made to the whole economy of a parish, as a tenth of garden produce was supposed to go to the support of the rector or vicar along with one in ten of the sheaves, fleeces, and lambs. The value of garden produce was much inferior to the field crops, as a parish with as much as 8 per cent of its tithe revenues coming from horticulture was quite unusual, and often a large proportion of that figure derived from flax and hemp.[14] If the proportion coming from fruit and vegetables is calculated, it often falls between 1 per cent and 6 per cent. The sums given can be tiny: at Deddington in 1432–3, out of the total annual tithe revenues of £18 the tithe on onions and garlic was stated to be worth 2*d.*, and it is only by stretching the definition of garden produce to include wax and honey from beehives (valued at 6*d.*) that gardens can be said to have made any significant mark on the tithe income from that parish at all.[15]

In some parishes fruit and vegetables can be calculated as producing valuable tithe revenues; for example, at Stoneham in 1341 the tithe of apples was said to be worth 13*s.* 4*d.*, which represents a total income from the fruit for those holding the trees and orchards of almost £7.[16] At Stokesay in 1252 garden tithes were valued at 10*s.*, the same as the tithe on wool, implying that horticulture generated an income for the parishioners of £5 per annum.[17] In an urban context, the 4 quarters 5½ bushels of onions and 1,100 heads of garlic collected in tithe from about 400 gardens in Warwick in 1465 were worth 17*s.* 3*d.*, and (multiplied by ten) represent for each small unit of production a mean output of perhaps 30 heads of garlic and a bushel of onions.[18] Tithe evidence cannot be taken at face value. In some parishes, such as Blunham in 1520, the decision was made to charge a standard 1*d.* for gardens, so the exact amount was not being calculated.[19] The assumption

[11] NA SC 6/992/20. [12] Amherst (1894); BL MS Sloane 686. [13] Dyer (1994*d*: 116–18).

[14] To the examples given in Dyer (1994*d*: 119–21) can be added Dinsdale, in Surtees (1816–40: iii. 239); Monk's Kirby, in Lincolnshire Archives, 2 Anc. 2/2/116.

[15] St George's Chapel, Windsor, XV.53.35.v (1).

[16] Vanderzee (1807: 126). [17] Rees (1985: 213–14). [18] Styles (1969: 80–2).

[19] Thompson (1990: 128–44). The custom was set out explicitly in the statement of tithe customs at Beckford, 1487–8: 'Everyone that is an householder that hath a garden ought to pay for his onions and all sorts of herbs and such like a penny at Easter': Gloucestershire RO, GDR, 40/T2.

behind such a small figure must have been that collecting tithes from many individual gardens would not justify the effort, and we may suspect that these gardens were more valuable assets than 1*d.* per annum would suggest.

The crops and plants in individual peasant gardens were sometimes assessed in terms of money when damage was done to them, and the tenant would bring a case of trespass before the courts, alleging damages to a certain value. In a typical case, animals invaded a garden and ate or spoilt vegetables at Hingham in Norfolk in 1449. The tenant claimed that the damage amounted to 6*s.* 8*d.*, but the jury settled for a more realistic sum of 2*s.* In similar circumstances at Haywood in Staffordshire in 1409 the produce was valued at 3*s.* 4*d.*, and servants at Marham in Norfolk in 1415 were said to have stolen fruit worth 5*s.*[20] These sums were by no means negligible, especially in view of the relative cheapness of fruit and vegetables, and are equivalent to a skilled worker's wages for a week or two. In major trespasses in Staffordshire involving damage to a number of houses at once, the number of fruit trees said to have been felled in each garden could be as many as ten at Agardsley in 1354, or as few as two or three in a town, Wolverhampton, in 1415.[21]

We can conclude that gardens were an integral part of the English economy in the later Middle Ages, and that they reflected the social hierarchy: some aristocratic houses had large, well-stocked, and professionally managed gardens, while almost all peasants and a considerable number of urban households had access to smaller parcels of garden ground. Gardens were numerous and commonplace, and although the quantity and value of their produce was much smaller than that of the fields and pastures, even the smaller ones contributed to the domestic economy.

Distribution of garden produce

Most gardens were worked by members of the household, and their produce was consumed directly, so that no money was used to hire labour, nor for sales and purchases, which has helped to ensure that gardening has left us with few historical records. Nonetheless fruit and vegetables had some value, and tenants were prepared to pay good money for a garden plot on the rare occasions that it was separately rented. A rood (a quarter-acre) of garden at East Bergholt, for example, was held in the late fourteenth century for 18*d.* per annum, which was far in excess of the normal rent for grain-growing land.[22] The same high values were reflected in entry fines and the purchase price of gardens or orchards.[23] Many of those who paid so much for gardens were buying the convenience of growing fruit and vegetables for their own kitchen, but garden produce was also traded.

[20] Norfolk RO, MCR/B/26; Staffordshire RO, D1734/2/1/427; Norfolk RO, HARE 2199, 194 x 4.

[21] Wrottesley (1891: 119; 1896: 58).

[22] Suffolk RO (Ipswich), HA6:51/4/4.7.

[23] For example at Blickling Hall, Norfolk RO, NRS, 10193; in 1483 Robert Aleyns sold a quarter-acre of orchard with a cottage for £2 13*s.* 4*d.*, which is six times the price of arable land in that year.

Commerce in fruit and vegetables was often conducted on a small scale and informally, with the result that it can only be glimpsed in the written sources. We hear of huxters in towns, many of them women, selling fresh fruit from baskets in the street.[24] More specialist male traders, known as leekmongers and garlicmongers, are occasionally mentioned.[25] Lords' officials and gardeners were able to buy seeds and plants all over the country. It was possible to obtain for planting at Eye (the Westminster Abbey manor in Pimlico and Mayfair) in 1327 the seeds of seventeen varieties of plant, including borage, hyssop, chervil, parsley, and spinach, as well as the ubiquitous onion and leeks. Leek seed was bought in the small market towns of Rotherham and Sheffield in the late fifteenth century.[26] Young trees and bushes were sold from plots of land set aside as nurseries, known as impyards.[27] Manorial producers regularly sold apples by the quarter and cider by the tun. Aristocratic and institutional households would buy apples, pears, strawberries, onions, garlic, and other vegetables, sometimes from middlemen in towns (London had its fruiterers) but often directly from the peasant producers. They often bought these in small quantities for immediate use and no great expense was involved: enough leeks for a household of more than forty people were bought for a few pence. Neither growers nor dealers in garden produce had any expectation of earning large profits.

Not all of the sources of supply of garden produce were local. Many of the fruits and nuts which were much enjoyed by the wealthier households were grown most successfully around the Mediterranean, and were therefore imported in dried or preserved form. These included the various types of dried grapes, such as raisins and currants, together with figs, dates, and almonds, which contemporaries categorized as spices, and which are not our main concern here. Regular imports were also made of onions, garlic, and cabbages from the near Continent. In the customs records and port books they are regularly recorded at ports such as Exeter, Hull, and Southampton, not as the main commodity, but as part of mixed cargoes.[28] For example, a ship from Waben in the Pas de Calais, the *Emmengard*, brought into Exeter in March 1321 a cargo mainly of dyestuffs (woad and weld) and potash, together with 121,000 onions.[29] Once unloaded at ports, these vegetables would be sent throughout the country. When the fifteenth-century aristocratic Stonor family of south Oxfordshire obtained supplies of garlic, they bought it from London, though it was collected from Henley-on-Thames, which had regular boat traffic with the capital.[30] The Waterton household at Methley in 1416–17 obtained garlic and onions from York.[31]

[24] Sharp (1982: 11); Kimball (1939: 37, 38, 43). [25] Fenwick (1998–2001: ii. 345).
[26] Westminster Abbey Muniments, 26873; Bodleian Library, MS DD Weld c.19/4/2–4.
[27] At Crowle an 'orchard' was called *Le Ympe Heye* in 1337: Worcester Cathedral Library, E13.
[28] e.g. Kowaleski (1995: 230, 243); Childs (1986: 50–1, 55–6); Foster (1963: 39, 67).
[29] Kowaleski (1993: 193). [30] NA C 47/37/7. [31] Woolgar (1992–3: ii, 511).

Eating and drinking garden produce

Aristocratic and monastic households were supplied with relatively small quantities of garden produce. At Glastonbury Abbey and Maxstoke Priory, we know how much came from the monastic garden, and the calculation can be made that garlic supplies were sufficient for a monk to have three cloves of garlic per day, and that there were enough apples for individuals to eat one weekly.[32] In fact fruit and vegetables would not have been served daily, but in larger quantities on occasion; most apples, for example, were eaten in the winter. More often the accounts will state at irregular intervals that a few pence have been spent on apples, or pears, or leeks, without details of the quantities. These references are sometimes so few that one might conclude that these foodstuffs scarcely figured in the household's diet at all. This would be misleading as the garden attached to the residence could have been the main source, but its contribution would not be mentioned. The Duke of York's household in 1409–10 at Cardiff and at Hanley Castle made occasional purchases of garden produce, but in the account we also find the statement that 8*d.* was paid in tithe for the onions from the garden, which implies that at least 10 bushels of onions were grown and presumably used without record in the castle kitchen.[33] This particular household was, in comparison with others at the time, a large-scale buyer of garden produce, with recorded purchases of apples and pears totalling 2,150, which may have been needed to supplement the fruit grown in the castle orchard.

The frequency of consumption of garden produce in an aristocratic household can be appreciated when a lord went on a journey, not staying on his manors, and buying all of his supplies. We can follow the Earl of March's progress in May and June 1378 as he travelled north from London, buying his food at such places as Ware, Royston, Huntingdon, and Stilton.[34] In thirteen days he made nine purchases of 'vegetables', 'herbs', onions, and garlic, so these were consumed regularly, if not every day. The quantities can be judged to have been modest as they cost between ½*d.* and 4½*d.* These items were bought as ingredients in meals, to which they contributed desirable flavours and textures. In these elite households fruits were especially highly regarded, and would have been the central element in a dish. They were eaten in the winter, and particularly at Christmas. The superior varieties of pear, such as warden, St Rule, and *jonett*, were luxury items, reserved for special occasions.[35]

All garden produce came to the educated medieval consumer with a health warning. Vegetables and salads were thought, according to the theory of the humours, to be cold and would be eaten in conjunction with foods with opposing qualities.[36] They

[32] Dyer (1994*d*: 128). [33] Northamptonshire RO, Westmorland (Apethorpe), 4. xx. 4., fo. 3[v].

[34] Woolgar (1992–3: i. 247–50).

[35] NA E 101/624/26 are the King's fruiterer's accounts of 1308, which include purchases of 900 pears and 1,700 apples for the coronation. For fruit as gifts, Ross (2003: 58).

[36] Scully (1995: 70–1); Albala (2002: 12, 70–1, 88–9).

were regarded as difficult to digest. Peaches were said to putrefy in the stomach if eaten at an early stage of the meal, and to demonstrate the point one English king died after eating this dangerous fruit.[37] Garden produce was also associated with the diets of the poor, and purchases of such vegetables as leeks were often concentrated in February and March, that is in the fasting season of Lent. When a treatise on household management recommended a frugal regime for a member of the lesser aristocracy, it suggested that ale and wine be supplemented with cider (it was said to be available without cost, as it came from the lord's own orchard). Really wealthy aristocrats would not usually have drunk anything but wine and ale.[38]

When we turn to the less privileged sections of society direct documentation for vegetable and fruit consumption diminishes, but they probably figured regularly in the diet. Whenever contemporary writers refer to the content of peasant meals, from Langland and Chaucer to the anonymous authors of shorter works, they mention vegetables.[39] Indirect indications in more objective sources have already been mentioned: every rural house, even the smallest cottage, was provided with at least a patch of garden. For a holding of less than 5 acres of arable land in the fields, a half-acre planted with fruit trees and vegetables could account for a considerable proportion of the resources of the household. When a peasant retired, it was not uncommon to reserve part of the garden for his or her use. Fruit trees, or a share of the fruit, could be specified as belonging to the former tenant's share of the holding. Quantities of cider might be mentioned as part of the retirement package, including an annual allowance of 120 gallons for a retired couple in Hampshire in 1457.[40]

Those earning wages expected to eat vegetables regularly. Leeks and other garden crops were an ingredient in the daily allowance of pottage given to farm servants working on demesnes. Reeves had to buy vegetables if the garden for some reason had not been cultivated.[41] Building workers in 1431 at Stratford-upon-Avon were served with vegetables (such as onions) every week, along with much greater quantities (by value) of bread, ale, and fish over the eight-week period for which records survive.[42]

The peasants and artisans probably shared some of the aristocratic attitudes towards this type of food. They enjoyed the strong flavours that garlic and mustard could bring to their meals, especially as they could not afford to flavour their meals with imported spices, and like the aristocracy they took pleasure in eating fruit. But they probably associated a diet with a high proportion of vegetables with poverty. Village by-laws allowed the village poor to pick green peas from the ends of the strips in the open fields.[43] Peas were mainly harvested after they had dried in the pod in August or September. The poor would have benefited from access to green peas in June and July when grain was most likely to be in

[37] Albala (2002: 109). [38] Myers (1959: 109).

[39] Pearsall (1979: 158–9); Benson (1988: 255, 258–9); Barr (1993: 94). [40] Dyer (1994*d*: 121).

[41] Dyer (1994*d*: 129). [42] Shakespeare Birthplace Trust RO, BRT 1/3/40.

[43] Ault (1972: 38–40).

short supply and highly priced. It should be added that everyone, including the wealthiest aristocrats, consumed dishes of fresh peas at that time of year, but the impression given by the by-laws is that the poor were being given access to an important source of sustenance, rather than a pleasant side dish. When workers' wages rose after the Black Death of 1348–9, they expressed their prejudices against the cheap foods that they had previously been given as part of their wages, such as cabbages, and as part of that movement, workers in Sussex had their allocation of cider replaced with ale.[44]

Peasants who lived near towns adopted a more commercial and specialized approach to their gardens. Individual gardens might be especially large and productive. The contents of a garden at Tooting, south of London, in 1397 were claimed to be worth 10*s*., and in 1340 in the suburbs of Oxford at Holywell the crops in a close, including vegetables and herbs, were said to be worth 40*s*.[45] The cultivators of these plots may have eaten more vegetables and fruit than their contemporaries out in the country, but of course selling their crops in the town provided their main motivation for production on a large scale. This is the main evidence, together with the many gardens cultivated by the townspeople themselves behind their houses or in separate areas of garden ground, for the urban consumption of garden produce.[46] The concentration of demand came partly from households, especially in the larger and more densely occupied towns, which lacked direct access to a plot. Larger towns also contained many poorer people and wage earners who, as we have seen, included a high proportion of vegetables in their diet, and a number of large and wealthy households, both those of merchants, and those of aristocrats in transit.

The significance of garden produce

Did gardens contribute significantly to standards of living? The suggestion has been made that in the period of high population in the late thirteenth and early fourteenth centuries, when grain prices were high at all times, and in a number of bad years rose 20 per cent higher than average, garden produce would have made a difference for the vulnerable smallholders.[47] There can be no doubt that the families of cottagers and peasants with a few acres of land would have appreciated the produce from their gardens in years of poor harvests, and that in the worst years cabbage and leeks figured among the 'famine foods' that helped to keep people alive. In normal years the addition of garden produce to a diet in which cereals and pulses predominated would have made dull food more palatable, and improved its nutritional quality. On the other hand, the diet of the whole population was based on the consumption of the main field crops, and serious shortage of supplies of basic bread grains and pottage corn could not

[44] Pearsall (1979: 159); Dyer (1994*b*: 96).
[45] Gomme (1909: 24); Merton College, Oxford, Muniments 4546.
[46] For example, Keene (1985: i. 151–3).
[47] Britton (1977: 157–9).

have been compensated by handfuls of vegetables. The people who suffered most hardship were the landless workers and the servants who would have been laid off in the worst years, and they did not have direct access to a garden.

We might also ask, in the case of the more affluent, whether the relatively restricted quantities of fresh fruit and vegetables included in their diet impaired their health. Their consumption must have fallen far below the intakes recommended by modern nutritionists, particularly at times of year when stocks of fruit from the previous autumn were exhausted and the spring vegetables had not yet begun to grow. Perhaps they received the minimum quantity of vitamins necessary to avoid scurvy, but when more detailed medical evidence becomes available in the early modern period many upper-class patients exhibited scorbutic symptoms.[48]

Garden production varied considerably not just between rich and poor, but also from one region to another, and over time. The size and productivity of gardens cannot be compared precisely, though the most systematic regional study yet undertaken, of manorial accounts in the counties around London, showed that there was a concentration of demesne gardens generating quite high revenues within a 30-mile radius of the capital.[49] We have already noted the concentration of gardening in the immediate vicinity not just of large towns, such as London and Oxford, but also in and near smaller towns such as Warwick.

The fragmentary evidence for peasant gardens suggests that they played an important role in East Anglia, where the plots attached to peasant houses were quite large, and holdings of field land relatively small.[50] Apple growing and cider drinking are well recorded in Hampshire and Sussex, and also in the west. A twelfth-century description of the vale of Gloucester waxes eloquent on the abundance of fruit: 'You can see the public roads clothed in apple trees . . . bearing fruits which far surpass others in taste and look.'[51] The people in the east, south-east, and west, in the districts of 'old enclosure' and a more wooded and pastoral countryside, may have practised gardening more than their counterparts in the belt of villages in the Midlands, the north-east, and central southern England. The latter depended on extensive cereal cultivation in open fields, combined with sheep grazing, where they had limited space or time for horticulture.[52]

Changes in gardening over time are not easily traced. The commercial growth which began with the foundation of towns in the tenth and eleventh centuries, and quickened in the twelfth and thirteenth centuries, undoubtedly stimulated demand for garden produce. Trade encouraged specialization in horticulture in some places. However it also led to the decline of English vineyards in the face of competition from the plentiful and high-quality wines from Gascony.[53]

[48] Lane (1996: 3). [49] Information from Margaret Murphy. [50] Dyer (1994*d*: 118).

[51] Hamilton (1870: 291). The English quotation is from William of Malmesbury, *The Deeds of the Bishops of England*, trans. D. Preest (Woodbridge, 2002), 197. [52] Thirsk (2000: 81–3, 116–17).

[53] James (1971: 9–10).

We might expect that these tendencies would go into reverse after 1350 or 1400, with the shrinkage of towns, the slowing down of commerce, and the scarcity of labour. Indeed gardens can be found reverting to grazing plots at this time. We cannot be sure about this period as one marking a decline in gardening, however, as even at the peak of commercial growth so much production had been for use rather than sale, and this must have continued. Also the commercial economy survived the shock, and the market moved in unexpected ways. The theory of 'alternative agriculture' predicts that when the market for the staple crops such as grain and wool declines, farmers will develop new products which give better returns, and the example from the later Middle Ages must be saffron, which needed intensive care to yield a very valuable spice and dyestuff.[54] Saffron gardens are found in that period over much of south-east England, with such a concentration in and around the town of Walden that its name was eventually changed to reflect its distinctive crop.

Finally, we must avoid the mistake of concentrating on the utility and commercial value of gardens and their products. Gardens were a great source of enjoyment and contributed to the quality of life. The aristocracy designed gardens for pleasure and held social gatherings and trysts in their enclosed spaces. In towns gardens might be called 'paradise', and guilds and fraternities would arrange for gardens to be laid out next to their halls, where the brethren could enjoy themselves.[55] Fruit was accorded a high status not fully reflected in its market price, and when the elite wished to acknowledge and honour their associates or superiors, they would send gifts of apples, pears, or cherries. The lower ranks of society were not excluded from this cultural regard for gardens and their produce. The people of Potterne gave fruit to their lord, the Bishop of Salisbury, in 1406.[56] A male and female servant in York in 1396 met in a garden, just like aristocratic lovers, to make a marriage contract.[57] We can sense the strong attachment of peasants to their gardens and their fruit from the retirement agreements registered in manorial courts, or the arrangements for widows made in wills. They often specified—as well as rooms which the old person would occupy, food and drink, and access to the hearth and kitchen—a share of the garden. For example William Spark of Elmley Castle surrendered 'a messuage with a garden adjacent' to Roger Hale in 1470 (the description of the holding is of interest, as the word 'garden' in the records of this manor replaced 'curtilage' in the mid-fifteenth century).[58] Spark clearly regarded his garden as an important asset, and the agreement included the provision that he should receive all of the fruit from the 'pear tree called a warden tree', and half of the pears from the 'pear genet tree'.

[54] Thirsk (1997: 16–17). [55] For example, Nightingale (1995: 417–20).
[56] Woolgar (1992–3: i. 418). [57] Goldberg (1992*b*: 115).
[58] Worcestershire RO, 899:95 BA 989/2/33.

Conclusion

Gardens were widely distributed through every social rank and every region in late medieval England. Every family with a small plot of land, which means a majority of the population, had the opportunity to grow and consume vegetables and fruit. Those who did not produce their own could buy garden produce from huxters or more specialist traders. Vegetable growing in England could not match the demand, resulting in an import trade. Although the rich may have eaten more garden produce than can be seen in their records, vegetables probably provided a higher proportion of the food of the peasants and wage earners. Horticulture was practised more intensively in town than in the country, and in woodland landscapes rather than in the open-field districts of the Midlands. It expanded under the stimulus of commerce in the thirteenth century, and did not always decline in the period of falling population after 1350.

Garden produce presents us with many paradoxes, as it was both cheap and highly regarded, commonplace yet not a major component in diet. The wealthy consumed it on a small scale to their nutritional disadvantage; the lower ranks of society ate vegetables, drank cider, and thought themselves deprived.

4

The Archaeology of Medieval Plant Foods

L. MOFFETT

Unlike the historical evidence for plant foods, which is concentrated in the late medieval period, remains from throughout the Middle Ages of the plants themselves are commonly found on archaeological sites in England. The majority of these remains are 'seeds' in the broadest sense of the term, but can also include whole fruits, parts of flowers, stems, and even roots and tubers. These can provide physical evidence of the plants used by people and sometimes this evidence can also suggest ways in which the plants might have been used or the methods of processing them for use. Dating of archaeobotanical remains is usually less precise than that of many historical documents. The information that can be deduced from archaeobotanical remains, however, is often very different from that in documents and the two sources can complement each other, to provide important contributions to the study of medieval agriculture, horticulture, and the use of wild plants. This chapter focuses on food plants alone, although mention will be made of straw and chaff remains from cereals.

Preservation of food plant remains

Preservation is a result of the interplay between human actions and natural environmental conditions and thus deserves some preliminary discussion. Most organic material decays quickly due to the action of micro-organisms such as bacteria and fungi. Plant material, therefore, will survive on archaeological sites only under particular conditions: understanding these is an important part of interpreting the evidence.

The most common means of preservation, especially of cereal remains, is by charring during exposure to fire. In an oxygen-rich fire, material tends to burn away leaving only ash. Under oxygen-poor (reducing) conditions, however, the plant material may survive as a carbon skeleton, preserving many of the morphological features of the material, though these are often distorted. Robust

I am grateful to Julie Jones, Angela Monckton, and Liz Pearson for permission to use their unpublished material. English Heritage supported this work.

plant material, such as cereal grains and dense seeds, tends to survive best. Oily seeds, light seeds, and light papery material, such as cereal chaff fragments, are less likely to survive, though are sometimes found.[1] Charred material is very stable and can survive for a long time if not subjected to mechanical damage. It is, however, physically fragile, so that trampling, freezing and thawing, and other physical processes may cause it to fracture.

Since charred material is so stable, it can be moved around an archaeological site by later activities such as digging. Backfilling a pit or a ditch with soil containing charred plant remains, for example, may result in the residue from several different activities becoming mixed in one feature. Material reworked in this fashion is open to misinterpretation, but often the reworking is apparent to a specialist. Different activities result in different types of charred assemblages. Interpretation is based on analogies with information derived from ethnographic studies of modern traditional farming societies,[2] from early writers on agriculture,[3] and from what we know of the biology of the crops themselves. All need to be used with caution and the more lines of evidence that can be drawn on the better.

Organic material can also be preserved by waterlogging in anoxic conditions, as in pits, ditches, and wells, where these have been dug below the water table, in low-lying waterfront deposits, and old river channels. Anoxic conditions prevent the actions of most micro-organisms and thus greatly slow the rate of decay. Sometimes organically rich material—not, technically, waterlogged—can survive in sealed conditions where it is protected from drying and oxygen. Two examples of this come from late medieval Worcester: the barrel latrine at Sidbury[4] and the buried stone floor at Fish Street.[5] In both cases, the organic richness of the material itself helped to maintain wet conditions; and because these features were well sealed and undisturbed, the conditions were also anoxic.

Waterlogging in general also preserves robust material well and can preserve more delicate remains than charring. Fruit stones and other food remains, including cereal bran, locally growing weeds, household rubbish, remains in faeces, flooring, bedding, and building materials are all often found in these deposits as are other organic remains, such as leather, textiles, wooden utensils, and hair.

Mineral replacement of organic material can take place in situations where there is a high presence of calcium phosphates or calcium carbonates, with sufficient water to dissolve these minerals and allow them to penetrate into the organic material. The most common place for this to occur is in latrines and other damp places where sewage was present. Mineral replacement tends to preserve robust material best, though sometimes the part that is replaced is the inside of the seed, forming a mineral 'cast', which can be very difficult to identify. Latrines are the main source of mineral-replaced remains. They are also likely to be better dated than many pits and ditches; but much of the material found in

[1] Boardman and Jones (1990).
[2] e.g. Hillman (1984); G. Jones (1984); Fenton (1978).
[3] Such as Markham (1668).
[4] Greig (1981).
[5] Miller, Darch, and Pearson (2002).

latrines is not the remains of food and careful interpretation is needed. Latrines are also relatively rare: they are found primarily on urban and high-status sites, limiting their usefulness as an indicator of diet in general.

Soot-covered thatch, also called smoke-blackened thatch, survives in a few buildings. This happens particularly with structures that had at one time been open halls, where smoke was able to percolate through the roof instead of being channelled up a chimney, and where the old thatch was never fully stripped off when new thatch was applied. The outer coating of soot prevents microbial decay. Although relatively rare, this material usually provides the best and most complete preservation of cereals, other field crops, and weeds, and occasionally other materials such as heather, which was used as an undercoat for the thatch. Most of the examples are late medieval and occur in southern Britain.[6]

Straw and chaff are sometimes found preserved in daub in late medieval buildings. There are few examples of this being studied in detail, but preservation in some cases appears to be excellent.[7]

Surveys

Several useful literature surveys cover medieval food plants.[8] Of particular note are Dickson's survey of archaeobotanical evidence for garden plants,[9] which encompasses many food and medicinal plants in Britain with site-specific references, though she does not include imported food plants or field crops; and that by Greig, who has surveyed the archaeobotanical record for both pollen and seeds of edible and useful plants from Britain and Europe from the eleventh to the eighteenth centuries and compared it with British documentary records.[10] The latter survey highlights some of the different biases in the archaeological and historical records. A similar pattern can be seen from Tables 4.1 and 4.2, which list food plants found on selected sites (or groups of sites from the same town) most of which date from before 1500. Most derive either from latrines or pits with possible sewage; those from Droitwich[11] and the Cowick Moat,[12] however, are from brine pit deposits. The range of garden plants that can be identified botanically is both wider and more specific than those which the historical records discussed in Chapter 3 can indicate. The archaeological specimens, however, do not give any representation of overall quantities available for consumption or of shifts in taste.

Cereals

Cereals are the most common food plant remains found in archaeological deposits of all periods, frequently as charred remains. This reflects the ubiquity of their presence and use, but it is also a result of the circumstances by which they

[6] Letts (1999). [7] Arthur (1960, 1961); Carruthers (1991). [8] e.g. Greig (1983, 1988*a*).
[9] Dickson (1995). [10] Greig (1996). [11] Greig (1997). [12] Hayfield and Greig (1989).

Table 4.1 Food plants other than fruits and nuts from selected sites, mostly pre-1500

	London	Leicester, Causeway Lane	Chester	Droitwich, Upwich	Beverley, Eastgate	Newcastle, Mansion House	Bristol, Redcliffe Backs	Hull, Mytongate, Qn. St.	Paisley Abbey	Worcester, Sidbury	Cowick Moat, Yorks.	Shrewsbury Abbey
Stone pine (*Pinus pinea*)												x
Opium poppy (*Papaver somniferum*)	x	x	x						x			
Garden orache? (*Atriplex* cf. *hortensis*)							x					
Beet (*Beta vulgaris*)							x					
Monks rhubarb (*Rumex pseudoalpinus*)									x			
Horseradish? (cf. *Armoracia rusticana*)									x			
Cabbages, etc. (*Brassica* cf. *oleracea/napus*)										x		
Black mustard (*Brassica* cf. *nigra*)	x	x								x		x
Brassica spp. and *Brassica sinapis*	x	x	x		x		x	x	x			x
Bean (*Vicia faba*)	x	x	x		x		x				x	x
Lentil? (cf. *Lens culinaris*)	x											
Pea (*Pisum sativum*)		x		x		x	x					
Flax (*Linum usitatissimum*)	x	x	x		x	x	x			x		
Chervil (*Chaeorphyllum aureum* and *C.* sp.)	x									x		
Coriander (*Coriandrum sativum*)	x									x		x
Alexanders (*Smyrnium olustratum*)	x											
Fennel (*Foeniculum vulgare*)	x		x		x		x			x		x
Dill (*Anethum graveolens*)	x				x	x	x					x
Celery (*Apium graveolens*)	x			x	x		x					x
Parsley (*Petroselinum crispum*)							x					
Carrot (wild ?) (*Daucus carota*)	x		x		x		x	x		x		
Borage (*Borago* sp.) [pollen]										x		
Vervain (*Verbena officinalis*)								x				
Hyssop (*Hyssopus officinalis*)												x
Marjoram (*Origanum vulgare*)								x				
Pennyroyal? (*Mentha* cf. *pulegium*)												x
Mint (cf. *Mentha* spp.)	x				x							
Mace (*Myristica fragrans*)									x			
Pot marigold (*Calendula* sp.)								x				
Garlic (*Allium sativum*)					x							
Leek (*Allium porrum*)		x	x									
Leek/onion/garlic (*Allium* sp.) [epidermis]			x						x			
Rivet wheat (*Triticum* cf. *turgidum*)		x		x		x	x					
Bread wheat (*Triticum aestivum*)	x	x	x		x	x	x					
Wheat (*Triticum* spp.)	x	x	x		x							x
Rye (*Secale cereale*)	x	x	x	x	x	x	x					x
Barley (*Hordeum vulgare*)	x	x	x	x	x	x	x		x			
Common oat (*Avena sativa* and *Avena* sp.)	x	x	x		x	x	x			x		x
Bristle oat (*Avena strigosa*)									x			
Cereal bran			x				x		x	x	x	x

Notes: The plant remains were preserved mainly by anoxic wet conditions, though some were mineral replaced. Most of the legumes and cereal remains (except bran) are charred.

x indicates presence. Rivet wheat includes *Triticum* cf. *turgidum* or *T. turgidum/durum*.

Table 4.1 *Continued*

Table Sources: London: tenth to fifteenth centuries, various sites, mostly pits and occupation layers, Jones, Straker, and Davis (1991); Giorgi (1997). Leicester, Causeway Lane: eleventh to mid-thirteenth centuries, pits and cesspits, Monkton (1999). Chester, 12 Watergate Street: mid-thirteenth century, rock-pit, probably with sewage, Greig (1988*b*). Droitwich, Upwich: thirteenth century, layers associated with brine pit reconstruction and late/post-medieval layer associated with brine well repair, Greig (1997). Eastgate, Beverley: twelfth to fourteenth centuries, cesspits and a layer, McKenna (1992). Newcastle, Mansion House: thirteenth to fifteenth centuries, waterfront deposits, Huntley (1995). Bristol, Redcliffe Backs: fourteenth century, waterfront deposits including sewage, Jones (2000). Hull, Mytongate and Queen Street: fourteenth- and fifteenth-century pits, cesspits, and layers, Miller, Williams, and Kenward (1993); McKenna (1993). Paisley Abbey: fifteenth-century drains, Dickson (1996). Worcester, Sidbury: fifteenth century, barrel latrine, Greig (1981). Cowick Moat, Yorkshire: late medieval, moat fill, Hayfield and Greig (1989). Shrewsbury Abbey, Queen Anne House: twelfth to sixteenth centuries, rubbish deposits and cesspit, Greig (2002).

are preserved. Cereals are annuals in the grass family which produce large seeds and a high yield of energy per unit of land. Wheat, rye, barley, and oats were the main cereals grown in medieval Britain. There is archaeobotanical evidence for two species of wheat and two species of oats, as well as at least two recognizable types of barley. Their use for food and drink, in bread, pottage, ale, and beer, has been discussed in Chapter 2, but grain was not the only important product. Cereal straw was used for animal fodder, bedding (both human and animal), building materials (daub, flooring, thatch, and insulation), and temper for ceramics, as well as for fuel or tinder.

It is important to recognize the uses of cereal straw and chaff because much of what we know about the species of cereals grown in the medieval period is based on the chaff remains, which can often be identified to species, whereas grains can often only be determined to genus. DNA studies may eventually broaden the range of characteristics that can be determined, but these studies are in their infancy and have so far been applied mainly to issues relating to the origins and distribution of domesticated crops.[13]

Certain characteristics of a crop will influence a farmer's decision about what to grow. Yield is one consideration; the quality of the grain may be another. The height and strength of straw may determine a cereal's suitability for thatch. The season of sowing may be important in terms of labour availability or crop rotation. Time taken to mature may determine suitability for sowing as a mix with another crop. Awns—the spiny 'beard' that protects the grain-sheath—provide important protection from birds, especially for free-threshing cereals such as bread wheat, rivet wheat, and rye, but they are also a nuisance at threshing time and can cause choking in animals if eaten in fodder.

Characteristics attributed to certain cereals, such as hardiness and suitability for soil type, are sometimes used in interpreting archaeobotanical material. Knowledge of these characteristics is based on modern cereals and needs to be used with this appreciation. Generally, however, where a particular attribute (such as tolerance of poor soils) is true of the whole species, and not just particular varieties, it is likely that this will have changed little through time. Table 4.3 lists the cereals so far identified from medieval archaeological sites in

[13] Brown (2001).

Table 4.2. Fruits and nuts from selected sites, mostly pre-1500

	London	Leicester, Causeway Lane	Chester	Droitwich, Upwich	Beverley, Eastgate	Newcastle, Mansion House	Bristol, Redcliffe Backs	Hull, Mytongate, Qn. St.	Paisley Abbey	Worcester, Sidbury	Cowick Moat, Yorks.	Shrewsbury Abbey
Black mulberry (*Morus nigra*)	x				x							
Fig (*Ficus carica*)	x	x	x	x	x	x	x	x	x	x	x	x
Walnut (*Juglans regia*)	x				x	x	x		x			x
Hazelnut (*Corylus avellana*)	x	x	x		x	x	x			x		x
Bilberry (*Vaccineum myrtillus*)			x						x	x		x
Gooseberry (*Ribes uva-crispa*)			x							x		x
Bramble (*Rubus fruticosus*)		x	x		x	x	x	x	x	x		
Raspberry (*Rubus idaeus* and *R.* cf. *idaeus*)					x		x		x			
Bramble/raspberry (*R. fruticosus/idaeus*)	x											
Dewberry (*Rubus caesius*)					x							
Strawberry (*Fragaria vesca*)	x		x	x	x		x			x		x
Dog rose-hip (*Rosa* cf. *canina*)	x											
Rose-hip (*Rosa* spp.)	x		x		x					x		
Peach (*Prunus persica*)					x							
Almond (*Prunus dulcis*)												x
Sloe (*Prunus spinosa*)	x	x	x		x		x			x		x
Bullace/damson (*Prunus domestica* spp.)		x	x				x		x	x		
Primitive plum (*Prunus domestica*)	x		x		x		x					x
Wild cherry (*Prunus avium*)	x						x					
Sour/morello cherry (*Prunus cerasius*)	x		x							x		x
Plum/cherry (*Prunus* sp.)	x				x	x						
Pear (*Pyrus communis*)			x									
Pear/quince (*Pyrus/Cydonia*)			x				x			x		x
Pear/apple (*Pyrus/Malus*)	x	x				x						
Apple (*Malus sylvestris*)		x	x	x	x		x		x	x	x	x
Rowan (*Sorbus aucuparia*)					x							
Whitebeam (*Sorbus aria* agg.)			x									
Wild service (*Sorbus torminalis*)			x									
Service tree (*Sorbus* sp.)										x		
Medlar (*Mespilus germanica*)												x
Hawthorn (*Crataegus monogyna*)	x		x		x		x					
Grape (*Vinis vitifera*)	x	x			x	x	x			x	x	x
Orange/lemon etc. (*Citrus* sp.)	x											
Olive (*Olea europaea*)						x						

Notes: As Table 4.1. Bullace/damson includes bullace and damson types (*Prunus domestica* subsp. *insititia*).
Sources: As Table 4.1.

Britain and gives a brief note of their economic characteristics;[14] Table 4.1 shows their presence on twelve sites.

Medieval cereal fields would have looked very different from their modern counterparts. Plant breeding and the setting of standards for the characteristics of variety and uniformity are very recent developments. Medieval cereals would have been much more genetically diverse, so that a single field—even of a single crop—might show (for example) variations in height, time of flowering, resistance to disease, and colour. Despite this internal diversity, there would still have been different races with characteristics in common that farmers would recognize: these are known today as landraces. Landraces offer a diversity of characteristics within a single crop, which reduces the risk of serious crop failure in unfavourable conditions. Under optimum conditions, this is generally at the expense of maximizing yields; but for farmers in traditional agricultural societies, the trade-off is well worthwhile, as some harvest is considerably better than none. This is also one of the main reasons for growing maslins and other mixed crops. A few late medieval and post-medieval historical records refer to particular cereal characterstics, such as season of sowing, ear shape, or grain colour,[15] which add to the picture of the diversity of medieval cereals, but cannot be correlated with varieties that are currently known.

Medieval fields can also be seen to have supported a considerable diversity of arable weeds. Many of these will have competed with the crop for water, light, and nutrients. Some, however, especially at field margins, may have been tolerated by farmers as a minor food source in their own right, especially for 'greens'. The various species of fat hen and goosefoot (*Chenopodium* spp.), for example, have edible leaves, and recent experimental work suggests the collection or cultivation of fat hen (*Chenopodium album*) in the late prehistoric period.[16] This may also have occurred in the medieval period, especially in times of food shortage.

Wheats Bread wheat is the commonest wheat found in the medieval period. It is found on sites throughout England and Wales and into lowland and coastal Scotland. Grain colour and flour quality vary greatly with different varieties as do milling and baking properties. Some bread wheats produce very hard grains with desirable bread-making qualities, while some have softer grains. Modern millers produce different flours from a mix of different types of grain depending on the properties desired. Other variable characteristics include the colour of the chaff, presence or absence of awns, yield, strength and length of the straw, and the length of the ear. Very short-eared bread wheats are called club wheats.

Rivet wheat (also known as cone, polled, or poulard wheat) and durum wheat (also called macaroni wheat) are biologically one species, but are sufficiently

[14] See also Chapter 2, especially Table 2.1.
[15] e.g. Skeat (1882); Plot (1686; 1705).
[16] Stokes and Rowley-Conwy (2002).

Table 4.3. Cereals found on archaeological sites in medieval Britain

Common name	Botanical name	Crop requirements	Notes
Rivet wheat	*Triticum turgidum*	Best on rich, well-drained soils. Can tolerate some cool conditions but needs good summers to ripen well. Winter sown (in Britain) and requires a long growing season.	Tall strong straw. Most varieties have long strong awns that discourage birds. Poorer flour for bread making than bread wheat. Resistant to smuts and other diseases.
Bread wheat	*Triticum aestivum*	Best on heavy and rich soils. Usually winter sown although some varieties can be spring sown.	The most variable and versatile wheat species. Some modern bread wheats have been bred to grow in fairly extreme conditions, but generally traditional varieties yield less and are less reliable than barley or oats on poor soils or cool uplands. Different varieties of wheat grow to different heights, but generally traditional varieties are taller than modern hybrids, most of which have been bred for short straw. Some varieties are awned and some not, but are generally not as well protected against birds as rivet wheat.
Rye	*Secale cereale*	Drought tolerant due to a deep root system and therefore often grown on sandy soils. Primarily winter sown, though spring-sown varieties exist.	Tall straw and generally with strong awns. The only cereal discussed here which is outcrossing and wind pollinated rather than self-pollinated (self-fertile). The most prone to infection with ergot (*Claviceps purpurea*), but otherwise fairly disease resistant.
Barley	*Hordeum vulgare*	Generally tolerates poorer soils than wheat and can have a shorter growing season. Winter-sown and spring-sown varieties grown.	Generally a weaker straw than wheat and rye and often shorter as well. Usually with strong awns. Both six-rowed and two-rowed varieties have been found on medieval sites. Naked (free-threshing) varieties of barley exist but have not so far been reported from medieval Britain.
Common oat	*Avena sativa*	Tolerates poor soils and short season growing conditions. Both spring- and winter-sown varieties exist.	The weakest straw but also the most nutritious if used for fodder. The grain is also the most nutritious for humans in terms of protein and calorific value, but oats yield less than other cereals grown in good conditions. Archaeological finds of naked oats have not been reported in medieval Britain, but would also be very difficult to identify.
Bristle oat	*Avena strigosa*	Tolerates the poorest conditions of all.	Small grained and low yielding, generally grown where no other crop is worthwhile.

distinct from each other in terms of agricultural characteristics, grain, and flour quality often to be discussed as different species, though in fact they interbreed. Neither rivet wheat nor durum wheat is grown in Britain today. Durum wheat, grown in warmer climates, is now used for making pasta and biscuits. Archaeobotanically it is not usually possible to separate rivet wheat from durum wheat unless whole ears are present. Archaeobotanical remains, therefore, are usually identified as being possibly of either, though whole ears recovered from soot-covered thatch confirm the presence of rivet wheat in Britain and have so far failed to show any evidence of durum wheat.[17] Rivet wheat is more likely in Britain on ecological grounds, as it is more tolerant of cool conditions than durum wheat, but it is close to its northern limit.

Rivet/durum wheat has so far been found mainly south of a line roughly from Chester to Ipswich, both in charred assemblages and in thatch.[18] It appears to have been a less common crop than bread wheat, but nevertheless was found across southern England throughout the medieval period and later. Until recently there was no well-dated evidence for it before the Conquest, but a radio-carbon date now suggests that it may have been present in the Late Saxon period.[19] Rivet wheat is often found mixed with bread wheat, though medieval cereal assemblages are so often of mixed cereals that it is difficult to say whether these two wheats were frequently grown together or merely processed, handled, and disposed of similarly.

Rye Rye is a very resilient crop because of its extensive root system. The baking qualities of rye flour are poorer than those of bread wheat, but mixing the flour with bread wheat flour will make a lighter loaf. Rye is more difficult for farmers to manipulate in terms of selection for improving the following year's crop because it is outcrossing. A farmer who selects particular ears for seed because the plant has desired characteristics will find that these characteristics do not necessarily come true.

Barley Barley with three fertile flowers at each node of the ear is called six-rowed barley, referring to the appearance of six rows of grains when the ear is viewed down its length from the top. In varieties where the ear is long and lax it may appear that there are only four rows of grains. This four-rowed barley was the type called *bere* or *bigge*. In many varieties only one of the flowers at each node is fertile, however, and this is called two-rowed barley. Most modern malting barleys are two-rowed, because these types are lower in protein and higher yielding than the six- or four-rowed types. The six-rowed type, however, is found most commonly on medieval sites, with two-rowed types identified much less frequently.

[17] Letts (1999: 35).
[18] Letts (1999); Moffett (1991*a*).
[19] Moffett (in preparation).

Barley has both hulled and naked forms. The hulled form appears to have been the one grown in the medieval period, as it still is in Britain and on the Continent. Naked barley is often very loosely held in the hulls making it susceptible to birds and fungal diseases. The tightly adhering hulls of hulled barley mean that it has to be processed to remove them before it can be used for human consumption. One means by which this was done traditionally was by pounding the grain in a mortar and pestle, a process known as hummelling.[20] Loose milling with the millstones set a grain's width apart will also remove the hulls. This processing, however, is not necessary for barley used for malting. Hulled barley and hulled oats are also better digested by horses if crushed to break the hulls.

Oats Oats are common from medieval archaeological sites, despite the fact that oats yield less than other cereals when grown on good soils and are more prone to lodging (a weakness at the base of the stem causing them to fall over). Archaeobotanical and historical evidence both confirm that oats were not restricted to areas of poorer soil. It is not always possible to tell whether the oat found in an archaeobotanical assemblage was a crop or a weed. Cultivated oat can only be distinguished from wild oat with certainty by the chaff parts, which are fragile and survive poorly. Large oat grains are often attributed to cultivated common oat, as the wild species tend to produce slightly smaller grains—though the majority of grains overlap in size between wild and cultivated oat. Wild oats (*Avena fatua* and *A. sterilis*) are successful crop weeds and may even have been tolerated by farmers in the past, especially when their crop yield might otherwise have been poor.

Bristle oat is a small-grained but very hardy cultivated oat. It is seldom grown except in conditions where nothing else will yield. Finds have been made on medieval sites in Scotland,[21] but this oat was not confined to the north: it has also been identified in medieval Stafford.[22] It is not entirely clear from archaeobotanical evidence whether bristle oat was a crop in its own right, or whether, especially in southern areas, it was a contaminant of common oat.

Mixed cereal crops Mixed crops, such as maslin (winter wheat and rye), dredge (spring barley and oat), and mixtil (winter wheat and barley), are frequently referred to in medieval documents, but rarely identified archaeobotanically, since it is generally difficult to say that the seeds from two crops found together were also grown together. Part of the reason for sowing mixed crops is that it buffers the risk. Should one crop fail or do poorly, the other may still give a decent yield. It has been argued that proportions of crop grains cannot be used to identify mixed crops since, even if they were sown at a ratio of 1 : 1 (which would not necessarily be the case), they were unlikely to yield at 1 : 1.[23] Ethnographic evidence from Greece has shown that post-harvest

[20] Fenton (1978). [21] Boyd (1986). [22] Moffett (1988). [23] Veen (1995).

crop-processing can also affect the ratios of the different cereals in a mixed crop.[24] Statistical analysis of the weeds associated with crops may determine whether the crops were grown together and, in a case study of prehistoric sites in north-east England, it has been demonstrated that two crops (emmer and hulled barley) were *not* grown together.[25] The difficulty of this approach is that it relies on different pure crops being grown under sufficiently different environmental circumstances to have different weed floras, and it can only demonstrate the negative—that two crops found together were not grown as a mixed crop.

Interpretations Seed assemblages are often very difficult to interpret. Typically an assemblage might include two or more cereal crops, some arable weed seeds, and a few fragments of cereal chaff. Sometimes one cereal clearly predominates, but often this is not the case. Occasionally a few legume seeds may be present. Although not all medieval assemblages look like this, this type of assemblage is frequent enough to present a challenge to archaeobotanists trying to reconstruct crop-related human activities from cereal remains.

Despite these difficulties there are a few cases where an assemblage has been thought possibly to have been deliberately sown as a mixed crop. Maslin is believed to have been found at St Mary's Priory, Coventry, where a fourteenth-century pit sample produced mainly grains of free-threshing wheat and rye in roughly equal amounts.[26] Small numbers of wheat and rye chaff fragments and weed seeds were also identified, and it was noteworthy that the wheat chaff suggested that both bread wheat and rivet wheat were present. Less certainly, dredge may be the crop in the malting kiln at Burton Dassett, a fifteenth-century deserted village in Warwickshire. Grains of barley and oats were present in roughly equal numbers and both cereals were identified as germinated. They were presumably charred in the process of roasting malt. Wheat was also present in the kiln, however, though in a smaller amount and none of the wheat grains could be identified as germinated. Preservation was poor and many cereal grains were unidentified. The assemblage is open to various interpretations: it is not known whether the cereal grains were all burned in the same firing, or whether they represent an accumulation resulting from several episodes of the kiln's use.[27]

Drying grain before storage has often been suggested as one way in which a number of different crops might accumulate and char in the same place—if grains leaked into the hotter portions of the drying kiln and if the kiln was not cleaned out between each crop. The grain would not necessarily need to be fully cleaned at this stage as any remaining contaminants such as weed seeds, bits of

[24] Jones and Halstead (1995). [25] Veen (1995). [26] Carruthers (2003).
[27] Moffett (1991*b*).

chaff, and small grit might be cleaned by hand just before the grain was used. It is far from clear, however, that farmers in late medieval England did ever dry their grain in kilns. Drying grain for storage is only necessary if the crop is damp or unripe. The labour involved is very considerable and it seems unlikely that farmers would dry their harvest in a kiln if air-drying either in the field or in a barn would suffice. Late medieval documents, though they make frequent reference to malting kilns, do not mention corn dryers.[28] Drying grain before milling, however, greatly improves both the ease with which the grain can be milled and the yield of flour. It is also said to improve the flavour. Grain drying before milling is known to have been practised in the wetter parts of the British Isles in the eighteenth to early twentieth centuries, especially after wet harvests.[29]

It has been suggested that some of the more mixed medieval assemblages, with many weeds, may be derived not from mixed crops but from thatch.[30] Although straw is threshed before being used for thatch, there are always a few grains remaining in the ears, and many weed plants have been found included in medieval thatch. If grains and weeds fell out of the thatch, they might be swept into the household hearth and there become charred. Any bits of straw that fell would likewise be swept into the hearth, but would be less likely to survive than the grains and dense weed seeds.

Malt The roasting of malt is another occasion on which grain is exposed to fire and the risk of charring. Malt is made from germinated grain, which is then lightly roasted to stop the sprouts from growing any further, but not heated high enough to kill the enzymes which are essential in brewing to convert the grain starch to sugar. The wet grain is heaped on a warm (generally heated) floor and turned regularly so that germination takes place evenly. Germinated grain with the sprouts (coleoptiles) still attached is distinctive even when charred. In theory, malted grain that had become accidentally charred should be easy to detect. Identification, however, is less straightforward. The coleoptiles detach easily from charred grain and if these are missing or broken it is impossible, from comparing the length of the coleoptiles, to tell whether the germination was even. In very poorly preserved charred grain, it may be impossible to determine whether the grain had sprouted or not. Medieval maltsters in any case may not have been overly concerned about the evenness of germination. Brewers may also have added unmalted grain to provide extra starch.[31] Malting and brewing were commonplace activities and, to some brewers at least, cheapness of production may have been more important than consistency and quality in the final product.

The twelfth- to thirteenth-century settlement associated with Boteler's Castle, near Alcester, had a malting kiln, which mainly contained oats and wheat, rather less rye, and hardly any barley.[32] Oats were insignificant in other features on the

[28] C. C. Dyer, personal communication. [29] Bowie (1979).
[30] De Moulins (forthcoming). [31] Corran (1975). [32] Moffett (1997).

site, where wheat was common, suggesting that oats may have been associated very specifically with the function of the kiln. This combination of oat and wheat is found widely in the samples from the suburb at St Mary's Gate, Derby, with some grains from one pit having germinated.[33] The 'corn dryer' had little in it, however, apart from arable weeds. Two possible malting kilns, one of fourteenth- and the other of fifteenth-century date, at Dean Court Farm at Cumnor, both failed to produce any convincing evidence for malting.[34] Perhaps charred grains should not always be expected in malting kilns if the malt roasting was done with care or the kiln was cleaned regularly. The cereals found in the kilns tend to confirm that each might be used for malting, either singly or in combination, much as the historical documentation for the later medieval period reviewed in Chapter 2 confirms.

Legumes

Legume crops occur widely in charred assemblages, though usually in small amounts (Table 4.1). Peas, beans, and vetch seem to have been exposed to fire far less frequently than cereals, though part of the reason for their lesser abundance may also be simply that they were produced in less quantity. Unless they are well preserved, the difference between peas and beans is not always easy to discern. Medieval beans were small and round, and once the external features of the seed (hilum and testa) are lost, the difference in shape and size is not always sufficient to separate them from peas.

Vetch (*Vicia sativa* subsp. *sativa*) was normally cultivated as a fodder crop that was probably employed as human food only in times of serious hardship. Like many legumes, the seeds are bitter and contain neurotoxins. Modern studies on pigs and chickens suggest that consuming too much vetch is likely to have ill effects.[35] Seeds have been found at several sites. The seeds are in general smaller than peas and beans, but overlap in size with the closely related wild subspecies (*Vicia sativa* subsp. *angustifolia*). Only seeds that are significantly larger than wild vetch can be reliably identified as cultivated.

Vegetables, herbs, fruit, and nuts

If cereal assemblages are often difficult to interpret, the frequency with which cereals appear to have been exposed to fire at least ensures their relative ubiquity and often their abundance. Remains of other food plants, however, seem to have been rarely exposed to fire. They are more likely to be found in wet, anoxic deposits, more rarely as mineral-replaced remains.

Vegetables, and some herbs, are particularly difficult to identify as these are often closely related to wild plants. Carrot, parsnip, celery, and the

[33] Monckton (2003). [34] Moffett (1994). [35] Enneking (accessed May 2004).

relatives of cabbage, for example, all grow wild in Britain in non-edible forms, but are the same species as the edible cultivated forms. The seeds of the wild and cultivated forms are indistinguishable. Seeds are the parts of vegetables most likely to survive, but vegetables grown for food are seldom allowed to run to seed beyond that needed for the next year's crop. Estate accounts suggest that seed was often purchased,[36] so it may be that gardeners growing vegetables for large landowners did not allow many of their vegetables to mature to provide seed for the following year. Seed remains of vegetables can therefore be expected to be rare and are often likely to be derived from wild relatives rather than crops.

Vegetative remains of vegetables are unusual, but finds have been made of epidermis of leaf tissue fragments from the onion family at York[37] and Chester.[38] It has not been possible to separate leek, onion, and garlic from the leaf tissue remains, but at Chester charred leek seeds were also found, and garlic cloves were found at Beverley.[39]

Seeds of herbs—where the seed was the part used—are found fairly often but seldom in abundance (Table 4.1). Dill, fennel, coriander, celery, black mustard, and opium poppy are not infrequent in cesspits. Opium poppy seeds may have been used as a flavouring for bread and cakes in addition to the use of the latex from the seed heads for medicine. Other herbs, such as savory, hyssop, and marjoram, are rarer.

Fruit stones and nutshells are much more frequently found than herbs and vegetables (Table 4.2). Being robust they survive well in anoxic deposits. Different types of primitive plums such as bullace, damson, and gages, as well as the wild sloes, can be distinguished from each other. Much depends on how well preserved the material is and how 'typical' the plum stones are. Plums all interbreed, so the form of their stones is not always distinctive. Stones of the larger, domesticated plums are found much less frequently. Walnut, wild and morello cherries, pear, apple, strawberry, grape, and fig are common. Peach, mulberry, gooseberry, medlar, and raspberry are less common, while the stone pine and almond found at Shrewsbury Abbey,[40] the olive stone from Newcastle,[41] and the single *Citrus* seed from Trig Lane in London[42] are very unusual for this period. Wild fruits and nuts would of course have been collected, including bilberry, bramble, hazel, rose-hips, sloe, rowan and service berries, and hawthorn.

Stone pine, olive, almond, *Citrus*, mace, and probably most of the figs and grapes would have been imported, the grapes in dried form as raisins or currants. Dried or preserved peaches and walnuts may also have been imported depending on the demand, as peaches in particular are labour intensive to produce in England.

[36] See also Chapter 3. [37] Tomlinson (1991). [38] Greig (1988*b*).
[39] McKenna (1992). [40] Greig (2002). [41] Huntley (1995). [42] Giorgi (1997).

Conclusion

Although the range of different food plants became greater in the post-medieval period,[43] even before 1500 the archaeological records confirm that some people had access to a substantial variety of food plants, including some imports. The twelve sites analysed in Tables 4.1 and 4.2 included some wealthy households, such as that which used the hunting lodge at Cowick Moat, as well as the abbeys; others in towns must have belonged to those of modest wealth. The finds from these last confirm not only that townsfolk ate the vegetables and fruits produced in their garden plots, discussed in Chapter 3, but also that they were able to buy vegetables, fruits (including dried fruits), and nuts in the local markets. Wheat, rye, barley, and oats were ubiquitous; both species and varieties, however, may have been more local. Rivet wheat, for example, was restricted by its growing range. Malt was made from all four grains.[44]

There are relatively few good synthetic studies of the archaeobotanical data which take into account the diverse range of contexts and other difficulties of interpretation. It is therefore not clear at present who had access to which foods and whether there were also regional or local differences as well as social differences in the use of these foodstuffs. A study of latrines in Amsterdam[45] has shown that using archaeobotanical remains to discern social differences is far from easy. The increasing amount of data available, especially for the late medieval period, offers considerable potential for integrated studies of archaeological and historical records of plants which may go beyond simple lists to a clearer appreciation of the role these plants played in people's lives.

[43] Greig (1996). [44] See also Chapter 2. [45] Paap (1984).

5

From *Cu* and *Sceap* to *Beffe* and *Motton*

The Management, Distribution, and Consumption of Cattle and Sheep in Medieval England

N. J. SYKES

Introduction

Throughout the medieval period cattle and sheep were, with pigs, the main meat-providing animals, but the value of their flesh was often outstripped by demand for their milk, manure, and traction or wool. Since cattle and sheep were often managed intensively for these secondary products, any study of beef and mutton consumption must be viewed against the backdrop of the medieval economy. Between the fifth and sixteenth centuries England witnessed considerable economic change which surely influenced dietary practices. As the documentary record does not cover the whole period or the activities of all social classes, animal bone studies provide one of the best opportunities to ascertain the scale of these influences. This chapter looks in detail at the archaeological evidence for cattle and sheep; Chapter 6 centres on the evidence for pigs, combining for the late medieval period in particular both historical and archaeological material; and Chapter 7 focuses on the late medieval historical record, especially for information that is more opaque in the archaeological material, such as the use of specialized meat products and offal, dietary change, quantities consumed, and some elements of dairy foods.

In this chapter, cattle and sheep representation, mortality profiles, and butchery patterns, based on a survey of over 300 archaeological assemblages, are examined (Table 5.1).[1] Relative frequencies of the two species have been calculated from bone fragment counts, a quantification technique that allows comparison between different assemblages. As the remains of larger animals tend to be more fragmented than those of small species, however, this method

[1] The data are presented in Sykes (in press *a*); examples of the principal sites are referred to in this chapter. The remains of sheep are difficult to separate from those of goats: in this chapter, both species are referred to as 'sheep' unless otherwise stated.

Table 5.1. Numbers of assemblages for each period and each site type analysed in Figs. 5.1–5.7

	Date (in centuries)	5th–7th	7th–9th	9th–11th	11th–12th	12th–14th	14th–16th
Site type							
Rural		19	15	8	9	9	3
Urban			18	35	45	43	21
High-status			14	18	27	26	9

Source: Sykes (in press *a*).

often suggests cattle to be better represented in death than was the case in life, perhaps the source of the frequently cited disparity between the archaeological evidence and that derived from historical sources.[2] In addition, fragment counts take no account of differences in carcass size;[3] thus even where sheep are numerically more abundant than cattle, the dietary contribution of beef may have surpassed that of mutton.

Economic change

After *c.* 400 the urbanism and large-scale production of Roman Britain was replaced by a localized economy based on mixed agriculture. Most early medieval assemblages show similar frequencies of species to the Roman period, with cattle being marginally better represented than sheep (Fig. 5.1). Production of prime beef and mutton was clearly central to the animal husbandry regime, as most fifth- to seventh-century assemblages, such as West Stow[4] and Barton Court Farm,[5] contain numerous remains from choice cattle and sheep, slaughtered between six months and two years of age. Many assemblages also contain sizeable numbers of animals aged under six months which, if not natural fatalities, hint at dairy production, with juveniles slaughtered to release milk for human consumption. Calves and lambs continue to be present in similar frequencies on seventh- to mid-ninth-century rural settlements, such as Pennyland and Hartigans.[6] Large numbers have also been recovered from the monastic sites of Flixborough[7] and Green Shiel (Lindisfarne),[8] although here their presence has been linked to vellum, rather than dairy, production. Despite the presence of these young individuals, overall there is a shift in cattle and sheep slaughter patterns from the Early to the Mid-Saxon period, with greater numbers of animals maintained beyond three years of age. Husbandry regime changes of this kind have been noted on several sites that span the Early to Mid-Saxon transition, such as Quarrington[9] and Eynsham Abbey.[10] Their coincidence with

[2] A. Grant (1988: 151).

[3] Harvey (1993: 228) suggests a single cattle carcass would have yielded 308 lb of meat compared with 31 lb from that of a sheep.

[4] Crabtree (1989).

[5] Wilson (1986).

[6] Williams (1993).

[7] Loveluck (2001: 114).

[8] O'Sullivan (2001: 42).

[9] Rackham (2003: 265, 268).

[10] Mulville (2003: 348).

Fig. 5.1 Variation over time in the representation of cattle and sheep: each bar shows the relative percentage of assemblages in which cattle are better represented than sheep (cattle dominated) or vice versa (sheep dominated)

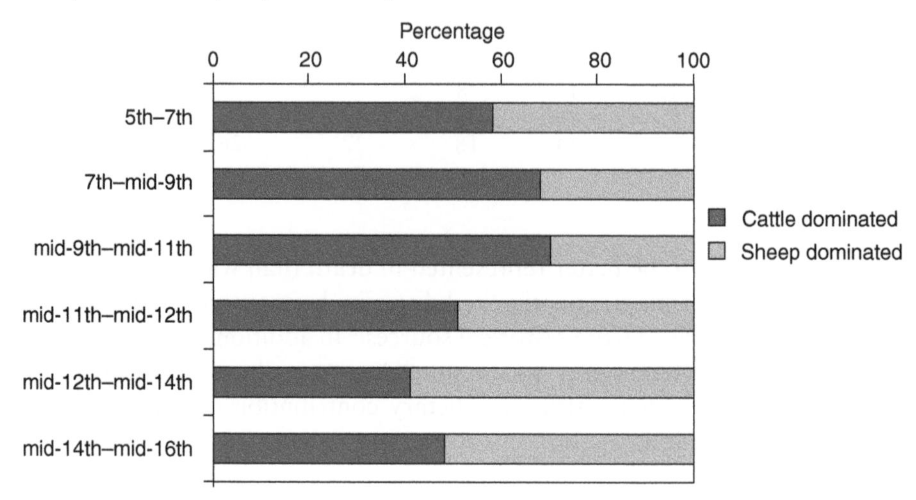

Note: For sample sizes, Table 5.1.
Source: Sykes (in press *a*).

the agricultural intensification of the Mid-Saxon period suggests that growing demands for traction, manure, and wool were satisfied at the expense of prime meat production.

Emphasis on agrarian output—and the need for draught cattle—increased throughout the Saxon period, and in most assemblages dating from the seventh to the mid-eleventh century, such as those from Walton, Aylesbury,[11] Flaxengate, Lincoln,[12] and the Cheddar palaces,[13] cattle are more numerous than sheep (Fig. 5.1). Further expansion of arable farming in the eleventh and twelfth centuries saw cattle kept to a greater age, presumably as plough animals, than in any preceding or succeeding period (Fig. 5.2). A grain-producing regime such as this would have struggled to maintain field fertility, and the need for high-quality manure may, in part, explain the post-Conquest rise in the frequency of sheep, whose dung has fertilizing qualities superior to that of other animals (Fig. 5.1).[14] Growth of the wool industry must, however, have been at the heart of this move towards sheep-farming. Wool production reached its height between the late twelfth and mid-fourteenth centuries,[15] and sheep—the majority of which were animals more than three years of age—are more frequent than cattle in most assemblages of this date (Figure 5.1).

Between the seventh and mid-twelfth centuries, the increasing number of calves and lambs raised to maturity would have restricted milk yields and may

[11] Noddle (1976). [12] O'Connor (1982). [13] Higgs and Greenwood (1979).
[14] McCormick (1991: 46). [15] Ryder (1983: 455–7).

Fig. 5.2 Percentage of cattle culled in each of seven age classes, mid-ninth to mid-eleventh and mid-eleventh to mid-twelfth centuries

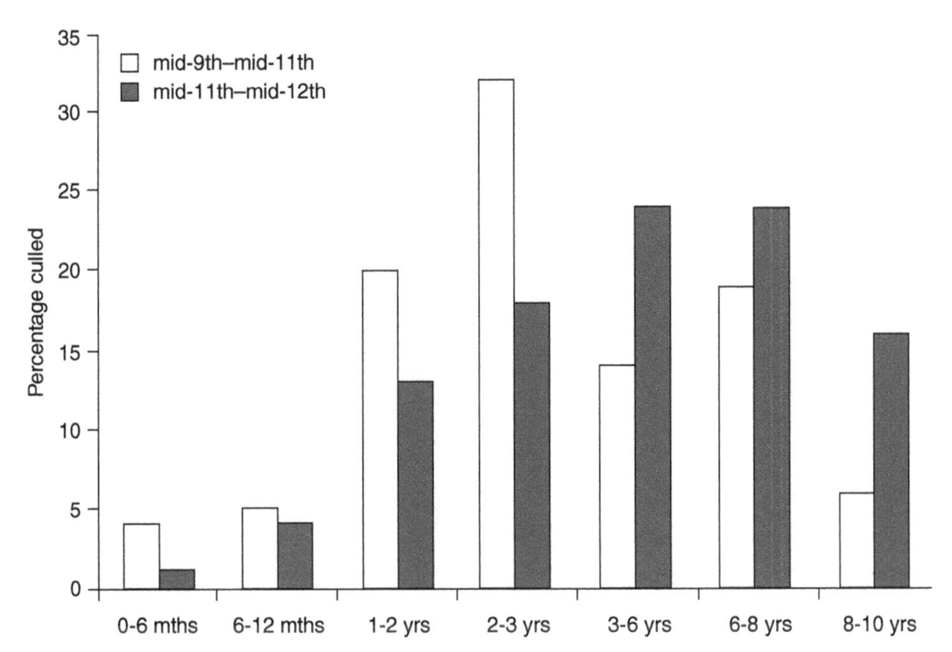

Note: For sample sizes, Table 5.1.
Source: Sykes (in press *a*).

have curbed the scale of dairy production, perhaps explaining the scarcity of dairy farms (*vaccaria*) mentioned in documentary sources such as Domesday Book.[16] Vaccaries appear more regularly in the historical record from the thirteenth century onwards[17] and a move towards specialized cattle dairying in the late medieval period is indicated by the documentary evidence outlined in Chapter 7. This is supported by the zooarchaeological evidence, with cattle cull patterns demonstrating an overall decrease in animal age (Fig. 5.3). Increased dairying was facilitated by the advent of the plough-horse, which gradually diminished the number of cattle required for traction, allowing more calves to be slaughtered for meat, in turn liberating milk for dairy production.[18] After the Black Death this economic regime became further defined, as population decline reduced the workforce for arable production, causing large swathes of land to be returned to pasture.[19] Cattle cull patterns for this period, which demonstrate a high frequency of both very young and very old animals (Fig. 5.3) reflect a situation where large herds of dairy cows were maintained to adulthood, while surplus bull-calves were slaughtered for meat. Growing demand for prime beef and veal in the late medieval period could explain a slight rise in the proportion

[16] Lennard (1959: 265). [17] Trow-Smith (1957: 107–8); Britnell (1993: 114).
[18] Langdon (1986); A. Grant (1988: 153–60); Albarella (1997: 22).
[19] Fryde (1996: 145); Britnell (1993: 156).

Fig. 5.3 Percentage of cattle culled in each of seven age classes, mid-twelfth to mid-fourteenth and mid-fourteenth to mid-sixteenth centuries

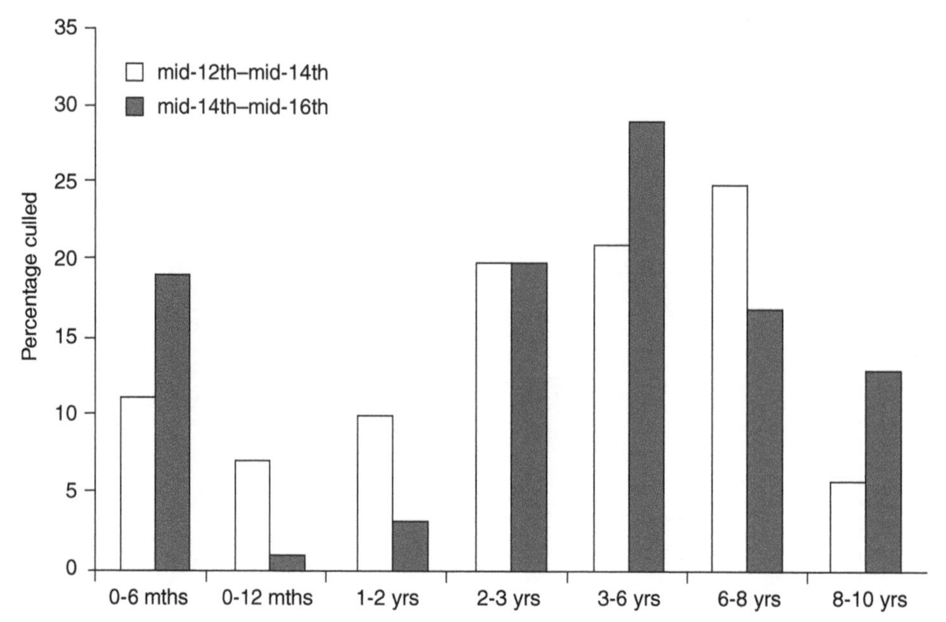

Note: For sample sizes, Table 5.1.
Source: Sykes (in press *a*).

of cattle (Fig. 5.1). It has been argued that this change reflects the murrain that swept the sheep flocks of the fourteenth and fifteenth centuries, reducing many by up to two-thirds.[20] Historical evidence indicates that wool prices soared as a result[21] and, with fewer sheep, farmers were encouraged to maintain their animals to a greater age than at earlier times (Fig. 5.4).

Social variation

Against this background of economic change it is possible to consider how patterns of beef and mutton consumption differed between social levels—whether practices consistently followed commercial trends or varied between groups. Historical evidence suggests that the upper echelons of medieval society ate more meat than the poor.[22] Animal bone studies cannot confirm this, but differences in meat consumption can be highlighted when the data are examined by site type.

Rural sites Zooarchaeological evidence is least abundant from farmsteads, hamlets, and villages, especially after the eleventh century. Assemblages from these settlements do, however, display some general characteristics relevant to all

[20] Lloyd (1977: 11); A. Grant (1988: 154).
[21] Grant (1988: 154); Ryder (1983: 455–7).
[22] Dyer (1983: 211); Salisbury (1994: 57).

Fig. 5.4 Percentage of sheep culled in each of eight age classes, mid-twelfth to mid-fourteenth and mid-fourteenth to mid-sixteenth centuries

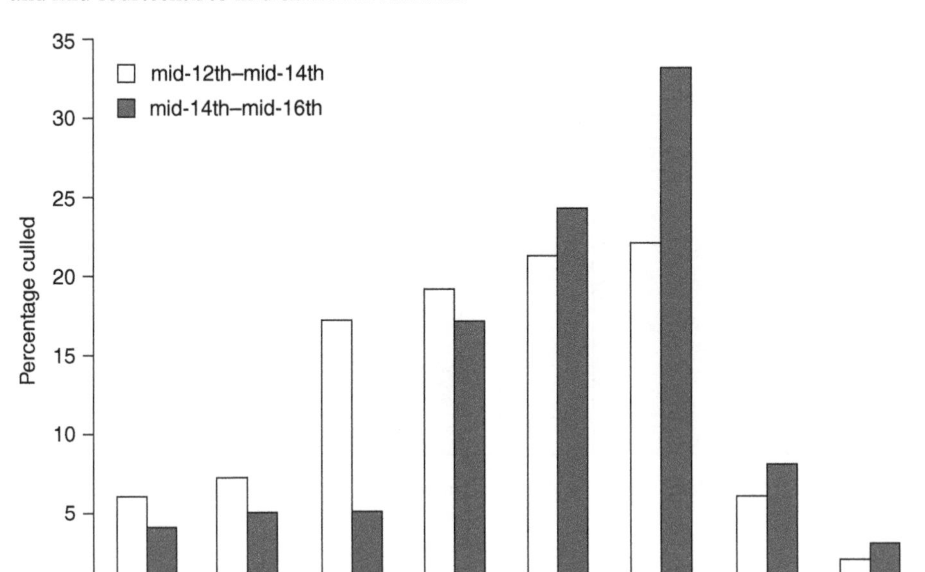

Note: For sample sizes, Table 5.1.
Source: Sykes (in press *a*).

periods. For instance, meat consumption within the rural environment was nearly always based more on mutton than in any other section of society (Fig. 5.5*a*). The assemblages from West Stow,[23] Wharram Percy,[24] and Marefair[25] all contain an abundance of sheep remains. Reasons for this are difficult to discern, but preferential export of cattle to consumer, in particular urban, sites would probably have left the rural population with comparatively more sheep to consume. Even though the peasant's diet was largely vegetarian,[26] when meat was consumed it derived predominantly from cattle and sheep: on average their remains account for 75 per cent to 88 per cent of the total assemblage from sites of this type (Table 5.2).

During the earliest part of the medieval period, production and consumption were closely linked.[27] The majority of fifth- and sixth-century sites considered in this survey (the most significant being West Stow[28] and Barton Court Farm[29]) were largely self-sufficient, with animals bred, butchered, and consumed on, or close to, the settlement. As society and settlement types diversified, the burden of feeding dependent urban and high-status populations fell upon the rural community. Mid- and Late Saxon farmers appear to have produced a surplus without adversely effecting their own practices of consumption: ageing data

[23] Crabtree (1990). [24] Stevens (1992). [25] Harman (1979*a*).
[26] Dyer (1983); Mennell (1985: 41). [27] Montanari (1999*a*: 168).
[28] Crabtree (1990). [29] Wilson (1984).

Fig. 5.5 Variation over time in the relative percentage of assemblages in which (*a*) sheep are better represented than cattle (sheep dominated) and (*b*) cattle are better represented than sheep (cattle dominated)

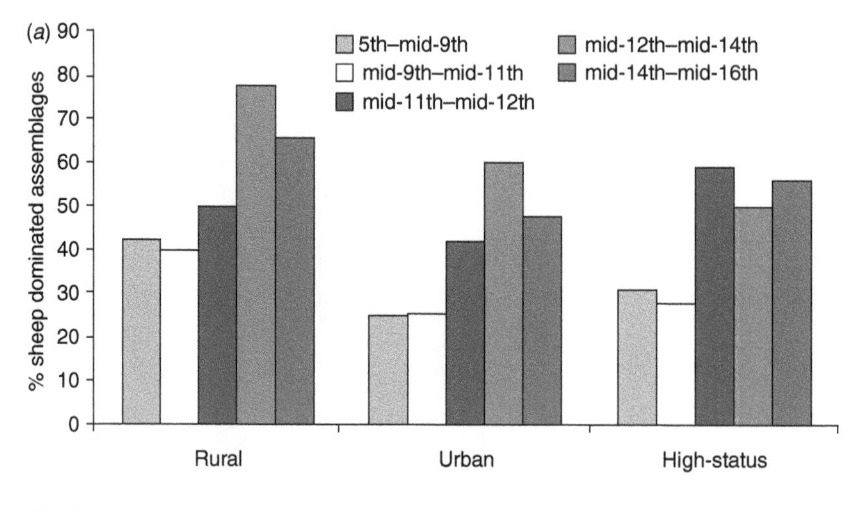

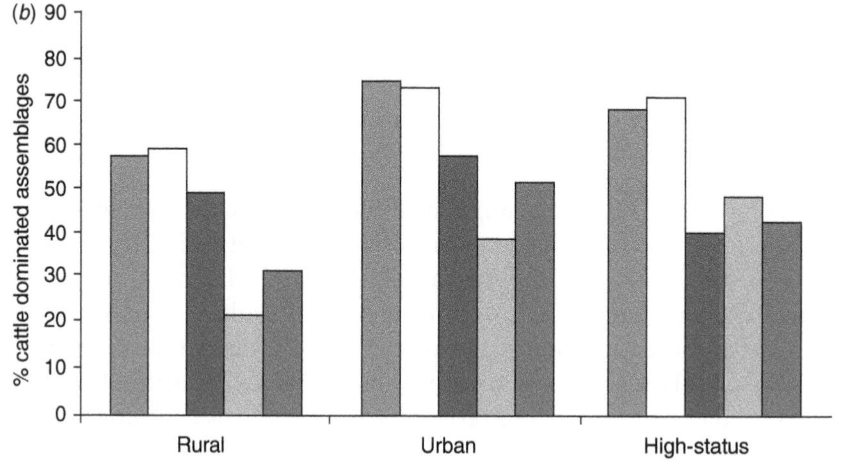

Note: For sample sizes, Table 5.1.
Source: Sykes (in press *a*).

indicated that the inhabitants of sites such as Yarnton[30] and Wharram Percy[31] continued to enjoy tender meat from lambs, calves, and prime-aged animals, sending their older stock for slaughter elsewhere. From the mid-eleventh century onwards, however, very young animals (those under six months of age) became less abundant on rural sites, such as West Cotton[32] and Marefair,[33] but better represented within assemblages from urban and high-status sites, such as Exeter[34] and Guildford Castle.[35] This suggests that growing consumer demand

[30] Mulville and Ayres (2004). [31] Stevens (1992). [32] Albarella and Davis (1994).
[33] Harman (1979*a*). [34] Maltby (1979). [35] Sykes (in press *b*).

Table 5.2. Average percentage contributions of cattle and sheep remains to vertebrate assemblages from sites of different type and date

	Rural	Urban	High-status
5th–7th century	80		
7th–mid-9th century	86	78	58
Mid-9th–mid-11th century	80	79	63
Mid-11th–mid-12th century	75	77	56
Mid-12th–mid-14th century	77	76	57
Mid-14th–mid-16th century	75	80	50

Note: For sample sizes, Table 5.1.
Source: Sykes (in press *a*).

for veal and lamb gradually induced farmers to market their surplus animals. Increasingly, the inhabitants of rural sites were left to eat up their old breeding stock; from the mid-fourteenth to the mid-sixteenth century, over 70 per cent of all the cattle and sheep they consumed were older than three years of age.

Urban sites Regardless of period, the majority of beef and mutton consumed within towns, cities, and *wics* would have been sent on the hoof from the rural hinterland,[36] confirmed by the general lack of evidence for on-site breeding in most urban assemblages. Practicalities of rural–urban provisioning probably explain the cattle-dominated character of most of these last (Fig. 5.5*b*). The species which provided the most meat per animal seem to have been favoured for town supply.[37]

Cattle outnumber sheep by two to one in many Mid-Saxon emporia, such as those from *Hamwic*[38] and Ipswich;[39] and it has been suggested that beef would have provided as much as 80 per cent of the meat consumed at Fishergate, York.[40] The restricted range of meats has been used to argue that the inhabitants of these early urban sites had no influence over what they received.[41] Lack of control has also been proposed based on the number of animals aged over six years within *wic* assemblages.[42] The bland composition of the meat diet in *wics* has frequently been cited as evidence that these Mid-Saxon emporia were maintained by a ruling elite, who forwarded to the towns lower-quality animals they received as food rent.[43] This theory finds little support from the zooarchaeological data. Compared with rural sites, the diversity of species in *wics* is less restricted (Table 5.2). Furthermore, ageing data indicate that prime-aged animals are better represented in *wic* assemblages than those from high-status sites (Fig. 5.6). The animals recovered from Ramsbury[44] were no younger than those from *Hamwic*, and many more prime cattle and sheep were present at Ipswich than the nearby high-status sites of Brandon and Wicken Bonhunt.[45] In

[36] O'Connor (1994: 145). [37] Zeder (1991: 38); Crabtree (1996: 64).
[38] Bourdillon and Coy (1980). [39] Jones and Serjeantson (1983).
[40] O'Connor (1994: 139). [41] Bourdillon (1994); Crabtree (1996); O'Connor (1994, 2001).
[42] Bourdillon (1994: 123). [43] O'Connor (2001). [44] Coy (1980: 47–9).
[45] Crabtree (1996: 65–7).

Fig. 5.6 Percentage of cattle culled in each of seven age classes, seventh to mid-ninth centuries

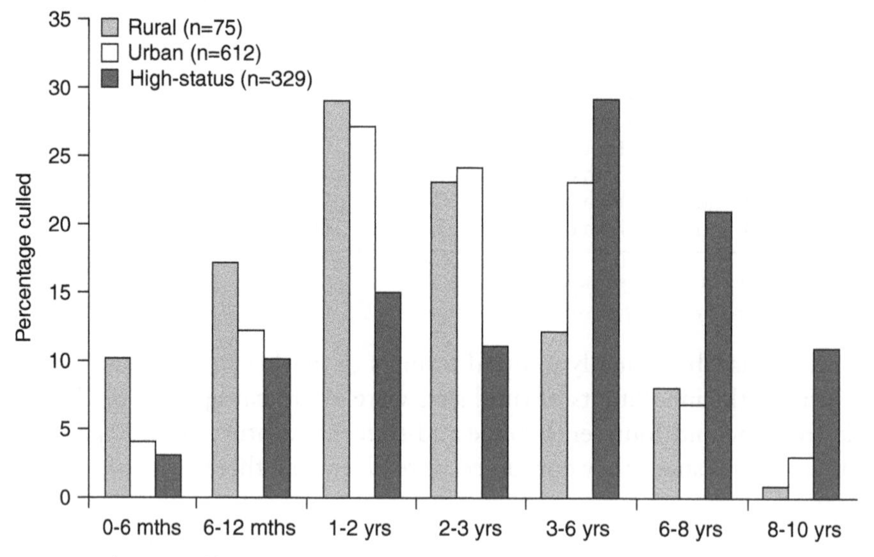

Note: For sample sizes, Table 5.1.
Source: Sykes (in press *a*).

the light of this evidence it might be argued that the urban population was provided with many of the better-quality animals, suggesting it had more control over its supplies than has previously been accepted. The low level of diversity of species, rather than representing lack of *control*, may simply reflect the lack of *choice* offered by an underdeveloped market.

Growth of commerce went some way to diversify the urban diet and from at least the late ninth century townsfolk were able to purchase varied meats from specialist traders, such as the fishermen and fowlers of Aelfric's *Colloquy*.[46] Although when aggregated the zooarchaeological data give little impression of this diversification (Table 5.2), assemblages from individual sites, such as Fishergate, York,[47] show a gradual proliferation in species other than cattle and sheep. As well as variety of foodstuffs, demands for tender meat were met increasingly from the mid-eleventh century. This is seen particularly at Flaxengate, Lincoln,[48] where lambs and calves are considerably more abundant in late eleventh- and twelfth-century deposits than in earlier phases. It was not, however, until after the Black Death, when the ratio of resources to people became high, that young animals, in particular calves, were sent to towns in large numbers. At first, consumer demand for veal did not conflict with the prevailing economy, since the burgeoning dairy industry produced a ready supply of surplus bull-calves. By the end of the medieval period, however,

[46] Garmonsway (1978). [47] O'Connor (1994: 145). [48] O'Connor (1982: 49).

urban demand was such that it appeared to threaten future stock, forcing the Parliament of 1532 to legislate against the sale of animals under two years of age.[49]

High-status sites While rural and urban assemblages demonstrate similar, economy-dictated, shifts in the frequency of cattle and sheep, assemblages from estate centres, religious houses, manor houses, and castles deviate from these trends (Fig. 5.5). This is particularly the case towards the later medieval period, suggesting that the diet of the elite was little influenced by the animal economy of the time. Distance from production appears to have been a mark of socio-economic status throughout the medieval period. As early as the seventh century it is evident that the social elite attempted to shun rural fare, consuming significantly less beef and mutton than any other section of society: in most cases almost half the animal bones deposited at high-status sites were from species other than cattle and sheep. The upper ranks of society were able to command greater variety in their diet, consuming larger quantities of pork,[50] poultry,[51] fish,[52] and game.[53]

Meat from animals central to the economy always seems to have been the least favoured, as not only did the elite eat less beef and mutton than other sectors of society but the ratio of beef to mutton also tended to differ from the norm. For the fifth to mid-ninth centuries this is apparent only when the data are viewed at a regional level (Fig. 5.7), when it can be seen that cattle are fewer in high-status assemblages from the north and east of the country, for instance, at Brandon[54] and the Saxon palaces at Northampton,[55] compared with rural and urban sites. By contrast, in the south and west of England, where sheep were generally more numerous, cattle were more common in all high-status assemblages, as at Ramsbury[56] and Portchester Castle.[57] Similar differences can be seen on high-status sites for the mid-twelfth to the mid-fourteenth centuries. In a period when sites across the country witness an increase in the frequency of sheep, high-status sites indicate a contrary trend (Fig. 5.5*a*). Conversely, in the centuries following the great murrain, when sheep populations were significantly lower and sheep ages significantly higher than the preceding period, the elite not only consumed more mutton[58] but also procured greater numbers of young and prime-meat animals than was the case in the mid-twelfth to mid-fourteenth centuries. In the mid-fourteenth to mid-sixteenth centuries, 5 per cent more individuals were slaughtered before four years of age than in the preceding period, a shift made more significant when viewed against the trends for rural and urban assemblages (Table 5.3).

[49] A. Grant (1988: 153). [50] See Chapter 6. [51] See Chapter 9. [52] See Chapter 8.
[53] See Chapter 11. [54] Crabtree (1996). [55] Harman (1985). [56] Coy (1980).
[57] Grant (1975).
[58] Increased mutton consumption is also indicated by the historical record: Woolgar (1999: 133 and table 8). See Chapter 7.

Fig. 5.7 Regional variation in the percentage of fifth- to mid-ninth-century assemblages in which cattle are better represented than sheep (cattle dominated)

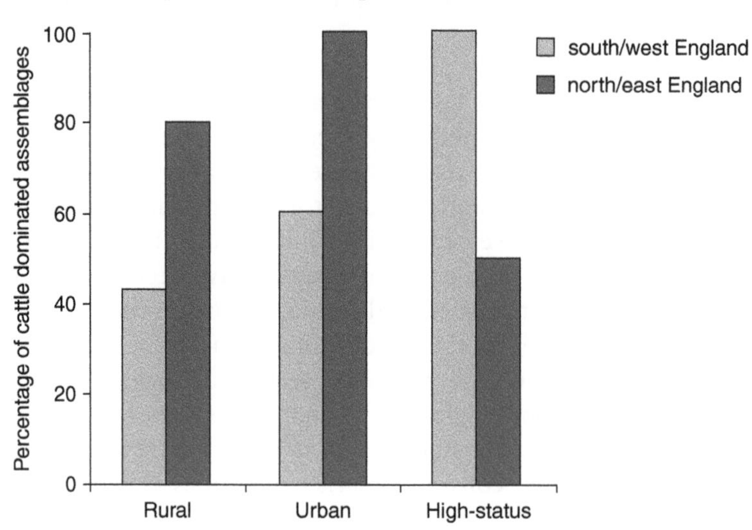

Note: For sample sizes, Table 5.1.
Source: Sykes (in press *a*).

If the upper echelons of society had control over their supplies, the ageing data for high-status sites, particularly for the earlier part of the period, are surprising. It is frequently stated that a wealthy diet is indicated by the remains of tender young animals,[59] but most of the cattle and sheep consumed at the Mid-Saxon settlements of Wicken Bonhunt,[60] Brandon,[61] and Lake End Road, Dorney,[62] were more than three years of age (Fig. 5.6). High-status sites dating to the Late Saxon period, such as Faccombe Netherton[63] and Eynsham Abbey,[64] also contain adult animals in abundance. Assumptions concerning meat values should be used cautiously, as there is no single preference applicable to all cultures in time and space; in many societies the fatty meat from mature animals is prized more highly than that from young individuals.[65] If, however, this was not the case in the early medieval period, the idea that the Saxon elite were consuming poor-quality beef and mutton is difficult to reconcile with current models of provisioning. It could suggest that lords and royalty had no choice over the animals they received in food rent or that they were content to take whatever they were given, especially since renders were frequently redistributed.[66] Redistribution through commensal hospitality, with the food rents eaten communally, may underlie the old age of the cattle and sheep represented in high-status assemblages. Communal consumption was central to the Anglo-Saxon elite, serving to

[59] Ashby (2002: 43); Grant (2002: 20). [60] Crabtree (1996: 65–7).
[61] Crabtree (1996: 65–7). [62] Powell (2002). [63] Sadler (1990).
[64] Mulville (2003); Ayres, Locker, and Serjeantson (2003). [65] Crabtree (1991: 174).
[66] Hagen (2002: 261–75).

Table 5.3. Inter-period and inter-site variation in the ages of (*a*) cattle and (*b*) sheep, shown as the relative percentage of mandibles in each age group

(*a*) Cattle	0–6 mths	6–12 mths	1–2 yrs	2–3 yrs	3–6 yrs	6–8 yrs	8–10 yrs
5th/6th century							
Rural (n=315)	14	18	15	19	9	13	2
7th–mid 9th century total%	**4**	**12**	**30**	**8**	**25**	**14**	**7**
Rural (n=66)	11	18	33	22	12	3	1
Urban (n=621)	4	13	26	24	22	8	3
High-status (n=338)	3	10	15	10	29	22	11
Mid-9th–mid-11th century total%	**4**	**5**	**20**	**31**	**15**	**19**	**6**
Rural (n=20)	25	40	10	5		15	5
Urban (n=406)	3	5	20	36	16	17	3
High-status (n=148)	4	1	10	21	15	24	15
Mid-11th–mid-12th century total%	**1**	**4**	**13**	**18**	**24**	**24**	**16**
Rural (n=12)			33	34	8	17	8
Urban (n=94)		8	17	23	29	22	1
High-status (n=204)	1	2	10	15	23	25	24
Mid-12th–mid-14th century total%	**11**	**7**	**10**	**20**	**21**	**25**	**6**
Rural (n=78)	4	22	5	16	15	37	1
Urban (n=168)	11	7	16	20	23	20	3
High-status (n=178)	14	2	7	22	21	24	10
Mid-14th–mid-16th century total%	**19**	**1**	**3**	**20**	**29**	**17**	**13**
Rural (n=20)	15		10	5	15	5	50
Urban (n=157)	26	1	2	17	26	23	5
High-status (n=117)	10	1	3	26	34	10	16

(*b*) Sheep	0–6 mths	6–12 mths	1–2 yrs	2–3 yrs	3–4 yrs	4–6 yrs	6–8 yrs	8–10 yrs
5th/6th century								
Rural (n=440)	10	30	22	8	15	13	1	1
7th–mid-9th century total%	**6**	**20**	**16**	**10**	**15**	**21**	**9**	**3**
Rural (n=232)	8	23	20	11	17	12	8	1
Urban (n=874)	2	14	20	22	14	21	6	1
High-status (n=402)	4	16	12	10	13	30	10	5
Mid 9th–mid-11th century total%	**6**	**10**	**20**	**20**	**20**	**16**	**7**	**1**
Rural (n=42)	10	24	21	14	17	14		
Urban (n=631)	6	9	23	22	21	15	3	1
High-status (n=272)	6	11	15	21	21	19	5	2
Mid-11th–mid-12th century total%	**5**	**6**	**24**	**18**	**16**	**19**	**9**	**3**
Rural (n=34)	3	9	18	29	17	12	6	6
Urban (n=639)	5	5	26	17	15	21	10	1
High-status (n=239)	7	8	20	18	17	14	8	8

Table 5.3. *Continued*

(*b*) Sheep	0–6 mths	6–12 mths	1–2 yrs	2–3 yrs	3–4 yrs	4–6 yrs	6–8 yrs	8–10 yrs
Mid-12th–mid-14th century total%	6	7	17	19	21	21	7	2
Rural (n=100)	1	16	22	15	16	18	11	1
Urban (n=845)	4	6	18	19	23	21	7	2
High-status (n=182)	12	6	8	20	16	26	8	3
Mid-14th–mid-16th century total%	4	5	5	17	23	35	8	3
Rural (n=54)		11	5	7	26	41	6	4
Urban (n=207)	4	3	3	13	19	45	10	3
High-status (n=198)	5	6	6	23	27	24	7	2

Note: For sample sizes, Table 5.1.
Source: Sykes (in press *a*).

define social and political relationships.[67] At a more practical level, prior to the existence of a true market system and in a period when pork was the only meat preserved regularly,[68] feasting was the easiest way of eating the large amount of perishable meat provided by a single beef carcass.[69] It seems probable that at feasts the old animals were used to feed the retainers, while the higher-status participants ate the more tender meat. The greater quantity of refuse created by the retinue would mask the food waste from the high table, producing an assemblage with many old animals.

Through the course of the medieval period aristocratic consumption was increasingly set apart from communal eating.[70] Assemblages from the mid-eleventh century onwards show a corresponding rise in the abundance of young animals. From the mid-twelfth to the mid-fourteenth centuries 14 per cent of the cattle and 12 per cent of the sheep consumed on high-status sites were slaughtered by six months of age to be eaten as veal and lamb. On some later medieval sites neonatal and even foetal animals have been reported. At Guildford Castle,[71] for instance, the remains of at least fifteen sheep or goats, killed within their first few weeks, were recovered from thirteenth-century deposits. On the basis of their skeletal representation and the contexts from which they derived, the remains were food waste rather than natural fatalities. A number of very young individuals has also been noted in the thirteenth- to fifteenth-century layers at Launceston Castle,[72] Dudley Castle,[73] and Eynsham Abbey,[74] but at these sites many were goat rather than sheep. Historical evidence suggests that lambs, kids, and calves were all popular at high-status feasts;[75] foetal and neonatal animals were also incorporated into the fast-day menu.[76] Although not permitted under

[67] Pollington (2003: 19–31); Magennis (1999: 17–28). [68] Hagen (1998: 39–41).
[69] McCormick (2002: 25). [70] Fleming (2000: 3–4); Mennell (1985: 57).
[71] Sykes (in press *b*). [72] Albarella and Davis (1996: 23). [73] Thomas (2002: 204).
[74] Ayres, Locker, and Serjeantson (2003: 394).
[75] Albarella and Davis (1996: 23); Woolgar (1999: 160); Dyer (2004). [76] Hagen (1998: 89).

the original Benedictine Rule, in the twelfth century dietary laws were manipulated to allow greater quantities of meat to be eaten in monasteries during periods of fast: offal was no longer considered meat. Salted, pre-cooked, or chopped flesh could also be eaten without breaking the Rule.[77] By the thirteenth century consumption of foetal and juvenile animals was also permitted on the basis that they came from the 'watery' environment of the uterus and could, therefore, be classified as fish.[78]

Butchery and cooking

Until the mid-tenth century, butchery on all sites appears to have been undertaken in a haphazard way. There is no clear archaeological evidence for specialist slaughter areas or butchers.[79] After this point, however, documentary records suggest that townsfolk were able to obtain beef and mutton from meat markets and fleshmongers. A charter of 932 mentions a cattle market (*hryþera ceap*) outside the city walls of Canterbury, and butchers' streets are recorded at both Winchester and York in the tenth century.[80] At about the same time, animal bone assemblages from urban sites show the emergence of standardized butchery techniques, with meat-cleavers as the main tool.[81] Rather than being butchered on the floor, as seems to have been the case in the preceding period, carcasses were suspended and split into equal sides, a technique suggesting greater professionalism and specialist premises with the apparatus for hoisting a carcass.[82] While evidence for professional butchery techniques is abundant for Late Saxon towns, these patterns do not appear in high-status assemblages until the late eleventh to twelfth centuries.[83] Interestingly, these standardized butchery patterns appear at the same time that cattle and sheep assemblages from high-status sites begin to include a high proportion of meat-bearing elements, perhaps suggesting that the elite was beginning to purchase ready-butchered joints from urban butchers.[84] The coincidence of these changes with the Conquest deserves comment. From this point, Anglo-Norman vocabulary was used for choice meat—*bœuf* (beef), *veau* (veal), and *mouton* (mutton)—while poorer cuts retained their English names (such as ox tail).[85] These shifts in semantics may reflect changes in the provisioning mechanism, away from food rents and towards purchases at market.

Those sections of society unable to buy fresh beef or mutton, and others who required provisions for consumption during lean times, would have relied upon supplies of preserved meat. Zooarchaeological evidence for preservation is difficult to discern, but one example is seen in high-status assemblages of the Saxon period which contain cattle shoulder blades trimmed in a manner

[77] Harvey (1993: 40). [78] Ervynck (1997*a*: 76). [79] Sykes (2001: 103).
[80] Hagen (2002: 315). [81] Sykes (2001: 104). [82] Grant (1987: 57).
[83] Sykes (2001: 110). [84] Sykes (2001: 100).
[85] Goody (1982: 136); Davidson (1999: 67, 604).

indicative of brining.[86] In the post-Conquest period butchery marks suggestive of hot-smoking become more abundant, especially on cattle bones.[87] This form of preservation produces a more flavoursome meat than brining and, although its life is shorter, it may have been preferred to salted beef. This type of butchery is less apparent in town assemblages,[88] supporting the argument that the urban population, having easy access to fleshmongers, had little need for preserved beef and mutton.[89]

One method of butchery common to all fifth to mid-eleventh-century assemblages, regardless of site type, is bone splitting, with long bones cleaved down the length of the marrow cavity. This ubiquitous pattern has been interpreted as reflecting a reliance on stews and soups.[90] A practical and economic method of cooking, boiling would have counteracted the taste of tainted meat, reduced the salty flavour and leathery texture of preserved beef and mutton,[91] and, most importantly, would have retained the meat juices and fat within the pot. Clear evidence for marrow extraction is less abundant in mid-eleventh- to mid-twelfth-century assemblages,[92] but it is unlikely that stews became unfashionable in Norman England. Studies of ceramics suggest a proliferation of sagging-based cook-pots in the twelfth century,[93] perhaps indicating a rise in the consumption of boiled meats. Beef and mutton of this date would have required lengthy cooking, as the ageing data suggest that animals were, on average, older than those eaten by the pre-Conquest population. Reasons for boiling beef and mutton went beyond practicality and taste, perhaps reflecting medieval humoral theory.[94] The theory of the humours also suggested that the flesh of young animals—veal, lamb, and kid—benefited from roasting.[95] In comparison to stewing, roasting is expensive of both fuel and labour, and consumption of roast meat must have been a sign of wealth. It may be no coincidence that, with the mid-twelfth-century rise in aristocratic consumption of juvenile animals, roasting dishes (or dripping pans) appear more regularly in the archaeological record.[96] Evidence suggests that percentages of burnt bone—probably associated with this process—also increase on some high-status sites.[97]

Conclusion

Cattle and sheep account for the majority of bone remains recovered from nearly all medieval sites, suggesting that between the fifth and mid-sixteenth centuries

[86] Sykes (2001: 237). Dobney, Jaques, and Irving (1995: 27) argue that shoulder joints were trimmed to allow the salt water to permeate the meat.

[87] Sykes (2001: 237). Both Lauwerier (1988) and Dobney, Jaques, and Irving (1995: 26–7) have cited as evidence for the hot-smoking process the meat-hook holes and shaving marks, produced when adhering dried meat was cut away from the bone.

[88] Sykes (2001: appendix Vb).

[89] Montanari (1999*a*: 249).

[90] Hagen (1998: 58).

[91] Montanari (1999*a*: 249).

[92] Sykes (2001: appendix Vb).

[93] Hinton (1990: 130).

[94] See Chapter 13. Scully (1995: 44).

[95] Scully (1995: 47).

[96] Brown (2002: 135–7).

[97] Sykes (2001: 235); Pinter-Bellows (2000: 168).

meat in diet, particularly that of the lower social classes, was centred on beef and mutton. Dietary contribution should not, however, be confused with the popularity of these meats, whose consumption was often dictated by economics rather than personal preference. This is demonstrated well by the trends observable in the assemblages from rural and urban sites, where cattle to sheep ratios and age profiles can be seen to chart shifts in trade and commerce: the early medieval need for traction and manure, the high medieval intensification of the wool industry, and the later medieval move towards prime beef and dairy production. In terms of the absolute quantities of meat in diet, however, the archaeological evidence is less revealing than the historical data, especially those for the late medieval period discussed in Chapter 7.

Power relationships between producers and consumers also influenced what was served at the table. Originally able to consume prime animals themselves, the inhabitants of rural settlements were gradually persuaded to send their choice stock to urban centres, where they could be redistributed, purchased from markets, or, later, bought from professional butchers. From the late eleventh century, many of the great households also began to shop within towns, purchasing ready-butchered joints of meat to supplement the food renders they received. But the elite could afford to be selective. Even in the Mid-Saxon period there is good evidence that the upper ranks of society avoided excessive consumption of beef and mutton, incorporating a greater variety of meats (pork, poultry, and game) into their diet. Through time, the pattern of aristocratic consumption became increasingly defined by its separation from that of the lower classes. In high-status assemblages, cattle to sheep ratios and the age at which animals were eaten were frequently contrary to the wider economic shifts, highlighting the desire and ability of the elite to set themselves apart. Differences in status were maintained not only by the type of meats eaten, but also by the method by which they were cooked: rather than ubiquitous stews, the elite, or the upper ranks in elite establishments, consumed greater quantities of roasted meats, sauced with the dripping;[98] and in the period for which we have good historical evidence, they consumed meat in much greater quantities.

Cattle and sheep were undoubtedly the mainstays of the medieval animal economy, but they represented far more than just traction, wool, meat, or milk. The mechanisms through which they were managed, distributed, processed, and eaten were often highly complex and linked inextricably to the socio-political structure of the time. By understanding the changing role of cattle and sheep in medieval England, we come closer to understanding medieval society itself.

[98] See Chapter 13.

6

Pig Husbandry and Pork Consumption in Medieval England

U. ALBARELLA

The tradition of keeping a family pig in a sty at the back of the house—widespread in the last century—is of long standing.[1] It was the product of changes in agriculture and land management through the last millennium, which caused a series of transformations in the manner of keeping swine in Britain. Several different forms of husbandry were practised over the centuries, but pigs always played an important role in economy and society, and, with their meat and lard, provided key components in medieval and modern diets. Although they were useful for the production of manure and had unselective eating habits that assisted in cleaning backyards and town streets, pigs were almost exclusively reared for their meat. Unlike cattle and sheep, pigs do not provide important secondary products, such as wool or traction. Pig milk was used in Hittite rituals,[2] but such exploitation is unknown in historic Europe. Their inability to provide as wide a range of products as other domesticates is compensated for by their productivity as a source of food, in terms of energy intake.[3] This productivity is partly a consequence of the fecundity of the species. Even when no selection of breeds had occurred, nine to twelve piglets—and sometimes more—could be produced every year.[4]

Pigs are adaptable animals, which can be reared even on poor-quality land.[5] In difficult times for cattle and sheep husbandry, pigs might continue to be a source of meat.[6] They were also an important source of fat, in an age when meat, produced from largely unimproved animals, tended to be much leaner. In addition,

This chapter was written while the author was based at the University of Durham, with the support of the Arts and Humanities Research Board. Marina Ciaraldi and Naomi Sykes provided valuable comments on an earlier draft. Becky Roseff introduced me to the work of Flora Thompson.

[1] Thompson (1939); Dyer (1998*a*: 196); Wiseman (2000: 46).
[2] Simoons (1994: 23).
[3] Campbell (2000: 165).
[4] Markham in Davis (2002: 47–60); Campbell (2000: 165); Kelly (2000: 81).
[5] Wiseman (2000: 37).
[6] Campbell (2000: 167).

pig meat was particularly suitable for long-term preservation,[7] a quality that cannot be overstated. In fact, the lower classes almost exclusively consumed pork in its preserved forms, bacon and ham.[8] Archaeologically the distinction between the use of pork in its fresh and preserved forms is hard to pinpoint, as animal bones normally derive from a diversity of activities, and only rarely can they be associated with specific butchery processes; but the distinction is present in the historical record for the later Middle Ages, and is discussed in Chapter 7.

There is a general consensus among historians that pork tended to be—after beef—the meat eaten most commonly in aristocratic households.[9] Peasants would rely on an almost exclusively vegetarian diet, but what little meat they could afford was mainly pork.[10] Even after the Black Death, when both historical and archaeological sources point towards an overall increase in meat consumption,[11] this situation probably did not change.

The archaeological evidence is not entirely comparable with the historical sources, as most assemblages of animal bones derive from urban sites, for which there is only a scanty documentary record. Whether sites are urban or rural, the archaeological record suggests that pigs tend to be the third most common species after cattle and sheep, with these last two varying in order of rank.[12] In eighty-seven of a sample of 112 medieval and post-medieval sites (and phases) from central England, pig remains represent less than 20 per cent of the total of cattle, sheep/goat, and pig remains.[13] The pattern is not dissimilar in other areas of the country.[14] Nevertheless, once the different weights of the three main domesticates are taken into account, there is little doubt that beef ranks as the meat most commonly consumed; but pork was likely to have been the second, at least in the early medieval period. This is a generalization, but it is nevertheless reassuring that historical and archaeological sources provide information that is—despite difficulties in comparability—by and large consistent.

Pigs: rise and decline

The heyday of swine husbandry probably belongs to the Saxon period, up to the Conquest. From the eleventh or twelfth century onwards there seems to have been a slow but steady decline in pork consumption, in comparison with other types of meat, and, apart from occasional local circumstances, pigs never regained the economic significance they had in Anglo-Saxon England.

[7] Woolgar (1999: 116); Wiseman (2000: 37).

[8] Dyer (1980: 328; 1998*a*: 116); Rixson (2000: 120).

[9] Dyer (1998*a*: 60); Salisbury (1994: 58); Campbell (2000: 103).

[10] Dyer (1998*a*: 154).

[11] Dyer (1998*a*: 158); Albarella (1997: 19–30); see also Chapter 7.

[12] Albarella (2005).

[13] This percentage is calculated by taking into account only sites where the total number of identified specimens (NISP) of the main three taxa is greater than 300.

[14] A. Grant (1988: 149–61); Albarella and Davis (1996); Albarella, Beech, and Mulville (1997).

Several authors—relying on documentary evidence—indicate that it was only after the eleventh century that swine husbandry became less widespread.[15] Trow-Smith goes as far as suggesting that pigs represented the 'hallmark of Saxon pastoral husbandry'. Since early Saxon literary sources are scarce and only a few assemblages of animal bones from Early and Mid-Saxon sites have been analysed, it is difficult to establish whether the importance of pigs emerged suddenly after the Roman period or whether there was a steady increase in their significance. The few sites which provide a chronological sequence within the Saxon period—such as West Stow,[16] Ipswich,[17] and St Peter's Street, Northampton[18]—do not offer a consistent pattern.

There is clearer evidence that pig bones are found more commonly on Saxon sites than on later ones. Table 6.1 compares pig frequencies with those of the sheep, as it has been suggested that the rise of sheep husbandry—among other factors—caused the demise of the pig.[19] Most sites register a decrease in pig frequencies after the Saxon period (Table 6.1, last column). Portchester Castle[20] and St Peter's Street, Northampton,[21] show an opposite trend, but the occurrence of exceptions is not surprising. Even in the later medieval period there were a few demesnes where pigs were the most common livestock.[22] Wherever possible Saxon (and Saxo-Norman) data have been compared with those from the eleventh to thirteenth centuries, rather than those from the fourteenth century onwards, demonstrating that a change in pig husbandry occurred relatively soon after the Conquest. Using more refined chronologies, Sykes has suggested that this may have happened in the second half of the twelfth century.[23]

The decline in pig husbandry continued, albeit gradually. Documentary evidence from Winchester indicates pork consumption going out of fashion in the later Middle Ages.[24] In the archaeological record, the gradual decrease in pig frequency during the Middle Ages (and even more so in post-medieval times) was highlighted by Grant[25] and has been confirmed more recently by further surveys.[26] As in the case of the transition between the Saxon period and the eleventh to thirteenth centuries, on most sites there is a further decrease in pig frequency between the eleventh to thirteenth centuries and assemblages from the fourteenth century to the end of the Middle Ages (Table 6.1, last column). More exceptions occur, however, suggesting that the phenomenon was less universal than the previous transition. Average percentages for all three main domesticates from a larger number of sites (including those which are not multi-period) demonstrate that while numbers of sheep increase steadily, those of pigs

[15] Trow-Smith (1957: 55); Harvey (1988: 130); Wiseman (2000: 39). [16] Crabtree (1989).
[17] Jones and Serjeantson (1983). [18] Harman (1979*b*). [19] Wiseman (2000: 39).
[20] Grant (1977). [21] Harman (1979*b*). [22] Campbell (2000: 167).
[23] Sykes (2001). [24] Dyer (1998*a*: 199–202); see also Chapter 7. [25] A. Grant (1988).
[26] Albarella and Davis (1996); Albarella, Beech, and Mulville (1997); Jones (2002).

Table 6.1. Number of identified specimens (NISP) of pig and sheep/goat at twenty-one multi-period Saxon and medieval sites in England

Site	Period	Pig n	Sheep/goat n	% Pig
Alms Lane (Norwich)	12th–14th century	159	482	25
	14th–early 16th century	113	376	23
Bedford Castle	11th–13th century	294	589	33
	12th–14th century	77	152	34
Berrington St. (Hereford)	Saxon	492	496	50
	11th–13th century	185	387	32
Burystead (Northants)	Late Saxon	171	365	32
	11th–early 16th century	79	199	28
Castle Mall (Norwich)	Late Saxon	277	236	54
	Saxo-Norman	216	208	51
	12th–14th century	62	133	32
	14th–early 16th century	122	308	28
Colchester	11th–12th century	188	178	51
	14th–16th century	264	1,042	20
Friar St. (Droitwich)	Saxo-Norman	93	103	47
	11th–13th century	110	159	41
	12th–14th century	292	367	44
Flaxengate (Lincoln)	Late Saxon	2,174	6,106	26
	Saxo-Norman	2,268	8,406	21
	12th–14th century	177	856	17
King's Lynn	11th–13th century	350	811	30
	12th–14th century	764	1,861	29
	14th–early 16th century	209	473	31
Launceston Castle (Cornwall)	12th–14th century	464	427	52
	14th–early 16th century	765	855	47
Lincoln (various sites)	Late Saxon	203	449	31
	12th–14th century	42	143	23
Lyveden (Northants)	12th–14th century	35	175	17
	14th–early 16th century	121	291	29
Portchester Castle (Inner Bailey)	Late Saxon	185	267	41
	12th–14th century	220	202	52
St Martin-at-Palace Plain (Norwich)	Saxo-Norman	1,140	1,102	51
	11th–13th century	1,433	1,801	44
	12th–early 16th century	312	310	50
St Peter's St. (Northampton)	Mid-Saxon	88	228	28
	Late Saxon	377	2,006	16
	12th–14th century	417	965	30
	14th–early 16th century	107	784	12
The Green (Northampton)	Late Saxon	137	452	23
	12th–14th century	309	2,661	10
	14th–early 16th century	133	679	16
The Shires, St Peter's Lane (Leicester)	11th–13th century	232	661	26
	12th–14th century	256	607	30
	14th–early 16th century	215	728	23
The Shires, Little Lane (Leicester)	11th–13th century	77	223	26
	12th–14th century	72	193	27
	14th–early 16th century	87	304	22
Thetford	Late Saxon	483	1,045	32
	Saxo-Norman	687	1,574	30
Walton (Aylesbury)	Saxon	331	511	39
	Saxo-Norman	396	883	31
	11th–early 16th century	292	847	26
West Cotton (Northants)	11th–13th century	318	531	37
	12th–14th century	230	826	22
	14th–early 16th century	35	309	10

Table 6.1. *Continued*

Notes: The percentage of pig of the total of pig and sheep/goat is also shown

Sources: Cartledge (1985); Grant (1979*a*); Noddle (1985*a*); Davis (1992); Albarella, Beech, and Mulville (1997); Luff (1993); Locker (1992*a*); O'Connor (1982); Noddle (1977); Albarella and Davis (1996); Dobney, Jaques, and Irving (1995); Grant (1971); Grant (1977); Cartledge (1988); Harman (1979*a*); Harman (1996); Gidney (1991*a*); Gidney (1991*b*); Jones (1993); Noddle (1976); Albarella and Davis (1994).

decrease, reinforcing the point that—though the main decline probably occurred early after the Conquest—the pig continued to lose ground to sheep husbandry in the later part of the period (Fig. 6.1).

These calculations are affected by a number of factors—above all, differential preservation and recovery between sites—but, given the large number of sites considered, general trends are apparent. For a phenomenon not to be obscured it has to be substantial: the gradual decrease in pig husbandry probably was. In addition, the direct comparison of pig and sheep bones, animals of similar size, minimizes the effects caused by bias in recovery. Although undoubtedly numbers of pigs decreased in relation to sheep, the archaeological evidence merely indicates a change in the relative importance of these two animals. Once we consider that, after the Black Death, there was an increase in pasture and that

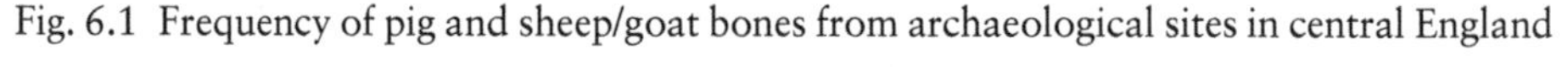
Fig. 6.1 Frequency of pig and sheep/goat bones from archaeological sites in central England

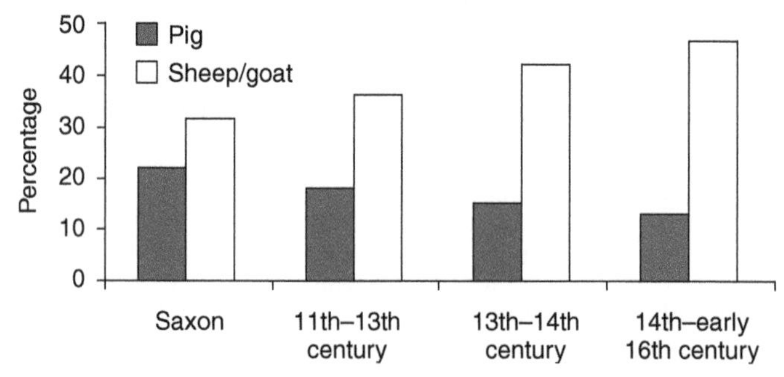

Notes: This is the average NISP (Number of Identified Specimens) of the percentage of the two taxa out of the three main species (cattle, sheep/goat, and pig) per site. Only sites where the total NISP for the three taxa was > 300 have been considered. Number of sites per period: Saxon, 33; eleventh to thirteenth centuries, 12; twelfth to fourteenth centuries, 19; fourteenth to early sixteenth centuries, 14. The Saxon phase also includes some Norman elements.

Sources and sites: As Table 6.1, with the addition of Bewell House, Hereford: Noddle (1985*b*); Black Lion Hill, Northampton: Harman and Baker (1985); Brackley: Jones, Levitan, Stevens, and Hocking (1985); Brook St., Warwick: Hamilton (1992); Causeway Lane, Leicester: Gidney (1999); Fishergate, Norwich: Jones (1994*b*); Flaxengate, Lincoln: O'Connor (1982); Full St., Derby: Patrick (1975); George St., Aylesbury: Jones (1983); Goltho: Jones and Ruben (1987); Great Linford: Burnett (1992); Ipswich: Crabtree (1994); Launditch Hundred: Ambros (1989); Lyveden: Grant (1975); Midland Road, Bedford: Grant (1979*a*); Mill Lane, Thetford: Albarella (2004); St John's Street, Bedford: Duke (1979), Grant (1979*b*); St Mary's St., Bedford: Grant (1979*a*); Thuxton: Cartledge (1989); Town Wall, Coventry: Noddle (1986); Walton Road, Aylesbury: Sadler (1991); West Parade, Lincoln: Scott (1986); Whitefriars St., Norwich: Cartledge (1983).

more animals were present in the countryside,[27] the absolute number of pigs may not necessarily have diminished, but simply become less in comparison with other species.

Regimes of husbandry

Medieval swine husbandry relied heavily on the exploitation of woodland areas, where pigs would have been taken seasonally to feed on roots, acorns, and beech mast (Plate 6.1).[28] The best demonstration of how widespread this practice was can be found in Domesday Book, which generally measured woodland in terms of the number of pigs that it could support.[29] The right to exploit demesne woodland for swine is known as pannage.[30] This practice probably dates back to the seventh century[31] and it survived at least until the end of the Middle Ages.[32] Apart from its economic impact—revenue obtained through the leasing of woodlands for pasturing pigs could be substantial[33]—it was an important element in the organization of society. Pannage was managed predominantly in a communal way:[34] swineherds collected pigs from different owners and drove them to woodland areas, where they might spend one or more nights in the company of the animals.[35]

Although the association between pigs and woods was particularly strong in the Saxon period and in the eleventh to thirteenth centuries,[36] it would be wrong to assume that it was the only strategy used to fatten the animals. Pannage only occurred in autumn and early winter,[37] the richest season for woodland products. Pigs had to be fed in other ways at different times of the year, as the *Seneschaucy* acknowledged.[38] Pigs were also kept—though probably in smaller numbers—in areas that were not so rich in woodland: the number of swine on some estates was higher than the number of animals local woodland might support.[39] In addition to natural resources pigs could be fed on cereals and legumes,[40] and occasionally grazed upon pasture.[41] Moreover, some might be housed or kept in yards.[42]

The system of pannage started to break up after the Conquest, mainly as a consequence of the gradual reduction of woodland.[43] This must have been an important factor in the relative decrease of pig husbandry in the later Middle Ages. There must have been geographic variation in the way this phenomenon occurred, as pannage continued almost unabated in regions rich in forest.[44] Where woodland was more depleted, intensive methods of swine husbandry

[27] Dyer (1980: 324); Overton and Campbell (1992). [28] Campbell (2000: 165).
[29] Williams and Martin (2002). [30] Harvey (1988: 127). [31] Trow-Smith (1957: 51).
[32] Overton (1996: 25). [33] Fryde (1996: 155); Wiseman (2000: 33).
[34] Harvey (1984: 228). [35] Kelly (2000: 82); Wiseman (2000: 33–4).
[36] Trow-Smith (1957: 53). [37] Wiseman (2000: 33). [38] Oschinsky (1971: 284–5).
[39] Williams and Martin (2002). [40] Kelly (2000: 83); Dyer (2003: 126).
[41] Trow-Smith (1957: 81). [42] Trow-Smith (1957: 53).
[43] A. Grant (1988); Campbell (2000: 166); Wiseman (2000: 40).
[44] Campbell (2000: 166); Dyer (2003: 17).

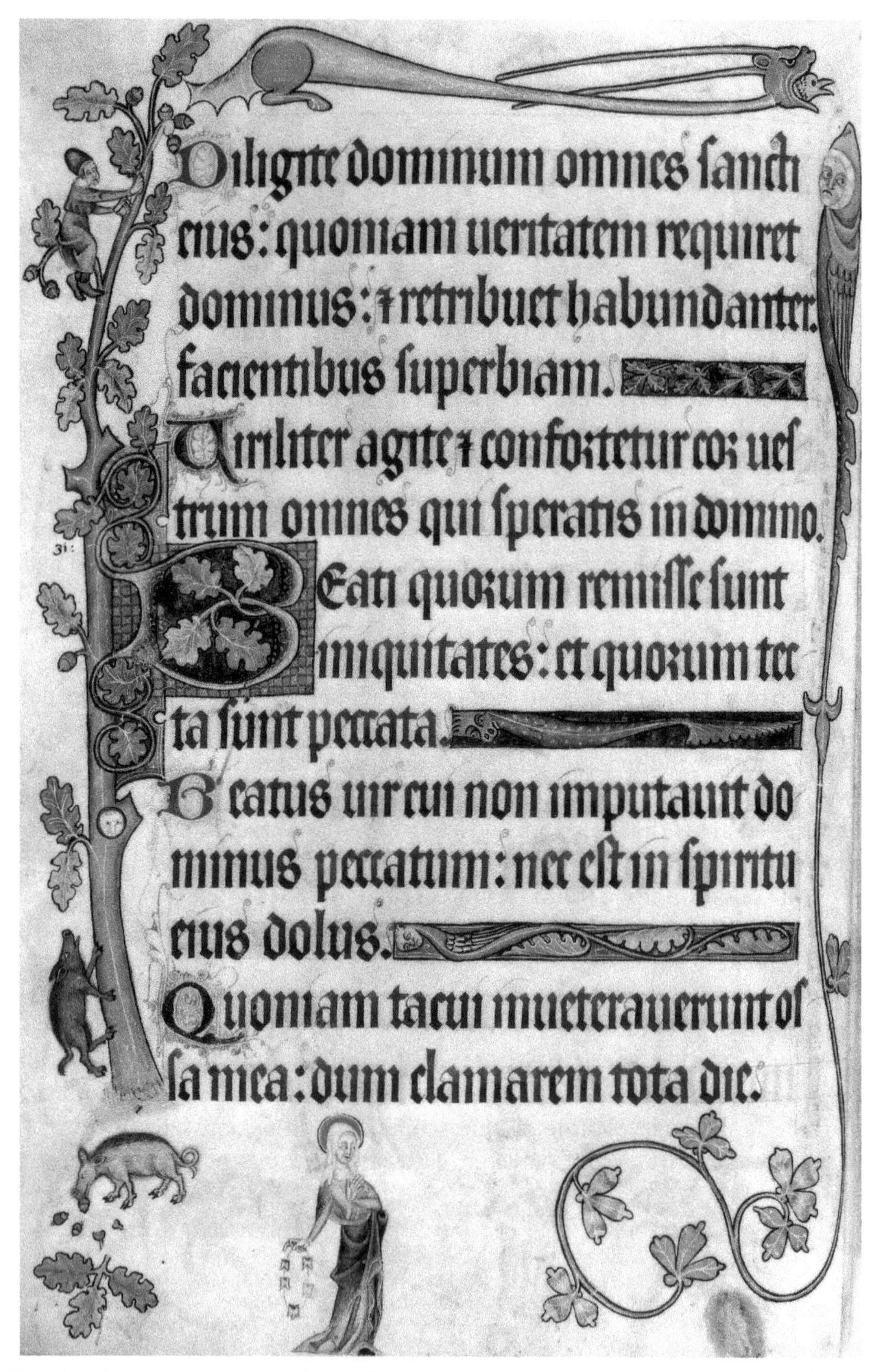

Diligite dominum omnes sancti
eius: quoniam ueritatem requiret
dominus: ⁊ retribuet habundanter.
facientibus superbiam.
Viriliter agite ⁊ confortetur cor ues
trum omnes qui speratis in domino.
31: Beati quorum remisse sunt
iniquitates: et quorum tec
ta sunt peccata.
Beatus uir cui non imputauit do
minus peccatum: nec est in spiritu
eius dolus.
Quoniam tacui inueterauerunt os
sa mea: dum clamarem tota die.

Plate 6.1 Pannage in practice: pigs feeding on acorns shaken down by their swineherd. From the Luttrell Psalter, *c*.1320–45, British Library, Add. MS 42130, fo. 59v. Photograph: © British Library.

became gradually more common. Sty feeding of animals, on legumes, cereals, house-waste, and even by-products of the dairy and brewery industries,[45] started replacing the traditional forms of free-range husbandry. This was mainly in response to the reduction of the season of pannage. While in Anglo-Saxon England pigs were allowed to roam in the woodland from August to December, after the Conquest this right was restricted to a period of six to eight weeks in October and November in many counties. In others, such as Lancashire, the length of the pannage season had not changed as late as the sixteenth century.[46]

These differences in the dietary regimes of pigs are difficult to detect archaeologically, but recent technological advances have started providing useful data. Isotopic analysis of pig bones from sites of different periods in Britain shows that omnivorous pigs were already to be found before the Conquest. They are more common on urban than rural sites,[47] possibly indicating that pigs kept in towns were fed on house-waste, rather than vegetable products alone. Work on tooth microwear has also shown that it is possible to differentiate between rooting and stall-fed pigs.[48] These criteria—initially devised on modern pigs with a known diet—have recently been applied to archaeological material from medieval York and other sites. Preliminary work suggests the presence of stall-fed animals in towns and foraging/rooting populations in rural contexts.[49]

Medieval urbanization brought other changes in pig keeping. A wealth of artistic, historical, and archaeological evidence demonstrates that pigs adapted very well to town environments.[50] Pigs, with their omnivorous habits, were occasionally encouraged to roam free and scavenge for food,[51] as an effective way to clean the streets; but this was not without its problems, and ordinances abound that ban or at least restrict the movements of pigs in towns.[52] There was also concern that they were a danger to children, and attacks were recorded.[53] Archaeological evidence from Norwich indicates that bones of neonatal pigs—suggestive of on-site breeding—increased in the sixteenth century, while new-born cattle and sheep were, by this period, no longer found. It is possible that an increase in urbanization at the end of the medieval period favoured the keeping of pigs, which did not need large areas of pasture to feed.[54]

Modifications in pig husbandry went hand in hand with changes in the general perception of the status of the pig and the consumption of its meat. By the time of the Domesday survey, pigs were regarded as animals of the poor, which may have affected the accuracy of their enumeration. It has even been suggested that there was a proportional relationship between numbers of pigs and numbers of poor.[55] The reputation these animals acquired may have derived from the

[45] Overton (1996: 25); Campbell (2000: 166); Rixson (2000: 120); Wiseman (2000: 41).

[46] Wiseman (2000: 33, 39–40).

[47] Müldner, Richards, Albarella, Dobney, Fuller, Jay, Pearson, Rowley-Conwy, and Schibler (in preparation).

[48] Ward and Mainland (1999).

[49] Wilkie, Mainland, Albarella, Dobney, and Rowley-Conwy (in preparation).

[50] Dyer (1998*a*: 196); Lilley (2002: 220); Dyer (2003: 199).

[51] Wiseman (2000: 42).

[52] Rixson (2000: 115).

[53] Smith (2000).

[54] Albarella (2005).

[55] Hallam (1988*d*).

increasingly poor quality of their meat, a consequence of inadequate feeding, based on industrial by-products.[56] This could certainly have been a factor, but the association between pigs and poverty probably relied more heavily on the fact that even less well-off peasants could keep one or two pigs at low cost and labour. It is possible that initially the low status of pigs resulted only from the way in which they were kept, but towards the end of the Middle Ages it extended to consumption—aristocratic households and townspeople would eat little pork by the sixteenth century.[57]

Social and geographic variation

There is a great degree of variation in the frequency of pig bones when town, village, and castle sites are compared, but, on average, they tend to be more abundant on high-status sites, followed by villages and then urban sites.[58] The contradiction with the preceding discussion is apparent rather than real, as historical and archaeological data describe different phenomena: the animal bones tell us about net consumption of meat, the documents inform us about supply and husbandry. Bones from archaeological sites do not necessarily come from animals that were raised locally, therefore the supposed greater emphasis in urban pig keeping that characterizes the later Middle Ages is not necessarily reflected in a higher consumption of pork there. Most of the meat consumed in towns was probably brought in and town-dwellers consumed less pork and mutton and more beef than their rural counterparts,[59] despite the fact that the number of pigs found in towns probably far exceeded the number of cattle.

The greater abundance of pig bones in village sites supports the view that pork was considered predominantly peasant food. According to the archaeological evidence, however, even on low-status sites beef seems to have been the meat that was most consumed, though not to the same extent as in towns. Despite problems of zooarchaeological quantification, there is a possibility that the consumption of beef by the peasantry has been underestimated by historians.[60]

The high levels of consumption of pork on high-status sites are more difficult to explain.[61] It is possible the decline in the status of pork only took place towards the very end of the medieval period, when pig meat had become less common in all sectors of society. There is also evidence that pig meat—especially from young animals—was considered appropriate for refined tastes,[62] and that supplies for castle garrisons were drawn from the peasantry.[63]

The few archaeological data that we have for monastic sites—also of a high status—confirm that pork played an important role in ecclesiastical diet. At the

[56] Rixson (2000: 120). [57] Woolgar (1999: 133); Smith (2000: 716). See also Chapter 7.
[58] Albarella and Davis (1996). [59] Albarella (2005).
[60] For a fuller discussion of this question, see Albarella (1999).
[61] See also Jones (2002: 157). [62] Dyer (1998*a*: 60). See also Chapter 7.
[63] Sykes (2001: 247).

Dominican friary in Chester,[64] Eynsham Abbey,[65] St Gregory's Priory in Canterbury[66] and Shrewsbury Abbey[67] pork was—in most phases—the second most frequently consumed meat; but at Evesham Abbey[68] and Austin Friars, Leicester,[69] it is less well represented. In the Misericord at Westminster Abbey, *c.*1495–*c.*1525, beef and veal constituted about 35 per cent of the meat consumed by the monks; mutton and lamb, 47 per cent; and pork, 14 per cent, more in line with the archaeological remains from Evesham and Leicester's Austin Friars.[70] It is possible that in earlier times more pork had been eaten: the trend towards a reduction in pork consumption throughout the Middle Ages has been observed on monastic sites as well.[71]

Pig husbandry and pork consumption also varied by geographic area. Where there was more woodland—as in the Weald, the south-west, and the north-west—there were more pigs (Plate 6.2).[72] Entries in the three counties covered by the Little Domesday show that pigs were—in comparison to sheep—more common in Essex than in Suffolk and Norfolk.[73] This probably emphasizes the effect that greater woodland resources could have on the success of pig husbandry. Unfortunately too few animal bone assemblages have been studied from Norman sites or from those up to the end of the thirteenth century in Essex and Suffolk to compare the archaeological and documentary evidence, but the archaeological data from Norfolk provide an interesting insight. In this county, evidence from sixteen assemblages—mainly urban—indicates that pig remains average 41 per cent of the remains of sheep/goat and pig. This contrasts with the 12 per cent suggested by Domesday Book. It is possible that Norfolk towns imported some of their pig meat from areas outside the borders of the county but, considering that nearby areas were also unlikely to have had a thriving pig husbandry, we must perhaps accept the possibility that numbers of pigs in Domesday Book are too low.[74]

Economic, rather then purely environmental factors, also played a role in the geographic distribution of pig husbandry. For instance, the East Midlands, despite poor woodland coverage, had many pigs.[75] Campbell suggests that the soil of this region was mainly dedicated to arable production, probably incompatible with extensive sheep and cattle husbandry. Therefore, in order to produce a sufficient supply of protein, there was specialization towards an intensive system of pig keeping.[76] The archaeological data are, however, once again at odds with the historical sources. A comparison of the frequency of pigs in Northamptonshire, Leicestershire, and East Anglia, on the one hand, and the West Midlands on the other, indicates that pigs were no more common in the

[64] Morris (1990). [65] Mulville (2003); Ayers, Locker, and Serjeantson (2003).
[66] Powell, Serjeantson, and Smith (2001). [67] Jones (2002). [68] Lovett (1990).
[69] Thawley (1981). [70] Harvey (1993: 53).
[71] Jones (2002); Ayers, Locker, and Serjeantson (2003). [72] Campbell (2000: 166).
[73] Compare Williams and Martin (2002). [74] Trow-Smith (1957: 80); Salisbury (1994: 28).
[75] Harvey (1988: 127); Mortimer in Davis (2002: 56). [76] Campbell (2000: 168).

Plate 6.2 Pigs and acorns, the rebus of John de Swinfield, d. 1311, precentor of Hereford Cathedral, from his tomb in the retrochoir there, early fourteenth century. The Swinfield family came from Kent, famed for the pannage its pigs had in the Weald. The arms on the saddles of the pigs are those of the Dean and Chapter of Hereford. Photograph: C. M. Woolgar.

East Midlands than elsewhere (Fig. 6.2). It is possible that the difference was too subtle to be highlighted by this rather crude comparison of archaeological data, but it is also possible that this is again a difference between data for consumption and information about production. Pig meat was not necessarily eaten where it was produced.

Slaughter, seasonality, trade

Pigs were slaughtered at almost any age,[77] but on average bones from pigs are from younger animals than those from sheep and cattle. This is not surprising: pigs,

[77] Trow-Smith (1957: 128).

Fig. 6.2 Frequency of pig bones from medieval sites in central England

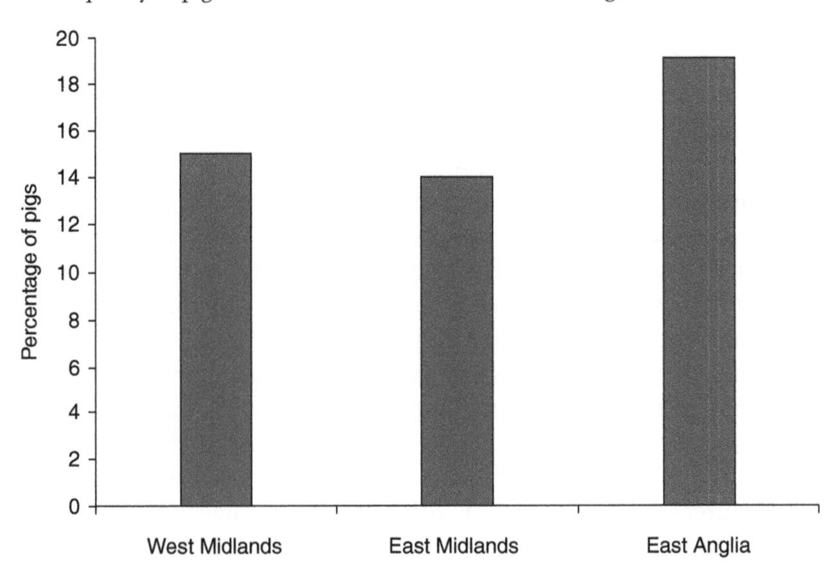

Note: This is the average NISP of the percentage of pigs out of the three main species (cattle, sheep/goat, and pig) per site.
Sources and sites: As Fig. 6.1.

unlike the other domesticates, were kept exclusively for meat, and consequently there was no point in keeping them alive beyond the point of maximum growth—apart from the few kept for breeding. Suckling pigs were occasionally consumed as a delicacy, but most pigs were fattened before slaughter.[78] Bones of very young pigs are found, but they are a minority. Some animals were killed in their first year, but pigs were believed to make the best porkers and baconers when they were rising two.[79] The archaeological evidence is consistent with this, as, in most sites, pig bones tend to belong to 'immature' and 'sub-adult' categories. There are a few exceptions, such as the twelfth- to fourteenth-century village of Thuxton[80] and the late medieval urban site at Towcester,[81] where first-year killings predominate. This suggests variations in husbandry and uses of pigs. Despite the claim that Saxon pigs were particularly slow growing and would not have been slaughtered until their third year,[82] there is little zooarchaeological evidence of a widespread change in kill-off patterns between the Saxon and succeeding periods. A tendency to kill animals at an earlier stage, however, has been detected for the early post-medieval period, and it is probably a consequence of the development of new, faster-growing breeds,[83] possibly associated with a further intensification of husbandry techniques.

[78] Kelly (2000: 85).
[79] Campbell (2000: 165); Davis (1794) in Davis (2002: 57). See also Chapter 7.
[80] Cartledge (1989). [81] Holmes (1992). [82] Trow-Smith (1957: 54).
[83] Albarella (1997).

Age at slaughter was partly determined by seasonal cycles. When there was only one farrow a year, this would generally occur in spring, whereas the most convenient time of year for slaughter was the late autumn or early winter, when pigs had been fattened in the woodlands, anticipating a shortage of natural resources to sustain them.[84] Many pigs born in spring would therefore have been killed in the early winter of the following year, at about one and half years old, which is consistent with documentary information for the thirteenth century.[85] Much of the flesh would have been preserved for later use. Seasonality can be difficult to detect archaeologically, but work on pig mortality, from medieval sites in Belgium, based on tooth wear, confirms a peak of killing in winter.[86] No corresponding data are available for English sites.

Throughout the Middle Ages pig husbandry was more rooted in self-sufficiency than in the case of other animals, to the extent that 'the amount of pork reaching the market was probably . . . small'.[87] Most castles, manors, priories, hamlets, and villages could afford a pigsty and swine could also be kept in urban environments—though much of the pork in towns would have been brought in from elsewhere. Demesne production was mainly for local consumption and most of the pork sold in the market was supplied by peasants.[88] Since manorial households rarely purchased pork in the market, most of the surplus must have been directed towards the urban market.[89]

On most medieval sites, pig bones derive from all parts of the body, supporting the view that complete carcasses, rather than selected joints of meat, were processed. Where there is a preponderance of particular parts of the body, these are almost invariably cranial elements, mainly teeth. This is much more likely to be the result of better preservation of the durable dental elements than of any pattern of butchery. Butchery patterns that reveal a systematic way of distributing the carcass, such as the longitudinal splitting of skulls and vertebrae, are almost exclusively found at urban sites, for example, at Aylesbury,[90] Leicester,[91] Thetford,[92] and Lincoln.[93]

Improvement

As long as pigs were managed within an extensive, woodland-based system of husbandry there was probably little opportunity or even motivation for improvement. With only limited control of breeding—domestic pigs could interbreed with wild stock during the pannage season—it was impractical to select new types and breeds. The increasing tendency, particularly after the Black

[84] Wiseman (2000: 35, 37); Dyer (1998*a*: 58). [85] Trow-Smith (1957: 126).
[86] Ervynck (1997*b*). [87] Farmer (1991*b*: 458).
[88] Thornton (1992: 35); Campbell (2000: 167). [89] Thornton (1992: 31); Farmer (1991*b*: 455).
[90] Jones (1983). [91] Gidney (1991*a*, 1991*b*). [92] Jones (1993).
[93] O'Connor (1982); Scott (1986).

Death, to enclose the animals and to maintain greater control of their life cycle may have generated the opportunity for change; but attempts to improve breeds may have occurred exactly at the point when the general economic importance of the pig went into decline.[94] There is little historical evidence for improvement during the medieval period and the earliest references to regional types are as late as the seventeenth century. Even in these examples, it is not clear whether the improvement was the result of genetics or nutrition.[95]

Nor does the zooarchaeological evidence provide much information about any increase in size before the end of the Middle Ages. It is not always clear whether an increase in size is a genuine phenomenon or the result of insufficient detail in the analysis of biometrical data. At Flaxengate, Lincoln, there was no change in the size of the pigs between Late Saxon and Saxo-Norman levels;[96] and the same is true for the data for the eleventh to fourteenth centuries at West Cotton[97] and Exeter.[98] It has been noted, however, that the pigs from Lincoln were smaller than contemporary animals from Exeter, which indicates some regional variation.[99] A hint that, in some areas, improvement may have taken place in the late medieval period is provided by the animal bones from Launceston Castle[100] and Castle Mall, Norwich.[101] At both sites pigs from the late medieval levels are slightly larger than those from earlier phases, though the increase is much smaller than that occurring in the post-medieval levels from the same sites. It is reflected in the bones rather than the teeth: since post-cranial bones can be modified by environmental factors, this slight change is more likely to be the consequence of an improvement in diet than a genetic difference.

Recent work at Dudley Castle has provided for the first time firm evidence that some degree of improvement occurred well before early modern times and the introduction of stock from Asia.[102] Increase in the size of both teeth and bones occurs at Dudley in the thirteenth and fourteenth centuries, perhaps a consequence of a change in husbandry, from an extensive, woodland-based system to the enclosure of pigs in sties. Since the increase affects teeth we can be reasonably confident that it is related to the creation or perhaps introduction of a genetically different animal. It is possible that the West Midlands was an area of experimentation in animal breeding, but it is unlikely to have been an isolated case. Larger animals were likely to grow faster and provide a greater output of meat in this timespan: an increase in size may be related to a tendency to slaughter animals at an earlier age.[103]

Meat gain could also be increased by improving birth and mortality rates, and the general health of stock. Birth rates could be improved by increasing the number of piglets per litter or the number of litters per year. In unimproved pigs,

[94] Wiseman (2000: 42). [95] Overton (1996: 115–16); Wiseman (2000: 47).
[96] O'Connor (1982). [97] Albarella and Davis (1994). [98] Maltby (1979).
[99] O'Connor (1982). [100] Albarella and Davis (1996).
[101] Albarella, Beech, and Mulville (1997). [102] Thomas (2002). [103] Albarella (1997).

litters were normally of eight or nine piglets, but by the fourteenth century, at Rimpton, there were litters of thirteen.[104] Early modern agricultural writers mention litters of up to twenty, but these were exceptional.[105] Farrowing traditionally occurred once a year, generally in spring,[106] but, writing in the late thirteenth century, Walter of Henley, recording the ideal rather than the usual, expected two litters a year.[107] At thirteenth-century Rimpton, piglet mortality was less than 10 per cent, a figure that compares positively with contemporary standards of intensive pig farming.[108]

Conclusion

Pigs played an important role in medieval economy and society and pork was—generally—the meat that was the second most commonly eaten during the Middle Ages. Peasants usually ate pork in its preserved form, while upper classes also consumed it fresh. Pig husbandry was prominent in the Saxon period, when there was sufficient woodland coverage in the country to allow widespread and extensive management of pigs, focused on the seasonal system of pannage. The gradual reduction in the forest led to a decline in pig production and consumption, and to a change in husbandry. Intensive systems of pig keeping, with pigs stalled or yarded, were present in the Saxon period. They became more common and, by the late Middle Ages, were the prevalent form of pig husbandry in many regions.

A consequence of these changes was that closer control of breeding became possible and this opened up the opportunity for improvement in meat output, the creation of regional types, and eventually distinct breeds. As faster-growing pigs were created, the age of slaughter might be advanced: many animals would consequently have been killed in their first rather than second year. Though most of the improvement took place in post-medieval times, archaeological data indicate that some changes occurred perhaps as early as the fourteenth century.

The medieval style of pig husbandry was less intensive than that practised in modern times, but it was not necessarily worse. Pig populations seem to have been healthy, the well-known fecundity of the species guaranteed good size litters, and the meat—as long as pigs were allowed to graze in woodlands—must have been of a good quality. The particular suitability of pork for long-term preservation sustained a supply of protein at times of the year when other types of meat were not readily available.

Pigs were food providers and all aspects of their management were directly or indirectly related to diet. It would, however, be unfair to regard them as

[104] Thornton (1992: 34); Kelly (2000: 81). [105] Markham (1657), in Davis (2002: 56).
[106] Kelly (2000: 81). [107] Oschinsky (1971: 334–5). [108] Thornton (1992: 35).

meat-producing machines; pigs were much more than that. They contributed substantially to the shaping of the medieval household and community, to its organization, settlement, movements, everyday activities, seasonal cycles, entertainment, and also feelings. It is a fair point that the role of the pig 'as a major contributor to the development of medieval society has rarely been acknowledged'.[109]

[109] Wiseman (2000: 39).

7

Meat and Dairy Products in Late Medieval England

C. M. WOOLGAR

The evidence for meat and dairy products reviewed in Chapters 5 and 6 can be supplemented for the late medieval period by documentary sources that allow us to address a series of questions about the contribution these foods made to diet that can be answered in no other way. First among these are the levels of consumption of meat and dairy foods: historical evidence offers the opportunity to quantify this, even on an individual basis, as well as providing more general indications of levels of food availability. Secondly, historical sources offer information about meat products that have left no more than a slight trace in the archaeological record, such as the use of sausages and offal, items of some consequence in a society which had restricted access to meat protein. Evidence for dairy foods is much more substantial in the historical sources than in the archaeological record and this permits us to consider questions of consumption, rather than production, and to encompass the contributions made by sheep and goats as well as cattle dairying.

In common with many foodstuffs, the emphasis with meat and dairy foods was not only on production for immediate consumption: meat, especially pork, but also beef and mutton, offered excellent opportunities, through salting, smoking, or a combination of drying and smoking of the main cuts of flesh, for preservation for the longer term. Other parts of the animal might be prepared for storage, as puddings, brawns, and tongues, or rendered for fats; and cheese and butter preserved milk fats. These were essential processes not only for maintaining these elements of protein in the food supply against the background of seasonal and dietary patterns, outlined in Chapters 13 and 14, but also for creating highly desirable—and hence marketable—products that were attractive for consumption at all levels. On these products and practices the historical sources have much to contribute.

Meat consumption in town and country

From the thirteenth century, levels of meat and dairy production and consumption among the urban population and the peasantry can be inferred from taxation assessments. These usually excluded day-to-day living necessities, such as items of daily food. Although the contents of larders and cellars were not included in the early thirteenth-century returns, by the end of the century some were described.[1] Two assessments for Colchester of 1295 and 1301 show a modest-sized borough with a population of 3,000 to 4,000, of which around 70 per cent were involved in agriculture.[2] In 1301, 389 households in the town and the villages of Lexden, Mile End, Greenstead, and West Donyland had a total taxable value of £518 1*s.* 4¾*d.* The assessments were commonly made at Michaelmas (29 September), when the surplus of the harvest might be easily estimated. At this point in the year, little meat, fresh or preserved, was listed in the town, as opposed to livestock. Many households in Colchester and the four villages were engaged in small-scale raising of animals. Where there were cows, most households usually had only one. There were very few sheep, but there is much more evidence for the presence of pigs. Excluding the six manorial households, 149 taxable households had pigs; in the vast majority of households where there were pigs or piglets, there was only one.

Does the presence of livestock reflect meat consumption, or a pattern of production? Pigs were possibly raised for sale. Their numbers echo the figures in Domesday for eastern England: tax assessments and the archaeological evidence suggest that comparable households elsewhere kept more sheep. The presence of carcasses, however, suggests that some at least were intended for local consumption. The value of a pig carcass was substantial, some reaching 4*s.*, as much as a beef carcass; and a piglet might be worth 1*s.* Regular meat eating on a modest scale is also suggested by the number of butchers in Colchester, a common feature of market towns. Six individuals described in this way held goods worth between 6*s.* 6*d.* and 26*s.*, with two with much more substantial valuations (for Colchester), at just over £3 and £7 15*s.* 2*d.* The total value of meat in their stocks ready for sale, or preserved, was 64*s.* 6*d.*, and just under half of that was in the hands of one man—a total of just under 2*d.* worth of meat for every household in the assessment. There were four further instances of preserved meat or a larder, with a value of 25*s.* in all. Livestock from the surrounding area brought into Colchester for sale may have allowed others access to meat.

An assessment for the ward of Henry de Gernemuta in the much wealthier borough of King's Lynn, *c.*1285–90, covers forty-three households with a total taxable value of £1,501 6*s.* 10¾*d.* Only seven sheep were among the goods of the ward; fifteen beef carcasses were noted, distributed between only ten households, and there were six cows in a further four households; but there were fifty-two pigs, one sow, forty-three piglets, and nineteen hams. The assessors valued meatstock at a similar level to Colchester, with piglets worth about 1*s.* each, each carcass of beef 3*s.* to 4*s.*, and hams at a considerable premium, some valued at more than 4*s.*[3] Even if a proportion of the pigs was destined for consumption

[1] Jurkowski, Smith, and Crook (1998: pp. xxix–xxx).

[2] Britnell (1986: 15–17); the assessments are printed in Blyke, Pridden, Strachey, and Upham (1783: i. 228–65).

[3] Owen (1984: 235–49).

elsewhere, the burgesses and innkeepers of Lynn far outstripped the inhabitants of Colchester in their carnivorous habits, particularly in their access to beef.

At the same time, the modest levels of stock and meat in Colchester surpassed those in the countryside. In Cuxham, an Oxfordshire village with a population probably in excess of 110 at the start of 1349,[4] the assessment of the sixteen taxable households in the village made on 12 March 1304 noted goods worth £13 10*s.* 7*d.* As this return was taken during Lent, one would not expect to find evidence of fresh meat. The only meat in store was ham: eleven were recorded, with an average value of just over 8*d.* each. Not only were these hams much poorer affairs than those in the towns, but there was no sign of preserved beef, mutton, or other meat in store. In terms of livestock, besides draught animals, ten households had a cow each and there were five calves; there were only six small pigs (*porculi*) and three piglets, although some of the peasants had geese and hens.[5] The pattern echoes the predominance of pig bones found on rural sites, indicated in Chapter 6, with the cows kept principally for a modest level of dairy production. The Merton College manor at Cuxham, which was taxed separately (and valued at £20 19*s.* 5*d.* in 1304), provided some meat for workers at harvest from its small group of pigs. In 1298–9, for example, five hams were consumed as harvest expenses, and the offal from the pigs was eaten by visitors to the manor. In 1317–18, four were consumed at harvest and one by visitors; the carcasses of six pigs were sent to the warden's hall in Oxford.[6] This pattern is familiar from that found in peasant maintenance agreements, with their evidence of the desperately low availability of meat in the countryside before the Black Death: even the middling and wealthy peasant households probably did not consume more than one or two pig carcasses a year.[7] Lists of inhabitants also suggest that at this period butchers were rarely based in the countryside. Slaughtering of animals might either be carried out on occasional visits, or be the part-time task of manorial workmen or others in the village.

After 1349, a very different situation prevailed, with much more livestock available per capita; and English evidence matches the large-scale rise in meat eating across Europe,[8] with a well-documented growth in the stocking of cattle, and an increase in both the meat trade and dairying.[9] References in sermons to the butcher's dog, with its bloody mouth, and the cries of compassion of swine and oxen for their fellows as their lives were brought to an end, indicate a much greater familiarity with meat processing.[10] Above all, there is much greater evidence for the trade in meat. Mutton was sold jointed, by the breast and side,[11] with the 'fatt chepe scholdirs' offered in Bread Street a particular attraction for Londoners of the 1350s and 1360s.[12] Butchers and their shambles were a prominent feature of almost every town, and trade in livestock supported their growth in

[4] P. D. A. Harvey (1965: 135). [5] Harvey (1976: 712–14).
[6] Harvey (1976: 313, 330, 332). [7] Dyer (1983: 205–7; 1998*a*: 154–6).
[8] Hilton (1983: 110–13); Dyer (1983); Stouff (1970). [9] Campbell (2000: 134–68); Dyer (2004).
[10] Owst (1933: 37–8). [11] Woolgar (1992–3: ii. 454). [12] Trigg (1990: 16).

the late medieval period.[13] The change in diet in the countryside was equally considerable: in 1256, the food allowances for the harvest workers of Sedgeford have been estimated at some 12,967 kcal per person, a ration clearly intended to feed several assistants as well as the principal worker. Of this total allocation, only 243 kcal came from meat. In 1424, on the same estate, out of the harvest worker's daily ration of 4,968 kcal, 1,169 kcal came from meat. In the thirteenth century, bacon was the main constituent of the meat ration, although beef was included towards the close of the century, and formed an increasing proportion, along with mutton and offal.[14]

Meat consumption: the upper classes and institutions

Before the Black Death, the contrast between the countryside and the practices of meat eating within aristocratic households could not have been greater. Seigneurial husbandry was geared to supplying the best-quality meats for the household, dispatching poorer stock to market.[15] Aside from small cash purchases of meat, the household of Thomas de Courtenay at South Pool, in the year from 29 September 1341, consumed forty-five and a half carcasses of beef, two calves, sixty-four sheep and three lambs, fifty-one pigs, one boar, and thirty-two piglets. With the exception of Lent, the household consumed one beef carcass a week, one or two pigs or the equivalent in bacon each week (probably as fresh meat from October to December; then, apart from the week of 14 April, as preserved meat); and piglets between the end of September and the end of December. Mutton was eaten between the end of October and the start of Lent, with one sheep in the week of 14 April, and three between the weeks of 28 July and 25 August. The beef came from demesne manors, aside from five oxen and two cows that were bought for the larder around the start of November.[16] There was a similar pattern of purchase in the household of John de Multon of Frampton. In 1343, the larder was stocked at Martinmas (11 November) with the meat of five cattle, forty sheep, and twenty-four pigs. In this year and in 1347–8, the purchase was supplemented by fresh meat, the latter account including regular purchases of veal from Martinmas through to Lent.[17]

Aristocratic households acted as leaders in fashions of consumption elsewhere, in monasteries, among the gentry, and upper echelons of the peasantry. Their already substantial consumption of meat can be shown, after the Black Death, to have become an extraordinary excess. By the fifteenth century, it is possible to estimate the quantities appearing in individual meals. On a day when meat was usually eaten, in October 1420, an ordinary member of the Earl of Warwick's household would have received approximately 1.84 lb of beef and

[13] Laughton and Dyer (2002). [14] Dyer (1994*b*: 81–7).
[15] Farmer (1991*a*: 386); Campbell (2000: 141). [16] Cornwall RO, AR12/25.
[17] Woolgar (1992–3: i. 233, 242–5). Compare the patterns in Laughton and Dyer (2002: 173–8).

1.28 lb of mutton (*c.*2,300 kcal) in the two meals of the day; a gentle member, however, would have had a further 2.5 lb of pork (largely piglets) and 2 lb of poultry at each of these meals, besides breakfast, bringing the total food set before him to nearly 13,000 kcal.[18] At the same time, these households had a predilection for younger, more tender and succulent meat. While pork may have been eaten comparatively young, from animals no more than one or two years old, even among the peasantry,[19] in the households of the high aristocracy, the upper clergy, and especially among women, there was a movement towards lighter meats in general, from younger animals, and towards poultry, birds, and game.[20] By the fifteenth century, there were specialists supplying younger meat, such as John Manchester, who supplied ready-butchered veal to the household of Sir Henry Stafford in 1469—a trend that is reflected in the archaeological evidence reviewed in Chapter 5.[21] The household of John Hales, Bishop of Coventry and Lichfield, preferred veal, piglets, and lambs in May and June 1461. The lamb, ranging from 10*d.* to 15*d.*, was more expensive than mutton, at half a carcass for 8*d*. Some beef was drawn from stock, but more was bought fresh in the market, especially veal, with one calf on most meat days.[22] The change in the great household's meat consumption overall is particularly striking between the fourteenth and sixteenth centuries. By 1500, beef consumption was dominant. Pork, in particular, was eschewed at this level and there was a common perception that it was peasant food, a reflection that is found in the archaeology discussed in Chapter 6 (Table 7.1).[23]

This pattern is mirrored in institutional records—many of which aped upper-class patterns of consumption—but with some variation. At Winchester College in 1556–7, thirty-six cattle and 500 sheep were acquired by the bursars to feed the College, but no pork was purchased, although some may have been reared at the College to be consumed there as ham and bacon.[24] A century earlier pork had been an important element in the diet, supplied to the College especially from its manor at Harmondsworth: in 1450–1 the bailiff delivered one boar and forty-seven pigs.[25] At the hospital of St Giles, Norwich, however, during the first thirty years of the sixteenth century, there was a trend away from beef consumption, some diminution in mutton, and a growth in the quantity of pork eaten.[26] The archaeological evidence reviewed in Chapter 6 confirms this variability in the taste for pork.

Meat products, sheep, and goats

If these were the general outlines of consumption, further patterns emerge from a closer examination of meat products. Prominent among them are the arrangements for preserving meat. In autumn, when pork was butchered, there was a brisk

[18] Woolgar (2001: 15–18). [19] Dyer (1998*a*: 58–60). See also Chapter 6.
[20] Woolgar (1999: 133–4). [21] Woolgar (2001: 15). [22] Woolgar (1992–3: ii. 452–63).
[23] Furnivall (1870: 272–4). [24] WCM 22213. [25] WCM 11504, m. 2^{d}.
[26] Rawcliffe (1999: 178, 186).

Table 7.1. Weights and proportions of meats consumed in aristocratic and ecclesiastical households, *c.*1350–*c.*1500

	Cattle	Veal	Mutton	Boars	Pigs	Piglets
Richard Turberville, 1358–9	10,400 73.9%		629 4.5%		3,037 21.6%	
John Dinham, 1372–3	4,798 72.5%		952 14.4%		869 13.1%	
William of Wykeham, Bishop of Winchester 1393	47.5%		13.8%		38.7%	
John de Vere, twelfth Earl of Oxford, 1431–2	11,772 72.8%		3,385 20.9%		302 1.9%	706 4.4%
Humphrey Stafford, first Duke of Buckingham, 1452–3	59,700 61.9%		35,315 36.6%		554 0.5%	932 1%
Sir William Skipwith, 1467–8	11,772 75.5%	150 1%	1,789 11.5%	140 0.9%	1,109 7.1%	617 4%
Richard Bell, Bishop of Carlisle, 1485–6	6,727 82.2%		1,451 17.8%			
Edward Stafford, third Duke of Buckingham, 1503–4	38,676 48.9%	13,310 16.8%	25,679 32.5%		592 0.8%	819 1%

Note: Meat weights in lb, based on Harvey (1993: 228–30).

Sources: Woolgar (1992–3: ii. 490–1, 497–8, 540–1, 554–5, 560); Harris and Thurgood (1984: 37–9); WCM 1; Staffs. RO D641/1/3/8, m. 1ʳ.

trade in chines, the backbone and immediately attached areas left over after the sides of meat had been taken for preservation.[27] The consumption of other by-products and offal concentrated particularly at these times of year. Evidence for sausage or pudding making is focused on the period of slaughter, particularly in December.[28] At this time of year sausages regularly appeared on the table at Merton College, Oxford, in the 1480s and 1490s, especially during the Christmas period, but occasionally at other feasts, such as the Purification (2 February).[29] The bill of fare at the three shepherds' fanciful meal in the Towneley First Shepherds' play, set at the first Christmas, included two *blodyngs* and a *leveryng*, that is, black puddings and a liver sausage, as well as an ox tail and meat from a sheep that had died of disease.[30] The records of Merton College contain useful information about offal and poorer cuts of meat, such as neck (*gullatts*),[31] tongue—consumed on a series of Tuesdays and Wednesdays in July 1492 and from September to November 1493[32]—and sheep heads.[33] These last must have been commonly available, from Sir Thomas More's reference to the rebels led by Jack Cade, who aimed 'to kyll up the clergie, and sel priestes heddes as good chepe as shepes heddes, thre for a peni, bie who would';[34] and sheep offal was a

[27] Woolgar (1999: 116–18). [28] Woolgar (1992–3: i. 125–6).
[29] Fletcher and Upton (1996: 16–17, 231, 235, 287–8, 383).
[30] Stevens and Cawley (1994: i. 114–15). [31] Fletcher and Upton (1996: 14, 17).
[32] Fletcher and Upton (1996: 471–5, 503–8; and intermittently before, e.g. 288).
[33] Fletcher and Upton (1996: 252). [34] Cited by Owst (1933: 290).

regular ingredient in monastic diet, in umbles.[35] Other by-products from butchery included calves' feet, for gelatine, calves' heads, for sweetbreads, and marrow bones.[36] Butchers were sometimes required to sell their offal to the urban poor—and the cheapness of this food must have made it attractive to collegiate caterers.[37] Fats were used for frying and suet was used in desserts like doucet, a sweet, custardy dish.[38]

Some meats were also particularly associated with events in the calendar. Lamb was highly prized as an Easter dish.[39] Collops—slices of pork—were the traditional fare on Collop Monday, two days before the start of Lent;[40] and Bishop Buckingham of Lincoln was scandalized that bacon and hard-boiled eggs were both provided and blessed in church by the rector of Nettleham on Easter morning before communion was taken.[41] One specialism was the fattening of boars in preparation for major feasts, particularly, but not exclusively, Christmas.[42]

The regional economies of England had different emphases in terms of livestock production, although these were not necessarily reflected in local consumption as livestock was usually moved before slaughter close to its place of consumption. Cattle breeding was a dominant feature of the pastoralism of the north and Wales. Goats were mentioned much less often, but they graced tables ranging from those of Bolton Priory and Bishop Mitford of Salisbury, to those of Merton College.[43] The rarity of goats is confirmed by the scant archaeological finds. The few remains that have been found have mostly been of kids.[44]

Dairy products

Dairy produce was among the most ubiquitous of foodstuffs in late medieval England. From Piers Plowman's two green cheeses, to the diet of kings, from the monastery to the tavern, it was highly esteemed and enjoyed. But if it was ubiquitous, it was not the most substantial element in diet, and its availability was closely linked to the scale of livestock husbandry. Milk, cheese, and butter had been consumed in the earlier Middle Ages, the products of both sheep and cattle, but it is not until the proliferation of the historical record in the thirteenth century that one can make a realistic assessment of their contribution to diet.

Milk, cream, cheese, and butter were the product of distinctive patterns of animal husbandry. Husbandry using sheep for milk was a major element in the economy in the thirteenth century, but it diminished subsequently with a concentration on wool production. Dairying with cattle was always important

[35] Harvey (1993: 219–20). [36] Fletcher and Upton (1996: 232, 315, 447).
[37] Dyer (1998*a*: 199). [38] Woolgar (1992–3: i. 229; ii. 453).
[39] Woolgar (1992–3: i. 224, 360). See also Chapter 14. [40] Woolgar (1992–3: ii. 437).
[41] Owen (1972: 141). [42] Woolgar (1992–3: ii. 500, 515).
[43] Kershaw and Smith (2000: 251); Woolgar (1992–3: i. 360); Fletcher and Upton (1996: 127); Dyer (2004).
[44] See Chapter 5.

Plate 7.1 A dairymaid (to judge by her headgear, a married woman) milking a cow, from a bestiary of *c.*1240–50, possibly produced at Salisbury, Bodleian Library, MS Bodl. 764, fo. 41^{v}. Photograph: ©Bodleian Library.

in some localities, but became more widespread in the fourteenth and fifteenth centuries (Plate 7.1). Goat keeping was important in this economy only to a minor degree. In the later Middle Ages, cream and butter were produced primarily from cows' milk, although a little butter was made from the milk of ewes;[45] cheese was made from the milk of cows, goats, and ewes. The commodities have different characteristics and durability. Milk is bulky to transport and goes off comparatively quickly. Converting it to butter and, especially, cheese offered better opportunities to preserve it. Although the period of production varied, the height of the dairying season encompassed the summer months when most stock were lactating and when milk was at its most vulnerable to heat. Thus, at Cuxham, in 1318, cheeses were made at the rate of one a day between 26 July and Michaelmas, with production diminishing into October, followed by a gap until June 1319. In 1351–2, the bulk of the cheese was produced between the end of April and mid-August. None was made between 1 November and 28 April.[46] Treatises on estate management indicated that the principal season of cheese making should end at Michaelmas, having run from a range of dates in April to May.[47] A longer season was possible for cows (but not for ewes), extending to year-round production, albeit with some fall-off in November and December, if there were supplementary stall feeding. Without that, the season might not last more than about five months.[48] The conversion of milk to cheese and butter therefore meant that dairy food was available for consumption at times of the year when comparatively little was produced.

Dairy produce was highly regarded and widely enjoyed. At one end of the scale was a premium product celebrated in royal and aristocratic circles, for

[45] Kershaw (1973*a*: 102). [46] Harvey (1976: 350, 522).
[47] Oschinsky (1971: 209, 283–5, 302). [48] Farmer (1991*a*: 402).

example, in the later Middle Ages, the cheeses of the priory of Llanthony by Gloucester.[49] Special coffers were made to carry the cheese of Edward I.[50] French cheeses were imported: the ship of Pevensey taken by Randulph de Oreford in 1242–3 carried goods from Rouen, including four cheeses.[51] And English cheeses were exported to France—particularly to some of the great abbeys of Normandy, as a part of their rents—and elsewhere on the Continent. The virgaters of Combe, in Hampshire, *c.*1230–47, owed carrying service taking cheese to Southampton at the time that the cheese was sent each year from England to the abbey of Bec, an obligation also of the tenants of Monxton, Quarley, and Brixton Deverill. The Prior of Ogbourne, whose duty it was to administer this part of the lands of Bec, owed to Bec each year a total of 32 weys of cheese (5,760 lb).[52] The virgaters of Minchinhampton were likewise obliged to carry their cheeses to Southampton, *c.*1170, for shipping to the abbey of the Holy Trinity at Caen, a practice that continued as late as 1320 and beyond.[53] Tolls on English cheese are recorded at Arras in 1036, and at Dieppe and Rouen in the fourteenth century.[54]

These exports are typical, however, of the business done by great estates, or specialists in the product. In terms of volume of trade, cheese as a premium product for export may have been exceptional. The centralized marketing of dairy produce, often in tandem with the wool crop, lessened in the fourteenth century, as the stock of manors was frequently leased to cowherds or dairymen. Without bargaining power, commercial contacts, and middlemen, the potential for marketing beyond a local level was diminished.[55] There is other evidence for the trade in cheese carried out locally, by small producers. At Exeter, the villeins from Pinhoe brought cheese and butter the 3 miles for sale at market.[56] The exemption of the town of King's Lynn, *c.*1284, from tolls on cheese, butter, fat, and vegetables, unless they were sold in volume, was a concession to this low-level trade. Orders made in a court leet at Lynn in the early fifteenth century reinforce this view of small producers.[57] Here female traders (traditional of dairying) carried on small-scale production from home. Where the trade was more substantial, male merchants, the cheesemongers, were involved. Both London and Norwich had cheesemongers by the late thirteenth century[58] and probably earlier, if the death of Elias le Pourtour, who fell dead in Bread Street in London, on 7 July 1227, carrying a load of cheese, is an indication.[59]

[49] NA E 101/414/16, fo. 29^{r}; Nicolas (1830: 14, 18, 33, 37, 44); Rhodes (2002: 149).
[50] Byerly and Byerly (1977: 39, no. 376). [51] Nicol (1991: 33).
[52] Chibnall (1951: pp. x–xi, 41–2); Morgan (1968: 42, 77–8). The weight of the wey is taken from the calculations in the Beaulieu Account Book, using 180 lb to the wey, or 15 stones of 12 lb each; but there was variation: Salzman (1931: 284 n. 6): a wey of cheese was usually 256 lb, but of cheese in Essex apparently 336 lb. [53] Chibnall (1982: 60, 109).
[54] Salzman (1931: 284); Morgan (1968: 42). [55] Farmer (1991*a*: 401–4).
[56] Kowaleski (1995: 286–7). [57] Owen (1984: 102, 422). [58] Carlin (1998: 30–1).
[59] Chew and Weinbaum (1970: 12–13).

The peasantry and dairy foods

In the late 1360s, Piers Plowman's two green cheeses were carefully husbanded against hunger and the dangers of excess. He had a modest amount of dairy products, oats, beans, and bran, parsley, leeks, and some vegetables, but no meat (or meat-producing animals) and no eggs. His small selection of livestock comprised a cow, for milk, with its calf, which he would no doubt sell, and a mare for a cart; but he relied on his cereal harvest, investing in the manuring of the fields, to stave off hunger in the longer term.[60] He was, however, much better endowed than many, particularly landless labourers. The direct link between dairy products and the amount of livestock continued to restrict the availability of dairy foods. The yield of a single cow on a demesne was unlikely to exceed 90 lb of cheese and butter a year. Peasant livestock may have been more productive, as it may have been better fed. Sheep might appear a less effective option in terms of dairy production on a small scale, needing about a dozen to equal the yield of a cow, but taxation returns, for example, for Blackbourne Hundred in 1283, indicate that many households did have a few ewes.[61] Between a quarter and a half of peasant households would not have owned livestock, and would not have had ready access to the daily average of ¼ lb of dairy produce per cow.[62]

Those peasants who were obliged to carry out harvest works for the lord were frequently entitled to cheese among the foodstuffs they were given for their labour. The boon workers at harvest at Bishopstone, a manor of the Bishop of Chichester, were to have on flesh days, at lunch, pottage, wheat bread, beef, and cheese; and at supper, bread, cheese, and ale; and on fish days, pottage, wheat bread, fish, and cheese, and ale.[63] But it was nonetheless not an excessive amount. Calculations of the diet of harvest workers of Norwich Cathedral Priory, *c.*1300, emphasize the cereal component of the diet. For every 2 lb of barley bread a harvest worker received 2 oz of cheese—that is, twice the daily allowance he might have had in a household of four, possessing one cow—1 oz of meat, and 4½ oz of fish.[64]

Monastic diet and dairy products

There is firmer evidence for the presence of dairy products in monastic diet. The observances of the different Orders placed different emphasis on elements in dietary regimes. Whereas meat, including the flesh of quadrupeds, had come to play a significant role in Benedictine diet in the later Middle Ages, the Cistercians aimed to adhere to the letter of their monastic rule and therefore to avoid flesh, lard, and, before the thirteenth century, the dietary supplements of pittances, discussed in Chapter 15. Cistercian diet focused on cereal products, vegetables, fish, dairy foods, and, exceptionally, honey—but excluded eggs and dairy foods in the seasons of

[60] Kane (1960: 341–2, ll. 264–75). [61] See Chapter 14. [62] Dyer (1983: 207–8).
[63] Homans (1942: 261). [64] Dyer (1998*b*: 59).

Plate 7.2 The refectory of Beaulieu Abbey, on the south side of the cloister, early thirteenth century, now Beaulieu church. Photograph: C. M. Woolgar.

abstinence, Lent and Advent.[65] From the second half of the fourteenth century, the differences in diet between the Orders were much less significant.[66]

Accounts for the Cistercian abbey of Beaulieu, for 1269–70, trace the production of dairy foods on its manors and granges.[67] On those at a distance from the house, in Oxfordshire, Wiltshire, and Cornwall, cheese was bought for harvest workers, or sold, or consumed on the grange. The cheese produced on the granges of the Great Close in the New Forest, those nearest to Beaulieu, at St Leonards, Bovery, Harford, and Otterwood, accounted for most of the 65 weys (11,700 lb) that the subcellarer received (Plate 7.2). This cheese was made from ewes' milk: it was not the total cheese production of these local granges, as the shepherds took a portion, and a further portion was consumed at the grange. A lesser beneficiary was the Abbey's porter, who drew from the granges all the cheeses there on the two Sundays before St John the Baptist (24 June) and the Sunday immediately after, at the height of the cheese-making season. Of the cheese received by the subcellarer, only 16 weys 4 stone (3,328 lb) were consumed as cheese, unmodified, by the convent. A further 2 weys were added to the convent's pottage. Seven weys were left at Michaelmas, approximately half a year's supply for the convent. Much, however, was wasted: 11 weys, about 15 per cent of the total, were lost as the cheese dried out. The records for butter show a similar pattern. Of 6 weys

[65] Knowles (1963: 210, 641). See also Chapter 13.
[66] Kershaw (1973*a*: 156–7). See also Chapter 13.
[67] Hockey (1975).

6 stone, 2 weys went on the expenses of the convent, 2 further weys for their pottage; the keeper of the guest-house received 6 stone; and just over 1 wey was lost as the butter melted. The accounts for clothing and shoes indicate that at this stage the Abbey had not less than seventy-three monks and sixty-seven or sixty-eight lay brothers.[68] Despite the quantities, therefore, the allowance of dairy foods remained modest: 3,688 lb of cheese and 720 lb of butter would have produced a ration on days when dairy produce was eaten per monk/lay brother of just over 2¼ oz of cheese and a little under ½ oz of butter.[69]

The presence of cheese, albeit in substantial quantities, does not imply that it comprised a substantial element in diet. At Bolton Priory, a house of Augustinian canons, it has been calculated that consumption, on average, between 1305 and 1315 was 115 stone (1,725 lb) of butter and 310 stone (4,650 lb) of cheese a year. Bolton was in an area of pastoral farming and from 1298 until the famine of 1316–17 rarely purchased dairy products, drawing them from its own livestock. The diet of its canons, however, was largely based on cereals, with the weight of cereals, meat, and dairy produce approximately in the ratio of 100 : 6 : 1.[70] At Westminster Abbey, the consumption of dairy foods ran by the end of the fifteenth century at perhaps as much as half a pint of milk a day for a monk between Easter and September; and, apart from Lent, Advent, and Fridays, 2½ oz of cheese a day, with butter used for cooking—quantities that were not significantly different from those at Beaulieu in the thirteenth century.[71]

Dairy food in high-status diets

In the diet of the nobility the amount of dairy products was, in the thirteenth and fourteenth centuries, comparatively limited. In 1341–2, the household of Thomas de Courtenay, at South Pool, with about twenty servants, consumed during the year 227 lb of cheese (that is, a little over 1 lb of cheese every meat day), 48 lb of butter (about 1 lb a week outside Lent), and 40 gallons of milk.[72] There was more dairy produce in the diet of the young, and it was used in some medicines.[73] Milk and cheese were stronger elements in consumption in households headed by women, although butter appears less often. While the household of Dame Alice de Bryene bought cheese for boon workers and other labourers in 1412–13 and 1418–19, in both these years there was a small account for milk and cream for the lady's chamber, in the first year at a total cost of 19*d*., in the second, 2*s*. 2*d*.[74] There were definite preferences for dairy products in the household of the Countess of Warwick in 1420–1. Along with her three daughters, six gentlewomen (one of whom had a child with her), and the female chamberers, the

[68] Hockey (1975: 17–18).

[69] Based on: monks, 73 × 238 days (i.e. not Fridays, the long Lent, or Advent); lay brothers, 18 at Beaulieu all year, × 238 days, and 50 on the granges returning to Beaulieu for 87 Sundays and feast days = 17,374 + 4,284 + 4,350 = 26,008 days. Hockey (1975: 16–23, 129–30).

[70] Kershaw (1973*a*: 155, 158). [71] Harvey (1993: 61–2, 70). [72] Cornwall RO, AR12/25.

[73] Frisk (1949: 67, 69). [74] Dale and Redstone (1931: 103, 118).

Countess focused dairy consumption on the days when some fish was eaten: milk, butter, and cheese appeared only on Wednesdays, Fridays, and Saturdays; and cheese mainly on Fridays, with a caesura only for Lent.[75]

Some aristocratic households maintained home farms, from which dairy foods might be drawn directly. As they may consequently have fallen outside accounting mechanisms, their consumption may be underestimated here. The account for the great household of Humphrey Stafford, first Duke of Buckingham, for 1452–3, at Maxstoke and Writtle, says little about dairy products, but does record six milk cows that were kept for the household.[76] Other great households relied on the market: butter and milk, together with eggs, were purchased each day for Sir Henry Stafford at Woking in 1469. This account displays another important feature, in the much larger quantities of butter that were purchased.[77]

While there had been some dishes that regularly used milk, butter, and cheese, the fifteenth century saw a much wider use of dairy products in aristocratic households, particularly as meals expanded and more effort was put into foods that one might class today as desserts. At the funeral feast of Bishop Mitford, on 7 June 1407, the dairy produce included 169 gallons of milk, 17 gallons of cream, and sixty-two cheeses.[78] The linen purchased for the Countess of Warwick in 1420–1 included 4 ells of linen (*spinall*) for jelly-cloths and cast-cream.[79] In the household of John Hales, Bishop of Coventry and Lichfield, in 1461, there were regular appearances of milk and cream, sometimes with honey, butter, eggs, and currants, for making doucet.[80] Cheese continued to appear by itself at the close of meals. John Russell's *Boke of Nurture* commended hard cheese almost certainly at the conclusion of the meal, along with wafers and hypocras, to aid the digestion.[81]

Conclusion

In the later Middle Ages, meat and dairy products supplied major elements of protein in diet, but access to these foodstuffs varied widely. Before the Black Death, historical sources show that the consumption of meat was severely restricted, principally constrained by the small amount of livestock relative to the human population, a pattern echoed in the cull patterns in the archaeological data. Even at this period, however, the consumption of large quantities of meat marked out upper-class eating habits. But meat and meat products were scarcely present in the diet of many before the Black Death, so any element of protein—and especially fat—derived from cheese, milk, and butter must have been a particularly significant addition to a diet largely based on cereals.

[75] Longleat MS Misc. IX, fos. 35^{r}–38^{r}.
[76] Harris and Thurgood (1984: 46).
[77] Westminster Abbey Muniments 12186, fos. 83^{r}–84^{v}.
[78] Woolgar (1992–3: i. 403).
[79] Longleat MS Misc IX, fo. 17^{r}.
[80] Woolgar (1992–3: ii. 452, 455, 458, 465, 467, 471, 473–4, 478); Harvey (1993: 61).
[81] Furnivall (1868: 123).

Both meat and dairy produce grew in importance in the second half of the fourteenth and fifteenth centuries, largely at the expense of the cereal component in diet. For a few, meat was the source of excessive amounts of protein, with wider use of selected dairy products in addition. If this was a model of consumption to which many might aspire, few would achieve it except on limited occasions, at boon feasts or as guests at Christmas celebrations. For many, the contribution of these foodstuffs to their diet cannot have exceeded a modest element, although by the fifteenth century that ration had grown significantly and less desirable foods, such as offal, might now be readily available to augment the diet of the poor. At the same time, as meat consumption became both more desirable and widespread, elements of distinction maintained its social cachet: eating young animals—piglets, calves, lambs, and kids—became a pronounced feature of the diet of the elite, supported by a specialist trade.

Meat and dairy produce were exceptionally valuable: the resource was carefully husbanded both to ensure that it was available for deferred consumption, at times when it might not be available fresh, and with the intention of creating an attractive foodstuff that might, in some circumstances, command a premium for its producer. Considerable investment was devoted to preserving meat, through salting or smoking; and a trade was well established in meat by-products, taking advantage of seasonal patterns of slaughter, and in dairy foods.

8

Fish Consumption in Medieval England

D. SERJEANTSON AND C. M. WOOLGAR

In the Middle Ages fish was, for some, a convenient and possibly free food. For others, it was at least another possibility in the face of the Church's restrictions on the consumption of meat, but it might offer more. A Wycliffite sermon, arguing that in the eyes of Christ fishing was a nobler activity than the genteel art of hunting, noted that fish were closer than meat to the food that men will have in Paradise.[1] As penitential food, there was an enormous demand for fish from many sections of society, particularly in Lent but also on the regular fast days throughout the year. These attitudes towards the eating of fish can be seen in historical sources for the later Middle Ages, but how far back into the early medieval period do they go?

The archaeological and historical evidence for the consumption of fish in medieval England is complementary in many ways. Most usefully, archaeology shows which fish were eaten in the earlier Middle Ages, before household accounts and other written evidence become available. Historical sources reflect particularly the experience of the households of the wealthy and the religious, providing important information about the dynamics of consumption, seasonality, and dietary impact.[2] Archaeology can widen the range of material to be considered, though unfortunately its potential is also very limited for revealing the eating habits of the peasantry. Much attention has been given to freshwater fish—the construction of fishponds was an important social statement, and some remain prominent in the landscape[3]—but marine fish and fisheries were much more important in dietary terms. Archaeology can reveal the origins of the production and trade in preserved marine fish—such as herring and cod—which lie in northern Europe in areas and centuries where little or no documentation survives, as well as their supply and consumption in England; written sources for the later Middle Ages supply evidence of the dynamics and volumes of these trades.

We are grateful to Alison Locker for her comments on a draft of this paper.

[1] Hudson and Gradon (1983–96: ii. 194–5; v. 208).

[2] Harvey (1993); Woolgar (1995; 1999; 2000).

[3] Aston and Bond (1988*a*); Dyer (1994*c*).

Plate 8.1 Eel fishermen with traps on the River Creuse, 1986. The traps are of a design that has not changed since the medieval period. Photograph: D. Serjeantson.

Historical research over the last three decades has produced a number of studies of fishing and of the supply and marketing of fish.[4] These have done much to establish the overall pattern of fishing and fish supply, and from them it is possible to make some general statements. Catches of fish were drawn from three principal environments: the sea, including estuaries, inshore, and deep waters; rivers and streams; and fishponds. Before the Norman Conquest, rivers and estuarine or inshore fishing were the most important sources. There is archaeological evidence for fish traps in rivers and estuaries, as well as weirs on the foreshore. Descriptions of river fisheries, especially in a monastic context, can be found in Anglo-Saxon charters possibly dating from the seventh century onwards, with large quantities of eels expected as renders in Domesday Book (Plate 8.1).[5] The layout of parishes in some coastal counties, with detached elements adjacent to the sea, giving access to this resource, probably preserves a very early arrangement.[6] There was a pre-Conquest herring fishery and Domesday records substantial renders of herring in East Anglia, Kent, Surrey, and Sussex.[7]

[4] Littler (1979); Childs and Kowaleski (2000); Kowaleski and Childs (2000); Woolgar (2000), Kowaleski (2000); Fox (2001); Locker (2001).

[5] Losco-Bradley and Salisbury (1988); Godbold and Turner (1994); Bond (1988: 69–112, especially 72–92).

[6] Fox (2001: 48–51).

[7] Campbell (2002).

Both historical and archaeological evidence suggests that two important changes probably came during the eleventh century, although the speed with which they took hold is unclear. Requirements for fasting appear in Anglo-Saxon contexts from an early period, and are clearly laid down in documents such as the Constitutions of Archbishop Oda of 942 × 946, the ecclesiastical section of Edgar's law code of 960, and one of Aelfric's pastoral letters, of 993 × *c.*995, addressed to Bishop Wulfsige III of Sherborne, which enjoined on everyone fasting on all Fridays throughout the year, except between Easter and Pentecost, Christmas and the octave of Twelfth Night, and other feasts celebrated by all. There may be an association between this emphasis on fasting and the spread of the Benedictine reforms of the tenth century.[8] What is less clear, however, is whether fasting necessarily implied eating fish instead of flesh, although it will have done so in some contexts. Another of Aelfric's letters, to Wulfstan, the Archbishop of York, *c.*1006, noted that the clergy might not hunt or hawk, but they could fish and thereby obtain food, citing the biblical fishermen Peter and Andrew as precedents. Here was a case that allowed the clergy to eat fish, particularly as opposed to meat, but it did not tie it to fasting, and other options for consumption on these occasions may have been a more virtuous solution.[9] At about the same time, in the decades around 1000, there is a remarkable growth in the presence of remains of herring and cod, especially on major urban sites, and it has been argued that they formed an integral part of a commercial revolution.[10] The progress of these changes may not have been straightforward and the impact made by cod and herring on diet in the countryside remained muted. In the years after the Conquest, some major monastic foundations had to persuade their monks to eat fish, rather than meat or meaty dishes, which suggests both a new zeal for clerical abstinence from flesh and that the place of fish in monastic diet might be opaque.[11] From this point on, however, there was to be greatly increased consumption of fish, with the development of marine fishing, especially beyond inshore waters.

Secondly, it is from this period as well that there was a major investment in the construction of ponds for freshwater fish, to supply the luxury end of the market. A poem, written shortly after 1200, on the martyrdom of St Edmund, King of East Anglia, killed by the Danes in 869, records that after the Angles and Saxons had conquered England they found a rich harvest of freshwater fish: this was one of the attractions of the land, or it was to an Anglo-Norman poet who imagined that it had also been so to his precursors as invaders.[12] The effect of these changes endured to the end of the Middle Ages, although fish eating declined in the fifteenth century. An Italian, probably a noble Venetian, visiting England in 1496–7, pointed to the abundance of fish drawn from rivers, springs, and streams, especially salmon, while at the same time noting the great preference for seafish.[13]

[8] Barrett, Locker, and Roberts (2004*a*: 629–30); Whitelock, Brett, and Brooke (1981: i. 73–4, 101–2, 225–6).

[9] Whitelock, Brett, and Brooke (1981: i. 300). For the association between virtue and diet, see Chapter 13.

[10] Barrett, Locker, and Roberts (2004*a*, 2004*b*).

[11] Woolgar (2000: 36).

[12] Grant (1978: 68–9).

[13] Sneyd (1847: 9).

Assessing the evidence

It is difficult to establish from historical sources how far fish was eaten before the Conquest, but there is very good evidence for it by the end of the twelfth century, particularly for the widespread consumption of marine fish in monasteries and great households. At the same time, there is a need to distinguish between the consumption of fresh fish and fish that had been preserved—and it was overwhelmingly marine fish that were preserved, along with salmon and eels, which migrated between fresh and salt water and which might be caught in either. Marine fish could have been transported fresh to all parts of England, since nowhere was further than two days' journey by packhorse from the coast. Preservation, however, was crucial to its widespread distribution and consumption at times when it might not be readily available fresh, or available in sufficient quantities to meet demand. A small group of species constituted the main staples of this market: herring, either red (smoked and cured) or white (salted or steeped in brine); and white-fleshed fish, such as cod, preserved either as salt fish or stockfish (originally wind-dried cod, but sometimes used to describe any white-fleshed fish that had been salted and dried). These fish may not always have been the most attractive in terms of consumption, but there is ample testimony in the later Middle Ages to their widespread presence. Margery Kempe expected that she would be eaten and gnawed by the people of the world 'as any raton knawyth þe stokfysch', and her Christ knew she would stick to him as closely 'as þe skyn of stokfysche clevyth to a mannys handys whan it is sothyn', imagery wholly consonant with the daily experience of a virtuous, fish-eating East Anglian.[14]

Excavated remains of fish are identified by the zooarchaeologist to species defined by modern taxonomy, while historical records refer to the purchase of fish described in terminology which is often now obsolete. Fresh and preserved fish of the cod family were known by a range of names, including stockfish, klipfish, green-fish, *milwell*, haberdine (originally dried fish imported from the region of Bayonne), *skreyth* (dried fish), dryling, cropling, or ling,[15] while the modern fish species to which these refer include cod, saithe, ling, haddock, and hake. The medieval buyer used terms reflecting the type of cure, the size or freshness of the fish; modern archaeozoologists may be able to identify species, but have developed only limited techniques for establishing whether bones are from fresh or preserved fish.

There are other constraints on the study of fish remains. Historical records document quantities of fish purchased or consumed;[16] but it is not possible to use excavated remains to establish absolute quantities of fish eaten, although some attempts can be made to consider relative amounts. There are several reasons for this. Only the most robust fish bones will survive as well as those of mammals and birds, and they tend to survive best in deposits such as garderobes (latrines)

[14] Meech and Allen (1940: 17, 91). [15] Wright (1996: 102–5). [16] Harvey (1993: 46–51).

and below-ground features like pits and cesspits. For this reason, archaeological evidence for fish consumption is biased towards wealthier settlements and religious houses where these features were constructed. During excavation, only large fish bones are visible to the naked eye, mainly those of the larger marine fish and a few others such as those of large pike and sturgeon, of which the dermal plates survive best. Most bones are very small and are only recovered when sediments are sieved using meshes of 2 mm or smaller.

Some fish, particularly the lampreys and members of the elasmobranchs (shark and ray families), have a cartilaginous skeleton which does not survive in the ground. The skates and rays, however, do have dermal structures (denticles) which are often found when deposits are sieved, but the only hard part of the former is a very small tooth which is rarely recovered. As the bones of salmon and trout are believed to survive less well than those of other fish families, salmonid bones may be absent or rare even when other fish bones are recovered.[17] Some bones can be identified only to family. Elements of cod, saithe, and some smaller fish of the cod family cannot be distinguished from each other and are usually identified in bone analyses as 'large gadid' and 'small gadid'. Salmon and trout remains are not usually separated, as only the premaxillary bone is distinct.

Comparisons usually focus on the relative numbers of bones and this is the only figure which can be compared across sites. It has been used in this chapter, but it is subject to the biases of comparing animals of different sizes and different classes. Conger eels, for instance, have twice as many vertebrae as the gadid fish, and common eels twice as many as herring, so comparisons of bone numbers may overemphasize the contribution of these species.

Archaeological data

The discussion of fish remains here is based on approximately 120 assemblages from about fifty sites. Most of the assemblages are from sieved samples, but some were collected by hand or sieved unsystematically. Some of the reports are unpublished and remain with excavation archives.[18] About twenty-five are substantial assemblages in which hundreds of fish bones have been identified, and it is on these that the discussion mainly relies. Earlier accounts of archaeological evidence for fish and fishing in England were limited in scope, even where they contain valuable insights.[19] A recent, more exhaustive survey, which is based on a slightly larger sample of assemblages, all sieved, has mainly been concerned with the trade in preserved herring and cod fishes.[20]

The assemblages discussed in this chapter are from a range of settlements, including castles, towns, religious establishments, and fishing villages. Most are

[17] Wheeler and Jones (1989: 14–26).

[18] Few of the assemblages from London have been published. We are grateful to Alison Locker for allowing us to use unpublished data.

[19] Coy (1989); Enghoff (2000); Locker (2001: 192–290).

[20] Barrett, Locker, and Roberts (2004*a*; 2004*b*).

single large samples or a related group of samples from a single household or area of a town. They comprise waste from different stages in the preparation and consumption of fish, and the disposal of the remains. The assumption has been made that in general they contain remains typical of the type of establishment from which they come, whether a castle or a suburb of a town, but any group may contain a restricted suite of remains from a particular activity. In order to give a broad overview of the main fish eaten, two comparisons are made. In the first the numbers of bones of the larger marine fish are compared (Figs. 8.1–8.4). This takes into account finds from both hand-collected and sieved deposits. The second comparison shows the four types of fish most commonly found, ranked one to four (Tables 8.1–8.5). The tables distinguish hand-collected and sieved assemblages.

Cod and other large, preserved fish Large fish, mostly of the cod family, provided, with herring, the bulk of fish eaten in the Middle Ages. Besides cod, the large marine fish included saithe or coalfish (especially in Scotland and the north of England), ling, and haddock, together with hake and conger eel. Large cod and other members of the cod family are mainly caught offshore, so fishing was only possible from boats which were substantial and seaworthy. The Vikings

Fig. 8.1 Relative numbers of bones of conger eel, cod, ling, and hake on selected sites, eleventh to twelfth centuries

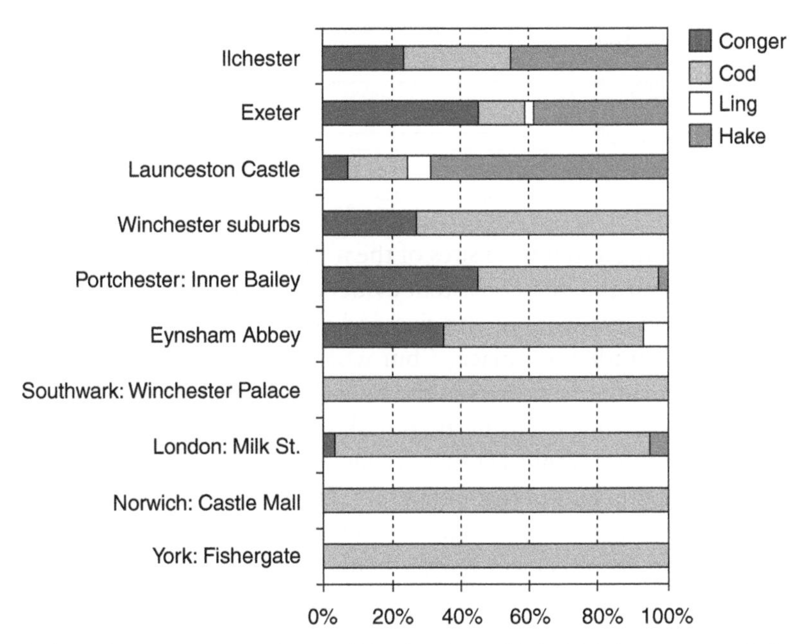

Source: Author's data (Serjeantson) from published and unpublished fish bone reports. References to main sites in text.

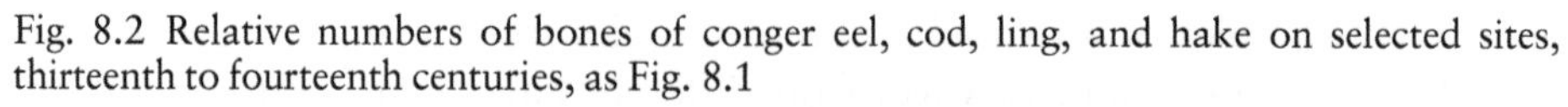
Fig. 8.2 Relative numbers of bones of conger eel, cod, ling, and hake on selected sites, thirteenth to fourteenth centuries, as Fig. 8.1

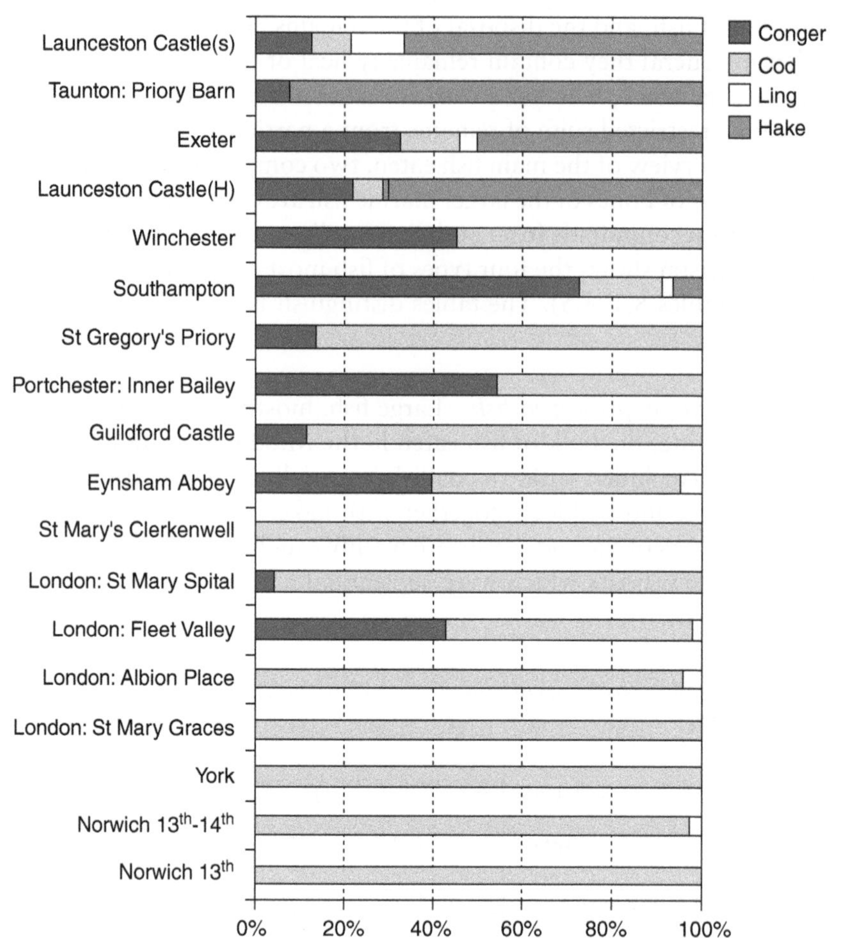

developed such boats, and it is on sites of the ninth century onwards in Scotland and Scandinavia that the first consistent evidence for the capture of large gadid fish is found.[21] The fish were of the size and type that were later traded: it is possible that some trade took place,[22] but work in Norway suggests that these fish were initially caught to provision the households of chiefs, and were traded only later. The earliest locations in the British Isles at which cod processing has been identified are the eleventh-century settlement of St Boniface in Orkney and, a little later, at Roberts Haven in Caithness.[23] The parts of the skeleton found at production sites are mainly those removed when the fish were preserved: bones from the head and from certain parts of the vertebral column. There is also evidence for the processing of large fish at a few sites in England: a predominance

[21] Barrett, Nicholson, and Cerón-Carrasco (1999: 355–6); Perdikaris (1997: 506; 1999).
[22] Colley (1984*a*: 127). [23] Barrett (1997: 620–8).

Fig. 8.3 Relative numbers of bones of conger eel, cod, ling, and hake on selected sites, late fourteenth to early sixteenth centuries, as Fig. 8.1

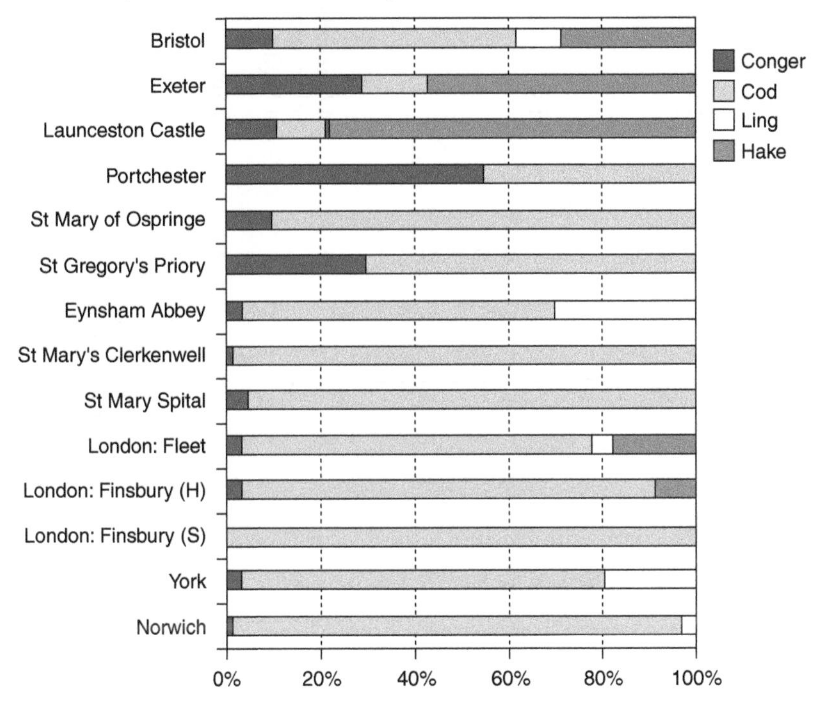

Notes: H, retrieved by hand; S, sieved.

Fig. 8.4 Relative numbers of bones of conger eel, cod, ling, and hake on selected sites, mid-sixteenth century, as Fig. 8.1

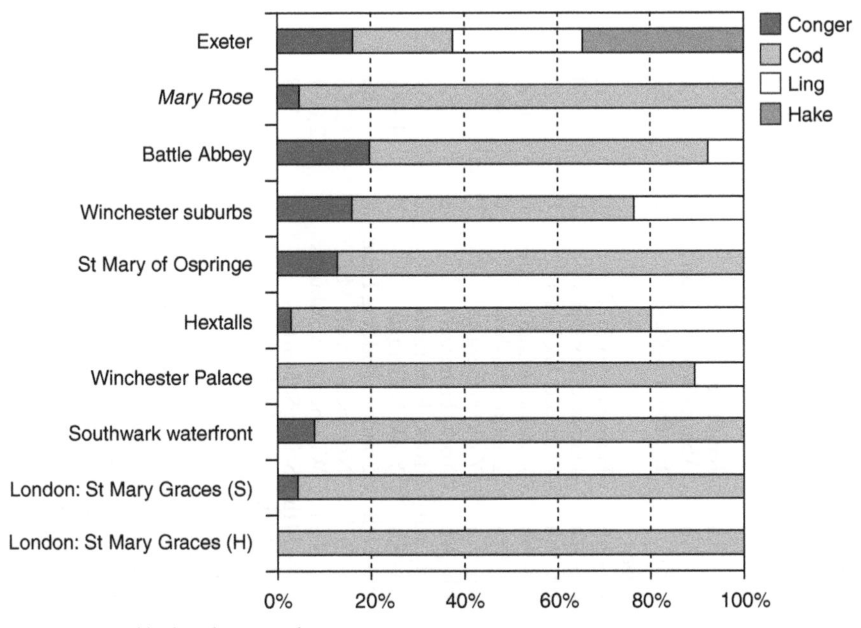

Notes: H, retrieved by hand; S, sieved.

Table 8.1. Fish from pre-Conquest sites: the four principal, large, marine fish, showing rank order

Date	Site	Type	n	Cod	Cong	Herr	Whit	Flat	Bass	Wras	Eel	Shad	Cypr	Pike	Other
Hand															
5th–7th	Bantham	F	—						2	4					1 Pollack 3 Scad
5th–7th	Abbots Worthy, Hants	V	21								1				
11th	Eynsham Abbey	R	21								1		3	2	
Sieved															
8th–9th	Hamwic: Six Dials	U	122			1		4	3		2				
8th–9th	Hamwic: Six Dials	U	150			2		3			1				
8th–9th	Hamwic: Six Dials	U	122			3		4			1				4 Salmonid
8th–9th	London: Maiden Lane	U	3,000			2		3			1		4		
8th–9th	London: Peabody	U	981			2					1	3	3		
8th–9th	York: Fishergate	U	276			2					1		3	4	
8th–9th	York: Fishergate	U	257	4		3					1		2		
Early 9th	York: Fishergate	U	539	4		2					1		3		
8th–9th	York: Fishergate	U	4,484			3		4			1		2		
9th–11th	London: Pudding Lane	U	185	1		4	2	3							
9th–11th	Norwich: Castle Mall	U	4,266	2		1	4				3				4 Mackerel
9th–10th	Winchester W suburbs	U	2,059			1		3			2				other species < 2
10th–11th	Scilly: St Martins	F	128		2					1	3				4 Mullet
10th–12th	Trowbridge	M	76										1		
10th–early 11th	Westminster Abbey	R	9,431			1	4				3		4		2 Smelt

Key to fishes: Cod (*Gadus morhua*); Cong, conger eel (*Conger conger*); Whit, whiting (*Merlangius merlangus*); Flat, plaice/flounder (*Pleuronectes platessa*/*Platichthysis flesus*); Herr, herring (*Clupea harengus*); Eel, freshwater eel (*Anguilla anguilla*); Cypr, carp family (Cyprinidae, excluding carp); Shad (*Alosa* spp.); Bass (*Dicentrarchus labrax*); Wras, wrasse species; Smel, smelt (*Osmerus eperlanus*); Pike (*Esox lucius*); Hake (*Merluccius merluccius*); Spra, sprat (*Sprattus sprattus*); Hadd, haddock (*Melanogrammus aeglefinus*); Gurn, gurnard species; Spar, sea bream species; Elas, dogfish/ray family, Elasmobranchii.

Site types: U, urban; R, religious house; V, village; M, manor/palace; C, castle; F, coastal fishing site. W, western; N, northern; E, eastern.

Sources: Author's database (Serjeantson), compiled from published and unpublished fish bone reports, and unpublished Museum of London excavation archives, as Fig. 8.1.

Table 8.2. Fish from eleventh- and twelfth-century sites, as Table 8.1

Date	Site	Type	n	Cod	Hake	Cong	Herr	Whit	Flat	Eel	Smel	Pike	Other
Hand													
11th–12th	Ilchester	U	54	2	1	3		4					
11th–12th	Exeter	U	414		3	1		2	4				4 = Wrasse
12th	Exeter	U	1,466		2	1		3	4				
11th–12th	Lewes	U	—	2		1			3				
11th–12th	Winchester N suburbs	U	251	4		2			1				2 = Rajidae
11th–13th	Portchester Inner Bailey	C	141	3		2			1				4 Bass
12th	Eynsham Abbey (pit)	R	60	4			1	4	3			2	
12th	Eynsham Abbey (kitchen)	R	196	4			1			2		3	
12th	Launceston Castle	C	71	2	1	3							4 Ling
Late 12th	Eynsham Abbey	R	36	2		3	3					1	
12th	London: Milk St.	U	251	4			1		3		2		
Sieved													
11th	York: Fishergate	U	1,674	4			2			1			3 Cyprinds
11th–12th	Winchester N suburbs	U	132	3			2			1			4 Mackerel
11th–12th	Winchester W suburbs	U	174				1		3	2			
11th–12th	Norwich: Castle Mall	U	1,643	2			1	4		3			
11th–12th	Southwark: Winchester Palace	M	221	2			1		3	4			
12th	Battle Abbey	R	45	2				2	1				4 Gurnard
12th	Eynsham Abbey (pit)	R	160				1	3	3			2	
12th	Eynsham Abbey (floor)	R	1,351				1			2		4	3 Stickleback
12th	Trowbridge	C	43				1			2			3 Cyprinid
12th	Norwich: Castle Mall	U	1,608	2			1	4	3				
12th	London: St Mary's Clerkenwell	R	44				1	2	3	4			
12th	London: Billingsgate	U	223	4			1	2	3	4			
12th	London: Milk St.	U	251	3			1		4		2		

Table 8.3 Fish from thirteenth- and fourteenth-century sites, as Table 8.1

Date	Site	Type	n	Cod	Hadd	Hake	Cong	Herr	Spra	Whit	Flat	Gurn	Spar	Eel	Smel	Cypr	Pike	Other
Hand																		
12th–13th	Taunton: Priory Barn	U	55			1	3			2	4							
L12th–13th	Eynsham Abbey	R	87				4	1						2			2	
M12th–M13th	Canterbury: St Gregory's Priory	R	28	2							1	3						
13th	Launceston Castle	C	1,284	3		1	2					4						
13th	Exeter	U	179			1	2			2	4		4					
13th	Exeter	U	897			1	2			3	4							
13th	Langport	U	36	3		4						1	2					
13th	London: Trig Lane	U	154	1							4							2 Ling 3 Elasmobranch
13th	London: Billingsgate	U	150	1				3		4	2							
13th–14th	Exeter	U	33	2		1	3			3	4							
13th–14th	Norwich: Castle Mall	U	202	1														Others <10
M13th–M14th	Canterbury: St Gregory's	R	119	2	3			1			3							
13th–15th	Eynsham Abbey	R	69	3				1			2			4				
E14th	Exeter	U	79	3		1	2				3	3	3					
14th	Portchester: Inner Bailey	C	107	3			1	4			2							
13th–14th	Okehampton Castle	C	*c*.3,000			2	4			1	3							
Sieved																		
12th–E13th	Launceston Castle	C	15			1												
12th–13th	London: Fleet Valley	U	320	4			3	1							2			
L12th–13th	Eynsham Abbey	R	354					1						2		4	3	
L12th–14th	Norwich: Castle Mall	U	1,427	2				1		3	4							
Early 13th	London: St Mary Spital	R	476					2		4	1					3		
13th	Launceston Castle	C	580			1	3			2		4						
13th	London: Albion Place	U	94	2						1	3	4						4 Elasmobranch
13th	London: Billingsgate	U	249	4				1		3	2							
13th	London: St Mary's Clerkenwell	R	215	3				1	4	2								
13th	London: St Mary Spital	R	82					2			1	3						
13th	Winchester W suburbs	U	549	4				2			1			3				
13th	Guildford Castle Palace	M	94		2					1	4			3				
13th–early 14th	York: Fishergate	R	1,043					1			4			2		3		
13th–14th	London: Fleet Valley	U	460					1		2	3				4			
13th–14th	Winchester N suburbs	U	67	3				1		4				2				
13th–14th	Winchester N suburbs	U	47		3			1			2			4				
13th–14th	Abingdon: Stert St.	U	191		4									2		1	3	
13th–14th	Norwich: Castle Mall	U	1,131	2				1	3	4								
L13th–14th	London: St Mary's Clerkenwell	R	707					1		2	4				3			
13th–15th	Eynsham Abbey	R	152					1		4				2		3		
13th–15th	Cleeve Abbey	R	41	4	2					1	2							
13th–15th	London: St Mary Graces	R	1,587					3	2					1	4			
13th–15th	Southampton fish market	U	1,177					1			3			2				4 Rajidae
14th	London: St Mary Spital	R	481					1			2			3	4			

Table 8.4. Fish from late fourteenth- to fifteenth-century sites, as Table 8.1

Date	Site	Type	n	Cod	Ling	Hadd	Hake	Cong	Herr	Whit	Flat	Gurn	Spar	Elas	Eel	Cypr	Other species
Hand																	
15th	Launceston Castle	C	476	2			1	3					4				
15th–early 16th	Canterbury: St Gregory's Priory	R	236	4					1	3	2						
Late 15th–early 16th	Eynsham Abbey	R	46	2					1	3	4						
14th	London: Customs House	U	257	1	2			3			4						
14th	London: Seal House	U	110	1						3	2	3					
14th	London: Trig Lane	U	69	1	3			4			2						4 Salmonid
15th	London: Swan Lane	U	80	1						2	4			3			
15th	London: Trig Lane	U	390	1					3	2				4			
15th–16th	London: Finsbury Pavement	U	39	1		4			3		2						
Sieved																	
14th–15th	Winchester N and E suburbs	U	25	3					1				4	2			4 Stickleback
14th–15th	London: St John's Priory	R	77	4					2	3	1						
14th–15th	London: St Mary's Clerkenwell	R	49	3					1	2							
14th–15th	Canterbury: St Gregory's Priory	R	2,616						1	2	3				4		
Late 14th–mid-16th	Norwich: Castle Mall	U	4,665	2					1		4			3			
15th	Launceston Castle	C	196				1	3		2			4				
15th	Winchester W suburbs	U	511	4				3	1					2			
15th	London: Fleet Valley	U	1,619						3	1	2		4				
15th–early 16th	London: Albion Place	U	101	4					2	1	3						
15th–16th	York: Fishergate	R	422			3			1	4				2			
15th–16th	London: St Mary's Clerkenwell	R	1,405						1	2	4			3			
15th–16th	London: Finsbury Pavement	U	190	2					1	3		4					
Late 15th–16th	Eynsham Abbey	R	4,299						1	3				2	4		

Table 8.5. Fish from early sixteenth-century sites, including some immediate post-Dissolution sites, as Table 8.1

Date	Site	Type	n	Cod	Ling	Cong	Herr	Whit	Flat	Gurn	Eel	Smel	Other
Hand													
16th	Exeter	U	1,440	4	2				3				1 Hake
16th	London: St Mary Graces	M	36	1				3		3	2		
16th	*Mary Rose*	S	4,193	1		2							
16th	Battle Abbey	U	150	3			2			4	1		
Sieved													
16th	St Mary of Ospringe	U	57				2		2		3		4 Mullet
16th	York: Fishergate	R	1,159				1				2		3 Cyprinid 4 Mackerel
16th	Hextalls, Bletchingley	M	1,278	2				1	3				4 Pike
16th	London: St Mary Graces	M	2,115				2	4			1	3	
16th	Southwark: Calvert's Buildings	U	177				3				1	4	2 Sprat
16th	Southwark: Winchester Palace	U	223	2			3	1			4		
16th	Southwark: waterfront	U	103	3			1		4				2 Haddock

of bones from the skull and the pre-caudal (proximal) end of the vertebral column indicate that cod and ling were processed in Hartlepool,[24] cod in York, and haddock in Lindisfarne and Jarrow.[25] The same bones in a waterfront site at The Parade, Plymouth, suggest the processing of hake.[26]

The optimum size of gadid fish for preservation is about 80–90 cm,[27] but many must have been used which fell outside this optimum range. Less is known about the technology for catching and preserving conger and hake. It is not a straightforward matter of relying on distinctions between the parts of the body found to delineate sites of production and consumption, because different bones were removed or left in according to the method of preservation used or local custom; but the preserved commodity may be identified from the parts of the skeleton present, combined with a more or less uniform size of fish. A very few sites—the Fleet prison is one example—have the signature for preserved cod, with exclusively appendicular and caudal vertebrae.[28]

From the later eleventh century, cod begins to appear among food remains. A recent survey has pinpointed the start of the trade in preserved cod to within a few decades of 1000.[29] From the eleventh century, it is the second most abundant species at some sites (Table 8.2). It is most frequent in the towns of eastern England, such as York,[30] Norwich,[31] and London (Figs. 8.1–8.4). From this time, regional variations can be seen in the large marine fish. At sites in the counties of south-western England, such as Ilchester,[32] Exeter,[33] Launceston Castle,[34] Taunton,[35] and Bristol,[36] hake was eaten in greater quantities than cod. In the south, conger eel was one of the main large seafish in Exeter and other settlements, such as Portchester Castle[37] and the suburbs of Winchester,[38] provisioned from the ports of the Channel coast. Even at Eynsham Abbey,[39] a much greater distance from the coast than the other three sites, conger eels feature among the food remains. These were not necessarily caught by local boats, as the Southampton port books show that large quantities were imported from the Channel Isles.[40] Though Londoners mainly ate cod, nearly 40 per cent of the large fish in one deposit at Fleet Valley was conger eel.

Hake is found mainly off the west coast of Britain, with a distribution which extends further south than cod, and conger is very abundant in the Channel. These local fisheries supplied much of the needs of the south and west, while stockfish from northern Europe and Iceland supplied the rest of the country. The consumption of large ling starts comparatively late. The large, preserved fish

[24] Locker (2001: 214–15). [25] Dobney (n.d.). [26] Locker (2001: 183).
[27] Locker (2001: 134). [28] Locker (2001: 214); Hamilton-Dyer (1995).
[29] Barrett, Locker, and Roberts (2004*a*: 622–3). [30] O'Connor (1991: 264).
[31] Locker (1997*a*: table 1). [32] Wheeler (1982: 284). [33] Wilkinson (1979: 76).
[34] Smith (1995: table 2). [35] Wheeler (1979*a*: 193). [36] Locker (2001: 200–2).
[37] Coy (1985: 258).
[38] Serjeantson and Smith (in press: table 5.32); Coy (in press: tables 3.15–3.19).
[39] Ayres, Locker, and Serjeantson (2003: 364, table 10.19). [40] Coy (1996: 58, table 2).

which were the subject of trade are present only from the fourteenth century onwards, and have only been found in significant quantities on a few sites from the fifteenth or sixteenth century (Figs. 8.3 and 8.4). Haddock, a smaller gadid fish that preserves well, is among the four most numerous species in a few of the assemblages from the thirteenth to the early sixteenth century (Tables 8.3 and 8.4). The method of curing in use today leaves only a few bones in the fish, so the presence of bones from the head as well as other parts of the body suggests that these fish were eaten fresh as well as preserved.

Herring The large-scale fishery for herring destined to be preserved and traded was believed to have started only in the tenth century,[41] but large quantities of herring from sites in north-west Europe in the sixth or seventh centuries suggest that, at least locally, herring fishing started earlier. Herring was of huge economic importance on the island of Bornholm in Denmark at this time, with more than 12,000 herring bones recovered in excavations at the site of Sorte Mulde[42] and large quantities at Eketorp in Sweden. Herring seems to have been processed there, since collections of complete heads were found. In England, too, there are early sites with herring. Some eighth- to tenth-century deposits—mostly in pits—in York and London[43] have yielded bones in quantities which suggest that fish, including herring, was eaten in these nascent towns (Table 8.1). One pit in *Hamwic* had a similar range of fish, but otherwise there is little evidence of fish consumption there.[44] The herring at these early sites is likely to have been caught locally, in the Ouse, the Humber, the Thames, and Southampton Water. Until they were fished intensively in later centuries, shoals of herring entered estuaries such as the Thames.[45]

By the eleventh century, herring are the fish most commonly found at nearly all inland sites as well as at sites within close reach of the coast (Tables 8.2–8.5). Showing fish by rank does not emphasize strongly enough how the numbers of herring bones can outnumber other species by an order of magnitude. At Castle Mall, Norwich,[46] herring bones are present in hundreds and cod, the second most frequent species, in tens. In the two largest samples from Eynsham Abbey, which date from the twelfth and fifteenth centuries respectively, herring bones are numbered in thousands and those of the next most frequent species, eel, in hundreds. Though other species predominate at a few fifteenth- and sixteenth-century sites (Tables 8.4 and 8.5), herring continue to rank first at many sites. As herring fisheries and trade became more efficient, they presumably became more readily available and cheaper than eels.

[41] Muus and Dahlstrom (1974: 210). [42] Enghoff (1999: 46, table 2).

[43] Locker (1988*a*: 428, table 123); Rackham (1994*b*: 131, table 12.2); Barrett, Locker, and Roberts (2004*a*: 622). [44] Coy (1996: 56).

[45] Wheeler (1979*b*: 70). Coy (1996: 56) believes the herring at *Hamwic* may have been imported rather than caught locally. [46] Locker (1997*a*: table 3).

Like other species, herring can be processed in different ways. Sometimes the whole head, and sometimes only a few bones from the back of the head and shoulder area are removed.[47] When vertebrae greatly outnumber bones from the head in an assemblage, it is likely that the assemblage consists of preserved fish. This was the case at Eynsham,[48] where approximately ten times as many vertebrae as cranial bones were present in the fifteenth-century samples. The processing of herring has not been identified at any archaeological site in England, and most sites do in fact have both cranial bones and vertebrae. In the case of herring, it is unlikely that many were eaten fresh at inland sites; it is more likely that the fish on those sites had been preserved.

When the number of bones is converted to the number of portions the fish would have yielded, we get a more realistic view of the relative quantities of food provided by cod, herring, and some other important species. Cod and other large, marine fish were nearly always more important in the diet than herring. The conversion has been made for a range of assemblages from London, Norwich, Huntingdon, Bristol, Winchester, and Eynsham. Eynsham and some of the London sites were monastic, but otherwise most of the deposits are from secular households, so this conclusion may particularly apply to towns and other secular households. In Norwich, for instance, cod made up over 50 per cent of the main food species as early as the eleventh century. From the twelfth century onwards, in all but five of nearly thirty assemblages, the archaeological finds suggest that cod (sometimes hake, as in Bristol) provided more food than herring. At the only small town studied, Huntingdon, herring predominated by both number and portion.[49]

Other marine fish After cod and herring, whiting is found in the largest quantities. It is among the four most frequent species found on sites in London, even before the tenth century, is more numerous than herring at one or two sites from the thirteenth and fourteenth centuries onwards, and is one of the four most frequent species in three-quarters of the assemblages. Flatfish ('butts'), mainly flounder and plaice, were also eaten in abundance from a very early date. In the thirteenth- and fourteenth-century deposits they outnumber even herring in the western—and wealthier—suburb of Winchester[50] and at two London ecclesiastical sites, St Mary Spital[51] and St John's Priory.[52] These fish had an enhanced cash value, and their predominance over herring and eel may reflect the relative wealth of these households.

[47] Enghoff (1996: 43–7). The absence of these elements among the remains from Trowbridge led Bourdillon (1993: 142) to suggest that the herring there were processed fish.

[48] Ayres, Locker, and Serjeantson (2003: 553, table A4.6).

[49] This calculation has been carried out by Locker (2001: 192–222, figs. 10.1–10.49) for a number of sites using the quantities per portion consumed by the monks of Westminster, as estimated by Harvey (1993: 225).

[50] Coy (in press: tables 3.15–3.19).

[51] Locker (1992*b*).

[52] Locker (1996).

Remains of fish of the shark and ray families are surprisingly common in view of the poor survival of the skeletons of these species: a few dermal denticles are found in most assemblages, and at some sites the elasmobranchs rank among the most frequent species, confirming their importance as a food fish. The other marine fish found regularly in unsieved assemblages are sea bass, the gurnards, and the wrasses, though their position is displaced by smaller fish in sieved assemblages. Gurnard is among the four main species found at Launceston Castle and even at some London sites in the deposits from the thirteenth and fourteenth centuries. Mullet and gurnard are also quite frequent in the assemblage from St Gregory's Priory, Canterbury,[53] ranking closely with cyprinids and eel, though after herring, whiting, and flatfish. Mackerel bones are common on one or two sites from the eleventh century onwards and are found in small numbers on many more, as are bass, turbot, and brill. Some fish found only occasionally in small numbers are probably incidental catches: these include the smaller members of the cod family and smaller flatfish.

Data from historical sources

In the later Middle Ages, the historical record confirms the enormous variety of fish that was available, in both fresh and preserved forms, beyond the staples of the cod family and herring. It was at its most extensive in the great households of the aristocracy (Tables 8.6 and 8.7), for example, in the household of Joan de Valence, Countess of Pembroke, in 1295–7, travelling between her manors in south-eastern England and the southern Midlands;[54] or in the household of Joan Holland, Duchess of Brittany, in 1377–8, residing largely at Castle Rising (Plate 8.2) and Swineshead.[55] Status brought with it a considerable emphasis on consumption of fresh fish, carried to the great household by a remarkable system of marketing. Residing at Potterne, on the northern edge of Salisbury Plain, in 1407, Bishop Mitford of Salisbury purchased his fresh marine fish at Warminster, much of which must have originated in catches in Devon and Cornwall. The Bishop's accounts record the expenses of the household's caterer and packhorse there;[56] and other households maintained horses specifically for the carriage of fresh fish. A horse was bought for bringing fish from Shoreham to the household of Thomas of Brotherton and Edmund of Woodstock, the young half-brothers of Edward II, in Sussex in June 1311.[57] Edward II's widow, Isabella, had a horse for carrying fish, listed along with twenty-four other horses and four mules in her stable after her death in 1358.[58]

The records of these households provide a useful picture for the later medieval period of what was available in the market and at what cost. In 1331–2, the clerk

[53] Powell, Serjeantson, and Smith (2001). [54] Woolgar (1999: 48–9); NA E 101/505/25-7.
[55] Woolgar (1995). [56] Woolgar (1992–3: i. 304, 307–8, 322).
[57] BL Add. MS 32050, fo. 6^{v}. [58] NA E 101/393/4, fo. 13^{r}.

Table 8.6. The marine fish, cetaceans, shellfish, and crustaceans eaten in two great households

	Joan de Valence, Countess of Pembroke		Joan Holland, Duchess of Brittany
	1295–6	1296–7	1377–8
Herring	xx	xx	xx
Herring (powdered)			xx
Herring (white)			xx
Herring (red)			xx
Aloses(shad)		x	
Sperling (smelts)	xx		xx
Capri marini (sprats)	xx		xx
Milwell (cod)	xx	xx	x
Codling			xx
Dried fish	x	xx	
Skreyth (dried fish)			x
Salt fish			xx
Stockfish	xx	xx	xx
Haddock	xx	x	xx
Hake		xx	
Ling			xx
Merlus (whiting)		x	
Halibut			x
Conger	xx	xx	xx
Bass			x
Dory	xx	x	
Gurnard	xx		xx
Mackerel	xx	x	x
Mullet	x	x	x
Red mullet			x
Sea bream		x	
Butts (flatfish)			xx
Flounders	x	x	xx
Sole	x		xx
Plaice	xx	xx	xx
Turbot	xx	x	xx
Ray	xx		x
Porpoise			xx
Whale			xx
Cockles			xx
Mussels	x		xx
Oysters	x		xx
Razors			x
Whelks	xx		xx
Crabs			xx
Crevices (crustacea)	x		x
Creye (crustacea)	x		
Chevres			x
Gricokes			x

Notes: x one to four appearances, xx five or more appearances in the account.
Sources: Joan de Valence, NA E101/505/25-7; Joan Holland, AD Loire-Atlantique E206/1 and 3.

Table 8.7. The freshwater fish eaten in two great households

	Joan de Valence, Countess of Pembroke		Joan Holland, Duchess of Brittany
	1295–6	1296–7	1377–8
Freshwater fish (unspecified)	xx	xx	
'Small fry'	xx	x	x
Bream	x		x [?]
Bream (small)			x
Chub		x	x
Dace	x		
Eels	xx	xx	xx
Lampreys	x	xx	x
Lamperns	xx	xx	xx
Perch	x	x	xx
Pike	xx	x	x
Pickerels (small pike)			xx
Roach	x		xx
Salmon	xx	xx	xx
Sturgeon			xx
Tench			xx
Trout	xx		xx

Notes and sources: As Table 8.6.

of the kitchen of Queen Philippa was charged at the audit for fish that he could not account for properly at the rate of 1,000 herring for 6*s.* 8*d.*, that is twelve and a half for 1*d.*; and cod at 4*d.* apiece.[59] The household of Thomas de Courtenay at South Pool in Devon made a series of large purchases around the start of Lent 1342, with white herring at 120 for 6*d.*, or twenty for 1*d.*; red herring at more than fifteen for 1*d.*; and 1,560 pilchards for 16*d.* White-fleshed fish was dominated by purchases of whiting, hake, and conger, with whiting (*merling*) at prices ranging from eighty for 15*d.* to 120 for 6*d.*; buckhorn (dried whiting), at 20*d.* for 120; 480 dried hake at six for 1*d.*; and conger at 3½*d.* each.[60]

Fishing was equally well established on the Bristol Channel coast of Somerset.[61] The Luttrell family of Dunster had in the 1420s two principal places of purchase, at nearby Minehead, where as lords of the manor the family was able to buy fish at a special, customary price; and at Exeter. Their purchases show a slightly different pattern of staple fish, with large quantities of cod (*milwell*) and ling: in 1429–30, Sir John Luttrell acquired 677 *milwell* and ling, bought at Minehead. Some was given away, including four to the reeve of

[59] JRULM, Latin MS 235, fo. 5[v].
[60] Cornwall RO AR12/25, m. 3[r]: all calculations use the long hundred of 120.
[61] Stevens (1985: 471).

Plate 8.2 Castle Rising from the north-west: the twelfth-century kitchen tower is in the north-west corner. In the fourteenth century, free-standing kitchen buildings were constructed in the inner bailey, remaining in use in the period that the castle was held by Joan Holland, Duchess of Brittany. Photograph: C. M. Woolgar.

Woolavington and twenty-four as part of his own sister's dowry;[62] Margaret Luttrell, in 1431–2, also gave salt ling or *milwell* to the reeve of Woolavington.[63] The accounts show substantial quantities of salted and dried hake, whiting (*scalpins*), stockfish, and herring (some given by the tenants of Minehead and therefore a local catch). In 1425–6, there was red herring at 6*s.* a mease of 600 fish, or a little over eight for 1*d.*; salted hake at about four for 3*d.*; and stockfish at about three for 2*d.*[64] There was some local salmon, with twelve from Watchet for 8*s.* 2*d.* in 1423–4;[65] and one for the household of Margaret Luttrell, bought in Bridgwater in 1431–2 for 18*d.*[66]

Both these households were based close to the sea and there would have been little additional cost for transport in their purchases. But even further inland, marine fish might be an accessible source of protein and vitamins. Manorial custumals demonstrate that fish, principally herring, but also whiting, cod, salmon, and eels, was commonly included in the meals given by lords to boon workers at harvest and at ploughing. Typically in the south and east of England, with some examples from further north, these show three or four herrings per worker per fish meal.[67] At Sedgeford, in Norfolk, it has been calculated that the fish in the diet of harvest workers in 1256 amounted to 694 kcal or 5 per cent of the daily allowance (which may have fed three or four people), drawn from herrings and *milwell*; and in 1424, from cod alone, to 135 kcal or 3 per cent of the daily allowance, probably for a single person. The amounts recorded in custumals for these limited occasions may not reflect—and may overestimate—day-to-day consumption, for at harvest there was merit in the lord using up preserved herring in anticipation of the imminent arrival of new stocks from that year's catch. So one may consider that the usual level of consumption of marine fish among these workers was at a lower level; and that others, not as privileged in their employment and circumstances, would have had considerably less: both find their reflection in the paucity of fish remains from rural sites.[68]

In coastal counties, where fishing itself was a part-time activity additional to agriculture, and where other commodities, such as the shellfish gathered by women and children, might supplement it further, there was potential for seafood to have a greater impact on diet.[69] It was here that fish might have most effect on a regional basis. In the south-west, for example, the change in the fifteenth century from an inshore fishing industry centred on shore-line 'cellar settlements', occasional and interim bases for practising this activity, to continuously occupied fishing villages, argues for the growth in availability of fish.[70]

In urban markets fish was available to a wide range of consumers. Evidence from London reveals both a highly specialized—and regulated—trade. London

[62] Somerset RO DD/L P37/11, m. 3^{d}. [63] Somerset RO DD/L P37/12, m. 2^{d}.
[64] Somerset RO DD/L P37/10C, m. 2^{r}. [65] Somerset RO DD/L P37/10B, m. 2^{r}.
[66] Somerset RO DD/L P37/12, m. 2^{r}. [67] Hallam (1988*d*: 833–40).
[68] Dyer (1994*b*: 83, 92–5). [69] Bailey (1990); Woolgar (1999: 119).
[70] Fox (2001: 145–9).

vocabulary is testimony to the range of fish found in the Thames and also that brought in for sale. Regulation in the fifteenth century forbade the salting for resale of fresh fish—flatfish, codling, whiting, mackerel, conger, herring, or other fresh fish; banned the sale of certain types of preserved fish, for example, which had been softened by soaking it in water (*wokedfish*); and placed strict limits on the length of time for which shellfish might be offered to sale.[71] The fishmongers of London had effectively taken over the supply of the royal household in the fifteenth century and it was on this marketplace that the kitchener of Westminster Abbey drew for the extensive provisions he required.[72] On Saturday, 1 May 1501, the household of Edward Stafford, third Duke of Buckingham, at Lambeth, had salt fish for breakfast; at lunch there were flounders, eels, pike, salt ling, soles, plaice, turbot, half a fresh conger, and mullet, supplied by two fishmongers, John Maston and John Bushe. For supper Bushe supplied a further salt ling, two plaice, and 100 prawns, while Maston supplied two further pikes.[73] Those staying at urban inns—and providing their own food—would expect to resort to local markets for their fish.[74]

Estuarine fish

As we have seen, fish was eaten in the early trading settlements of York, London, and Southampton. In addition to those species already discussed can be added some which are rarely eaten, indeed rarely caught, today: smelt and shad, smaller relations of the herring,[75] of which there are two local species, twaite shad and allis shad. These spend part of their life in the sea and part in fresh water. Shad is one of the four most frequent finds in one of the deposits from the Anglo-Saxon trading port in London, and smelt is second after herring in the pre-Conquest sample from Westminster Abbey.[76] Smelt continues to be one of the most frequent species in some London deposits even into the sixteenth century.

Freshwater fish

Eels and salmon spend part of their life cycle at sea, but are caught in rivers or estuaries, and are discussed here with freshwater fish. Before the Conquest, eels are outstandingly more common than other fish (Table 8.1). Their dominance is replaced by herring in the eleventh century, as at Westminster Abbey, and from that time, the bones often continue to be the next most abundant after those of herring (Tables 8.2–8.4). They fall into two general sizes, as exemplified by those from Eynsham Abbey.[77] The eels from the Late Saxon and Norman deposits were usually approximately 30–40 cm long. Some from the fourteenth to the sixteenth centuries, however, were over 50 cm. These were large females which

[71] Wright (1996: 105, 107). [72] Harvey (1993: 46–7). [73] NA E 101/546/18, fos. 4^{v}–5^{r}.
[74] Kristol (1995: 11–12, 43–4). [75] Wheeler (1978: 70–2, 160–1).
[76] Locker (1997*b*: 111–13). [77] Ayres, Locker, and Serjeantson (2003: 381, fig. 10.13).

could have weighed up to about 10 kg.[78] They must have been a luxury food. These may have been raised in the abbey fishponds or elsewhere locally, or might even have been specialist imports, such as those known to have come into London from Flanders.[79] They must have constituted some of the rich food served on feast days or at the abbot's table. Few assemblages have more than a handful of bones of salmon or trout, though salmonids were numerous in the pit at *Hamwic* in which fish bones were plentiful, and also at the later London waterfront site at Trig Lane.[80]

Pike is often found. It is one of the most numerous species in many sites earlier than the eleventh century. After that time, marine fish become more important elsewhere, but at Eynsham and at Stert Street, Abingdon,[81] pike continue in quantity. It was not until a century later that marine fish predominated at these inland sites as well. Just as with eels, both large and small pike were eaten. The pike from the Late Saxon period at Eynsham Abbey were quite large, and would have been a luxury fish, while those from the twelfth century are uniformly 30 cm or smaller. These small pike would have been classed as pickerels, and the purchase of pickerels for the abbey is documented.[82]

The other freshwater fish, the cyprinids, are found from the early period. Few parts of the skeleton of the cyprinids—dace, roach, and other members of the carp family—can be identified to species, and they are grouped together here. The remains of carp, however, are distinctive. Cyprinids are particularly common in the eighth- to ninth-century *wics*, suggesting that the capture of these freshwater fish was a habit which began well before the construction of fishponds following the Conquest. The numbers did not notably increase later, but the tables here show numbers in relation to other fish, and consumption of fish of all kinds increased from the eleventh century onwards. They continue to be found particularly on inland sites such as Eynsham Abbey.[83] In continental Europe, in households distant from the sea and supplies of marine fish, they predominated as the fish eaten on fast days.[84] Stickleback remains were unexpectedly found on the floor of the kitchen at Eynsham Abbey (Plate 1.1) and may be the gut contents of pike.[85]

Fishponds are difficult to date archaeologically and there is very little evidence for them from Anglo-Saxon England. A very few are recorded in Domesday Book.[86] There was, however, large-scale investment in the twelfth century in the construction and repair of fishponds, for example, at royal manors such as Brigstock in 1129–30 and 1160–1.[87] The Bishops of Winchester had large ponds by the end of the twelfth century, which later totalled some 400 acres.[88] Monastic Orders, first the Benedictines and subsequently other Orders, invested in ponds,

[78] Wheeler (1978: 62). [79] Woolgar (2000: 36). [80] Locker (1986).
[81] Wheeler (1979*c*: 21–3). [82] Ayres, Locker, and Serjeantson (2003: 380, fig. 10.12).
[83] Ayres, Locker, and Serjeantson (2003: 549–53, tables A4.2, A4.5, and A4.6).
[84] Hoffman (1994: 402). [85] Ayres, Locker, and Serjeantson (2003: 387).
[86] Bond and Chambers (1988: 356). [87] Hunter (1833: 88); Anon. (1885: 32); Steane (1988).
[88] Roberts (1986).

Plate 8.3 The fish house of the Abbot of Glastonbury at Meare, from the south-east. Possibly constructed by Abbot Sodbury (1322–35), the lower floor was probably used for preparing fish and storing fishing equipment; the upper storey had residential accommodation. A further, two-storey part of the building at the west end was demolished in the nineteenth century. Adjacent were three fishponds and Meare Pool, which continued to produce pike, tench, roach, eels, and other fish for the Abbey into the early sixteenth century. Photograph: C. M. Woolgar.

the later Orders employing more sophisticated water management systems (Plate 8.3).[89] It seems probable that fishponds proliferated in central and northern France at about the same time as they did in England and they may thus reflect a more widespread change in taste in fish consumption.[90] Fishponds were intended to provide bream and roach as well as pike, and some of the cyprinid remains must be of fish produced there. These freshwater fish were exclusively luxury items; and even where there were very extensive fishponds, they made a minimal contribution to diet.[91] Fishponds might, however, serve as an investment and they are found in villages, presumably with the intention that the fish be sold.[92]

Other cyprinids were probably caught in rivers and natural ponds. Fishing in streams and rivers with hook and line was frequently the pastime of boys, some quite young, while their fathers and elder brothers were engaged in more complicated operations with nets and fish traps.[93] The river fisheries and the Fens were potentially rich sources of food. They might supply the luxury market,

[89] Currie (1988: 270); Aston and Bond (1988*c*: 444); Rippon (2004).
[90] Querrien (2003: 412–18). [91] Harvey (1993: 47); Dyer (1994*c*).
[92] Aston and Bond (1988*b*: 431).
[93] Hanawalt (1986: 53, 132, 158); Hunnisett (1961: 48, 87–8).

for example, with specialist pike fisheries;[94] and weirs on major rivers, such as the Severn, produced salmon.[95] General river fishing might produce other valuable fish: for example, in February 1316 a fisherman of London presented Edward II with a sturgeon that his men had caught in the Thames at Woolwich.[96] Given this value, fish of all sorts, not just freshwater fish, were employed as gifts. Edward III sent his wife, in April 1349, a present of four lampreys.[97] This is one of many references to lampreys, but we should not expect to find remains, although a tooth of a lampern has been identified at York.[98] In 1344–7, while Edward III was principally abroad, the keeper of the wardrobe of his household recorded a range of gifts of fish, totalling six salmon, thirty-nine large pike, ninety-nine other pike, seventy-nine carp, 124 bream, 1,383 perch and roach, two eels, and 100 flounders, all of which were consumed in the household.[99] The carp recorded here were almost certainly given and consumed on the Continent. The case for the introduction of carp to England in the fifteenth century is supported by the absence of archaeological finds before that date; and even in the sixteenth century it must have been a rare novelty. An early site where remains have been found is Nonsuch Palace,[100] where they are in late seventeenth-century deposits, coinciding with the period Nonsuch was used by the Berkeley family.

Fish and diet

The best evidence for the contribution made by fish to diet comes from the great households and monasteries of late medieval England. Although there was a great variety of fish eaten in the household of Bishop Mitford of Salisbury, underlying consumption focused on herring, whiting, and eels in October 1406; and in March 1407, on herring, cod, and whiting. As the number of meals served is known, it is possible to calculate a minimum calorific value for the fish content of these meals (Table 8.8). Additional (or alternative) fish would have been available to the gentle members of the household. This household placed a noteworthy emphasis on herring. There was considerable variation between households, but herring became less popular in the fifteenth century.[101] The very ample allowance of fish in Mitford's household would have meant that even in Lent, or on days when only one meal was eaten (as was often the case on Fridays), the average level would have satisfied present-day calorific recommendations for consumption. The level of provision also compares well with other aristocratic households of the fifteenth century. The volume of herring gave the diet in Mitford's household more energy than that in the household of the Earl of Warwick in 1420–1 (541 kcal in October 1420 and 988 kcal in Lent 1421) or in that of Henry Stafford in July and August 1469 (613 kcal).[102]

[94] Lucas (1998: 42–3). [95] Pannett (1988). [96] NA E 101/376/7, fo. 42v.
[97] NA E 36/205, fo. 7r. [98] Jones (1988). [99] NA E 101/390/12, fo. 51r.
[100] Locker (2005). [101] Harvey (1993: 49).
[102] Woolgar (2001: 16–17); Martin (2002) uses greater weights for the fish.

Table 8.8. Minimum calorific value for the fish content of the meals of the household of Bishop Mitford of Salisbury

Fish	kcal per meal October 1406	kcal per meal March 1407
Herring	611	1,331
Whiting	290	86
Cod	20	137
Eels	313	11
Total	1,234	1,565

Notes: Based on October 1406: 642 fish meals; 1.82 lb/0.82 kg fish per meal (0.78 lb/0.353 kg white fish; 1.04 lb/0.472 kg fatty fish). March 1407: 920 fish meals; 2.11 lb/0.96 kg fish per meal (0.62 lb/0.282 kg white fish; 1.49 lb/0.676 kg fatty fish). Figures calculated using Harvey (1993: 226–7), and McCance and Widdowson (1998).

Source: Woolgar (1992–3: i, 261–430).

Monastic and institutional records provide some evidence for the decline in fish eating in the later Middle Ages, particularly relative to the consumption of meat. At Bolton Priory, in 1305–6, the bulk purchases of the kitchen amounted to £39 4*s*. 3*d*., of which fish constituted £29 0*s*. 2½*d*.; in 1306–7, the fish came to £25 9*s*. 6*d*. out of £45 15*s*. 0¾*d*.; but in 1377–8, fish bulk purchases ran to £24 9*s*., while those of meat reached £49 6*s*. 6*d*.[103] At the Hospital of St Giles, Norwich, before the Black Death the expenditure on bulk purchases of fish was usually substantially greater than that on meat; by the 1370s, there were signs that the balance was changing—and the balance was not just in price, but in quantity. Whereas the bulk purchase of fish in 1374–5 amounted to £26, it only came to £10 or more in eight of the years for which there are accounts between 1465–6 to 1526–7—although in all these cases more fish would have been consumed, accounted for in the larger sum expended on diets, that is, the day-by-day purchases.[104]

Records have shown that the rural poor in the later Middle Ages did eat some fish. We might expect locally caught freshwater fish and cheap, preserved fish to be present, but up to now the archaeological evidence has been very scant. Typically, very few fish bones are found on inland, rural settlements. This might be dismissed as a consequence of poor recovery, but that explanation can be ruled out when the small and delicate bones of rodents and amphibians are retrieved, as at Eckweek and Middleton Stoney castle.[105] Part of the reason for the absence of fish bones is the poor survival of remains of small fish, but, as we have seen, bones of cod and other large, marine fish survive reasonably well—as well, for instance, as those of domestic chicken. These should therefore be recovered, if present, but few have been found at rural settlements. We have to conclude on

[103] Kershaw and Smith (2000: 196, 214, 232, 558).

[104] Rawcliffe (1999: 181–3) (table B, for 'fish' read 'grain', and for 'meat' read 'fish').

[105] Davis (1991: table 5); Levitan (1984: 119, 121).

present evidence that the larger marine fish were rarely eaten in inland villages: the results of isotopic analysis of human bones for this period, reported in Chapter 16, accord with this view. The few coastal settlements that have been analysed do, however, include fish bones among the food remains, mostly from species found close to the shore.[106]

Fish consumption in early towns is notably varied. Remains were found in only three pits at *Hamwic*—despite the fact that dozens were sampled[107]—suggesting that, in the eighth century, the taste for fish was confined mainly to the trading ports of the eastern seaboard. Fish, mainly eel and herring, was eaten in towns from the ninth century onwards. From the eleventh and twelfth centuries and after, the bulk of the fish remains in the towns is of the large, marine fish, as well as herring and eel. At Castle Mall, Norwich, 'other' fish make up 14 per cent of the sample, and include twenty different species, mostly marine.[108] In the thirteenth- and fourteenth-century Winchester suburbs, fish other than herring, eel, conger, and cod make up 20 per cent of the sample, which includes altogether twenty identified species.[109] One interpretation is that the fish most frequently found were preserved, and those retrieved in small quantities were fresh; but this hypothesis cannot be proved.

At the upper end of the social scale, from excavations at castles in their heyday of use, palaces, certain wealthy manors, and many of the religious houses, typically a range of about twenty to twenty-five species of fish have been found in large samples. The fish remains therefore do not necessarily suggest a more diverse selection of fish than is found in towns. It is more difficult to say whether or not the archaeological evidence supports the notion that religious households ate more fish than other groups. Some of the largest samples of excavated fish bones have come from religious households: the excavations at Eynsham Abbey and St Gregory's Priory happily encountered some deposits where fish bones had survived well in protected areas beneath and on kitchen floors, but the quantity of remains cannot be taken as an index on its own. One source of evidence, the ratio of fish bones to bird bones, can point to levels of relative consumption. In the two large deposits at St Gregory's Priory, the ratio of fish to birds from the kitchen floor deposits is more than 9 : 1; and from the refectory it is 8.5 : 1. If the same calculation is applied to the food remains from a pit of household refuse at Hextalls, Surrey, a manor with royal connections, the result is a ratio of only 3 : 2.[110] The smaller quantity of fish in the secular household is plausible, but the contexts of the finds were different and the sieve-mesh sizes were different, so the two groups may not be comparable.

The archaeology of some wealthy sites suggests that there were some places where fish was avoided. At least three sites where fish might be expected had few, even though large quantities of bones of small birds were recovered. At

106 e.g. Bantham Ham, see Coy (1981: 107–10). 107 Coy (1996: 56).
108 Locker (1997*a*: table 1). 109 Serjeantson and Smith (in press: table 5.32).
110 Bullock (1998).

Carisbrooke Castle, Middleton Stoney, and Faccombe Netherton, a manor belonging to the Bishop of Winchester,[111] the ratios of fish to birds are particularly low: 1 : 29, 1 : 49, and 1 : 11.5. Studies of consumption in aristocratic and gentry diet make it clear that patterns of eating fresh fish varied more between individual households than did patterns of meat consumption: taste may govern fish consumption at these sites.[112]

As well as the prospects of status and Paradise, fish eating might bring medical benefits. Although saints routinely cured those who had eaten bad fish[113] or who had fish bones stuck in their throats,[114] there are examples of individuals eating fish with a therapeutic end in view. Throughout his illness in the summer of 1501, Henry Stafford followed a regime of fasting and fish eating. To assist him he had a variety of crustacea, marine, and freshwater fish. Pike were supplied by Robert Almayn, a fishmonger who seems to have specialized in this species.[115] But fish eating fell from medical favour: Renaissance dietaries frequently warn against it, compounding their strictures particularly where poor-quality water was also involved—a combination that condemned many of the freshwater fish that had hitherto been highly prized.[116]

Conclusion

Archaeological finds establish the patchy nature of the consumption of fish in the Middle Saxon period and that major changes in fish consumption took place from the late tenth century onwards. They strengthen the evidence for the predominance of marine fish in the diet from this time: if the English preferred marine to freshwater fish it was because these were readily available, both fresh and preserved. Before this pattern of consumption was established in the eleventh and twelfth centuries, fish, when it was eaten, was drawn from a narrower range of habitats and species, and came from more local sources. The species and quantities found in later medieval deposits are substantially those which documentary sources lead us to expect, but with a few surprises. The consumption of eel and herring, which is found more widely and overtakes eel in abundance, is possibly higher than might have been anticipated; but the range of other fish matches those which feature regularly in the accounts. It is reassuring that once effective methods are adopted to recover fish bones, the important role of herring and eels in medieval English diet is confirmed. The large, marine fish—cod, hake, and conger eel—are present in quantity from the twelfth and possibly the eleventh century, with the regional variation in their supply and consumption more marked than might have been expected. Pike and other freshwater fish

[111] Numbers (fish) in Smith (1994) and (birds) in Serjeantson (2000: 182, table 32); Levitan (1984: 134–5); Sadler (1990: 500, 506).

[112] Dyer (1998*a*: 27–38); Woolgar (1995).

[113] Robertson (1875–85: ii. 92).

[114] Foreville and Keir (1987: 285); Grosjean (1935: 130–1).

[115] NA E 101/546/18, fos. 22^r, 50^r, 56^r, 89^v.

[116] Albala (2002: 44, 122).

were eaten in quantity at inland sites until the thirteenth century. Cyprinids featured in diet at least from the middle of the first millennium; they continue to be found in good quantities until the later Middle Ages, by which time most came from managed fisheries.

Whether or not the desire to eat fish was driven by religious observance is hard to establish when the remains are studied in isolation from other sources of evidence. The taste for eels, herring, and other locally caught fish, evident especially in eastern England in the Saxon period, developed in north-west Europe beyond the influence of Christian teaching and Mediterranean taste. The Viking cod fisheries equally developed beyond these influences. The main changes in fish consumption came at a time when the Church began to place greater emphasis on the avoidance of meat, and fish may have become a food of choice as fasting became part of the way of life of the widening Christian world and an emblem of virtue. The fisheries for cod and herring which had developed in the north of Europe were able to fulfil the requirement for fish as a substitute for meat. The second important impulse for the development of the trade in preserved fish must have been growing urbanization.[117] Town-dwellers had to be fed, and fish was an ideal source of food, readily preserved and transported. While some of the most abundant evidence for fish eating is from religious houses, the laity also ate fish extensively, as the remains from many deposits in towns show.

The rise and decline of fish eating were the products of a complex mixture of taste, of social and religious standing, of availability in the face of seasonal and regional diversity, and many other factors besides. Fish made a major contribution to the diet of the aristocracy, in the monasteries, and in some parts of the countryside and in some towns, certainly the larger conurbations. Elsewhere, the benefits of fish eating were probably nugatory, perhaps non-existent. The pattern of decline in the mid- to late fifteenth century is partial. Indeed, the archaeological evidence for diminishing levels of fish eating in the later medieval period and after the Reformation is not particularly clear. In London, for example, fish continues to be present in deposits from the seventeenth century and later.[118] The Reformation, with the abandonment of the Lenten fast, took away one of the cultural reasons for the consumption of fish. Although the noble great households of the seventeenth century consumed fish, they did so in nothing like the variety or quantity of former periods, even in households in maritime counties. Nonetheless fish long continued to be an acceptable gift, welcome at very many tables, and it may have sustained its significance in other sectors.[119]

[117] Barrett, Locker, and Roberts (2004*a*: 630–1).

[118] Locker, personal communication. Locker (2001) cites many assemblages from the sixteenth and seventeenth centuries from London, Bristol, and elsewhere.

[119] Munby (1986); Gray (1995–6); Starkey, Reid, and Ashcroft (2000: 45–110).

9

Birds: Food and a Mark of Status

D. SERJEANTSON

In England, the eating of chickens and geese began only during the Roman period. With few exceptions, the practice of eating wild birds is found for the first time in the early Middle Ages. Thereafter, every section of society ate domestic fowls (chickens) and geese, but eating other birds was restricted by custom, by decree, and by expense, and was a mark of an individual's standing.[1] The consumption of birds was also linked to religious observance: the meat of birds was permitted where that of four-footed animals was forbidden.[2]

The bones are a better source for the range of species eaten than the historical record, not least because the names given to birds have been surprisingly fluid over the past millennium. Some medieval designations were used for more than one species or for groups of birds, and some correlations are uncertain (Table 9.1).[3] Archaeological remains provide evidence for the relative numbers of the different species which were eaten. They cannot indicate where poultry and gamebirds were procured—this is an area in which written evidence can be especially informative[4]—but they can suggest how they were raised. It is clear from the very many illustrations in which kings and noblemen are seen carrying a hawk that the possession of a bird for use in falconry was an important sign of status, and archaeological evidence has made an important contribution to understanding the origin and development of medieval falconry.

As with other foodstuffs, archaeology provides more information on the patterns of consumption of the wealthy and least about the peasantry. Rural settlements produce few or no deposits, such as pits or wells, in which bird bones might be preserved effectively—but it is here, as Chapter 10 shows, that historical information for the later Middle Ages is particularly important. Bird bones

[1] See also Chapter 13. [2] Harvey (1993: 39).

[3] As Gurney (1921: 28) observed in reference to the birds listed in Aelfric the Grammarian's *Nomina avium*. Fisher (1966: 184) recorded that, of the fifty birds referred to in the works of Shakespeare, 'some are, as usual, indeterminable between close species'; and Bourne (1981: 332) noted of the birds eaten in the feast hosted by Henry VIII in Calais in 1532, 'the identity of most species is clear enough, but some cause problems'.

[4] See Chapter 10.

Table 9.1. Birds from medieval sites in southern England and recorded in medieval documents

Scientific name	Modern/medieval name	Abundance
Introductions		
Pavo cristatus	Peafowl	xxx
Phasiansus sp.	Pheasant	x
Numidia meleagris	Guinea fowl	0
Native		
Cygnus olor	Swan, mute swan	xxx
Pluvialis apricaria	Plover	xxx
Perdix perdix	Partridge	xxx
Gallinago gallinago	Snipe, 'snite'	xxx
Scolopax rusticola	Woodcock, 'cock'	xxx
Turdus spp.	Thrush, fieldfare, etc., 'small bird'	xxx
Numenius arquata	Curlew, 'prane'	xx
Larus spp.	Gull, 'sea pie'	xx
Grus grus	Crane	xx
Phalacrocorax carbo	Cormorant	xx
Ardea cinerea	Heron, heronsew (young heron)	xx
Fulica atra	Coot	xx
Vanellus vanellus	Lapwing, 'peewit', 'wype', 'plover', 'upupa'	xx
Limosa spp.	Godwit, 'brewe', 'prane'	xx
Tringa/Philomachus spp.	Ruff, rees, 'snite' [redshank, other waders]	xx
Gavia spp.	[Divers]	x
Podiceps spp.	[Grebe]	x
Puffinus puffinus	'Puffin' [Shearwater]	x
Sula bassana	Gannet	x
Phalacrocorax aristotelis	'Cormorant' [Shag]	x
Botaurus stellaris	Bittern, 'bittour'	x
Ciconia ciconia	Stork	x
Platalea leucorodia	'Shovelard', 'popeler' [spoonbill]	x
Coturnix coturnix	Quail	x
Rallus aquaticus	[Water rail]	x
Otis tarda	Bustard	x
Gallinula chloropus	Moorhen, 'heath hen'	x
Haematopus ostralegus	[Oystercatcher]	x
Calidris spp.	Knot, dunlin, 'stint'	x
Charadrius spp.	Dotterel, sandpiper, 'stint'	x
Numenius phaeops	Whimbrel, 'brewe'	x
Sterna spp.	Tern	x
Uria aalge	Guillemot	x
Alca torda	[Razorbill]	x
Fratercula arctica	[Puffin]	x
Fringillidae/Emberizidae	Small bird [finch, bunting]	x
Alauda arvensis	'Lark' [skylark]	x
Passer domesticus	Sparrow	x
Egretta sp.	Egret	0
Hirundo rustica	Swallow	0

Notes: The scientific name, the English medieval or modern name, and the archaeological abundance are shown.

Relative abundance: xxx, more than ten sites; xx, more than five sites; x, one to five sites; '0', bird recorded in documents, but not archaeologically. Remains of birds not eaten, such as hawks and commensals, are excluded.

are much more common at castles, manorial sites, and religious houses. In towns survival is typically mixed. In the northern and eastern suburbs of Winchester, for instance, about twenty closely dated groups of bones were studied, and of these fewer than half a dozen have more than a handful of bird bones.[5] Most assemblages discussed here are those with substantial groups of well-dated bird bones, often from protected contexts.

Sieving archaeological deposits has an impact on the retrieval of bird bones. The bones of birds of the size of chickens and larger can usually be seen and retrieved by hand except in very difficult sediments, but those of smaller birds are much less likely to be recovered. To retrieve bones of small birds consistently it is necessary to sieve deposits using a mesh of 5 mm or smaller. A number of the assemblages show this clearly. The percentage of wild birds is greater in the sieved samples from Launceston Castle than in the material retrieved by hand,[6] despite the otherwise good quality of recovery at that site. At the manor of Hextalls, in an early sixteenth-century pit filled with debris including remains from what appears to have been a single visit or single occasion of feasting, the small songbirds, thrushes, and smaller passerines make up 38 per cent of the bones from the sample sieved to 6 mm, but only 9 per cent from the material retrieved by hand.[7]

Some limitations are particular to bird bones. Not all can be identified to species; members of the same family can have skeletons which are similar in shape and size, which means that some can be identified only to family and size. It is not always possible to distinguish between wild and domestic strains of geese and ducks—not surprisingly as they must have interbred in the past as they do today. The woodpigeon can be distinguished from other pigeons, but wild and domestic rock doves cannot be separated. It is also difficult to separate the bones of the closely related chickens and pheasants. Most zooarchaeologists distinguish the tarsometatarsi and femora of the two species[8] but there is no consensus on whether other elements can be identified. It is possible that some bones of pheasants have been missed—and also that some bones of domestic fowl have been misidentified as pheasant.

We cannot take it for granted that all the bird bones recovered from archaeological sites were food remains. Cut marks made during the preparation and consumption of the birds can indicate that birds were eaten, but most bones show no butchery marks. Few species were *not* eaten in medieval England: birds which modern taste would regard as inedible, such as herons, cormorants, and seabirds, were eaten by all social classes, although in the late Middle Ages the consumption of some species might be reserved, at least in theory, for particular groups.[9] The exceptions, which were not commonly eaten, were birds of prey,

[5] Serjeantson and Smith (in press). [6] Albarella and Davis (1996).
[7] Serjeantson (2001*a*: 266). [8] e.g Albarella and Davis (1996).
[9] See also Chapter 10; Woolgar (1999: 133–4).

Fig. 9.1 Percentage of chickens, geese, and other birds: average of all sites. 'Other' birds include all ducks and pigeons as well as wild birds

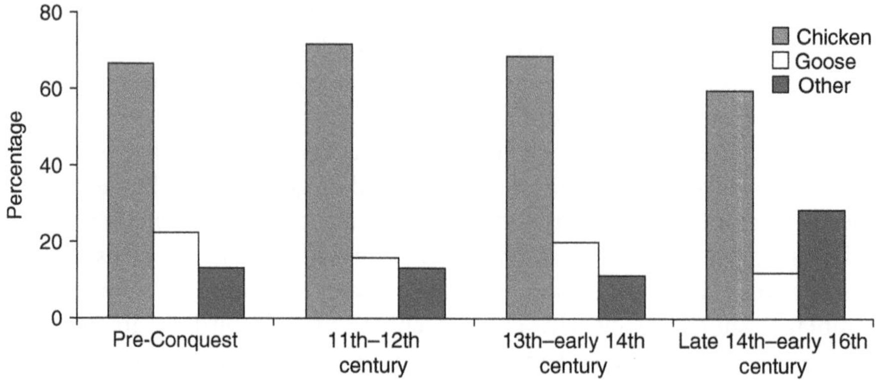

Note: Counts are based on number of identified specimens (NISP).

both falconers' birds and other raptors such as kites and buzzards, and members of the crow family. The circumstances of the finds often confirm this distaste. Corvids were scavengers in villages and towns,[10] which must account for the presence of their bones in archaeological deposits.

Methods of comparison

The percentage of chickens, geese, and 'other' birds from just under fifty assemblages in England dating from the seventh to the early sixteenth centuries is compared in Figs. 9.1–9.5. The calculation uses the number of identified specimens (NISP); this is the least unsatisfactory method, as it is the only count which can be used in comparisons between different assemblages. The 'other' birds are all other species which were probably eaten and they include all ducks (including possible domestic ducks) and pigeons.[11]

In the Middle Ages, chickens and other birds were classed by age and sex, and the distinctions made in the kitchen are sometimes visible in archaeological material. The bones of immature birds can be recognized from their porous character and lack of fusion of some elements, so pullets, goslings, and cygnets, for instance, can be distinguished from adult birds. The presence or absence of a spur on the tarsometatarsus is a guide (though not a wholly reliable one) to the sex of chickens: this bone with a spur is from a cock and without, it is from a hen. Bones with an area of scarring where the spur is about to attach are from an immature male, and so probably from capons. The number found, however, is too small to account for the many capons listed in the documents, so other

[10] Mulkeen and O'Connor (1997: 443).

[11] Except for the West Cotton assemblage, where the pigeon bones appear to come from a dovecot: Albarella and Davis (1994).

Fig. 9.2 Percentage of chickens, geese, and other birds: pre-Conquest. Left to right: manors, religious houses, towns, village

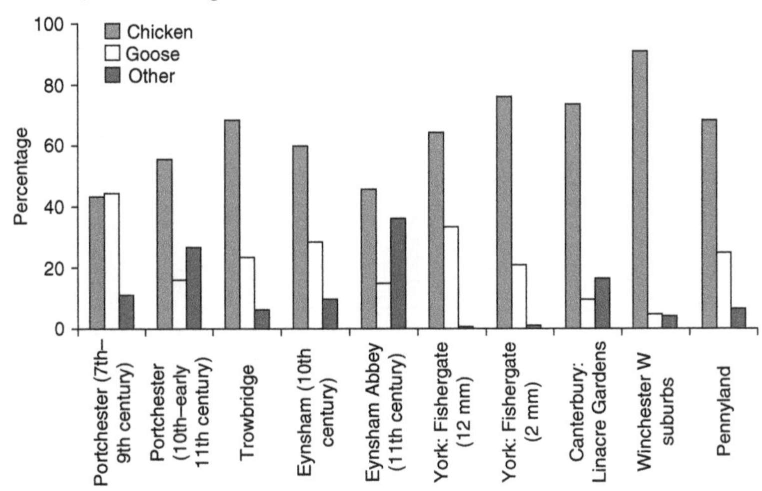

Notes: mm, sieve mesh size; W, western.

Sources: Eastham (1976: 291); Bourdillon (1993: 137, table 22); Mulville (2003: 344, table 10.1); Ayres, Locker, and Serjeantson (2003: 365, table 10.20); O'Connor (1991: 261, table 71); Driver (1990: 244); Coy (in press: tables A3.29–A3.35); Ashdown (1993: 154).

Fig. 9.3 Percentage of chickens, geese, and other birds: late eleventh to twelfth centuries. Left to right: castles, manors, religious houses, towns, village

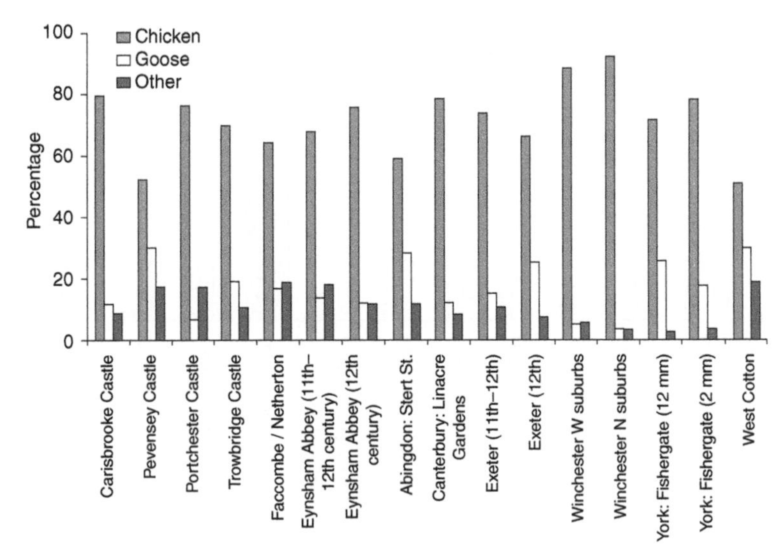

Notes: W, western; N, northern.

Sources: Serjeantson (2000: 182, table 32, and 183, table 33); Powell and Serjeantson (in press: table 15); Eastham (1977: 236, table VIII); Bourdillon (1993: 137, table 22); Sadler (1990: 500–4); Ayres, Locker, and Serjeantson (2003: 365, table 10.20); Bramwell and Wilson (1979: 21); Driver (1990: 244); Maltby (1979: 204–6); Coy (in press: tables A3.36–A3.46); Serjeantson and Smith (in press); O'Connor (1991: 261); Albarella and Davis (1994: table 1).

Fig. 9.4 Percentage of chickens, geese, and other birds: thirteenth to mid-fourteenth centuries. Left to right: castles, religious houses, towns, village

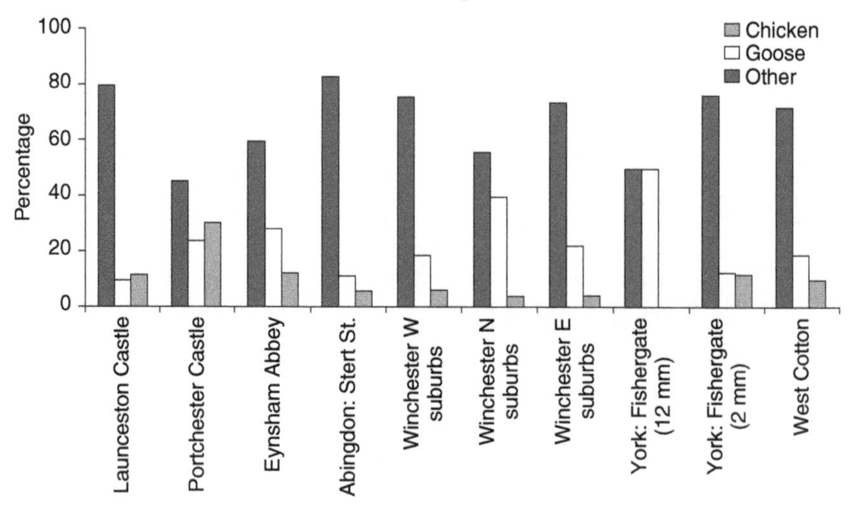

Notes: W, western; N, northern; E, eastern; mm, sieve mesh size.

Sources: Albarella and Davis (1996: 65–6, table 1); Eastham (1985: 263, table XXVIII); Ayres, Locker, and Serjeantson (2003: 365); Bramwell and Wilson (1979: 20); Coy (in press: tables A3.39–A3.46); Serjeantson and Smith (in press); O'Connor (1991: 261); Albarella and Davis (1994: table 1).

Fig. 9.5 Percentage of chickens, geese, and other birds, mid-fourteenth to mid-sixteenth centuries. Left to right: castles, manor, religious houses, towns, village

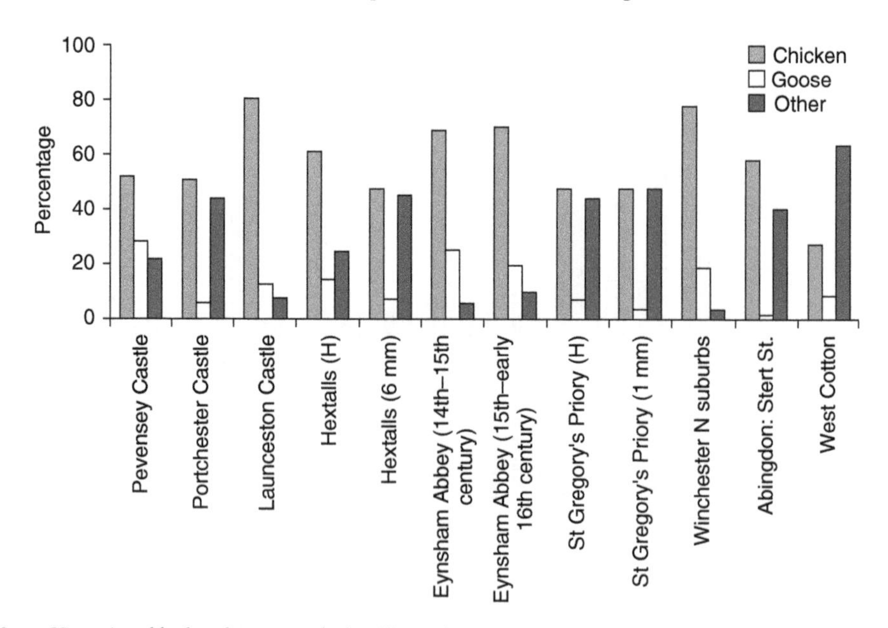

Notes: H, retrieved by hand; mm, mesh size; N, northern.

Sources: Powell and Serjeantson (in press: table 1); Eastham (1985: 264, table XXIX); Albarella and Davis (1996: 65–6, table 1); Bourdillon (1998: 142, table 6.3); Ayres, Locker, and Serjeantson (2003: 365, table 10.20); Powell, Serjeantson, and Smith (2001: 318, table 40); Serjeantson and Smith (in press); Bramwell and Wilson (1979: 20); Albarella and Davis (1994: table 1).

mature birds must have been fattened as well to be eaten as capons. Hens in lay can be identified from the presence of medullary bone, most often seen in the femur. As a good hen would lay more than 100 eggs in a year,[12] medullary bone might be present for one third of the year. Bones of very young chicks, too small to be edible, are presumed to be natural casualties, and the presence of these is usually taken to indicate that fowls were raised at the settlement.[13] Eggshell is sometimes noted in deposits or found when sediments are sieved, but is difficult to identify to species and to quantify in any systematic fashion.

Before the Conquest

Numbers of bird bones are never very frequent on rural settlements before the Conquest: they typically make up less than 10 per cent of identified bones. At the Anglo-Saxon village at Yarnton, for instance, they constitute 3 per cent to 4 per cent of identified bones. A rare exception is the eighth-century settlement at Eynsham. In one pit, 20 per cent of the identified bones are of birds. The site was a thegnly manor at this time, and may already have been a minster.[14]

At this time the domestic birds raised were chickens, geese, and possibly ducks, with chickens outnumbering geese at most sites (Fig. 9.1). More than 80 per cent of all birds from the wealthy western suburbs of Winchester were chickens, but fewer than 50 per cent at the Anglo-Saxon settlement at Portchester (Fig. 9.2). The chickens at *Hamwic*, based on the tarsometatarsi, were in the ratio of one cock to four hens, as might be expected in a typical barnyard flock. Four possible capons were also identified there.[15] In the tenth- to eleventh-century suburbs of Winchester the ratio of hens was higher, suggesting eggs were already important.[16] They certainly were in the Late Saxon period at Eynsham Abbey: there eggshell was recovered from a deposit thought to have been the kitchen floor.[17] Of the chickens eaten in the Late Saxon period at Eynsham Abbey and St Albans Abbey,[18] 30 per cent and 40 per cent respectively were immature; but more than 90 per cent from the poorer Winchester suburbs were mature birds. Even at this time, it appears that wealthier establishments were able to choose to eat more palatable birds.

Most of the goose bones are from domestic birds, but the large number from Anglo-Saxon Portchester may include wild birds caught in Portsmouth Harbour as well as domestic geese. There are surprisingly few geese in the western suburbs of pre-Conquest Winchester; this was a wealthy suburb from the first, and geese may not have been favoured for food by the nobility. Small pigeons as well as woodpigeons have been identified at some sites, possibly already kept under

[12] Although the *Husbandrie* gives 115 days, in reality 100 days constituted a good laying season: Chapter 10.

[13] Coy (1989: 32).

[14] Mulville (2003: 355, table 10.11).

[15] Bourdillon and Coy (1980: 116, fig. 17.20).

[16] Serjeantson and Smith (in press: fig. 5.41).

[17] Keevill (1996).

[18] Author's unpublished data.

domestic conditions. The first references to pheasant occur in the eleventh century. It must have been a desirable addition to a banquet as much for its plumage as for its food value,[19] but confirmed finds of pheasants are very rare. One bone found in Anglian deposits in York was identified as coming from a pheasant. It was a coracoid, which is difficult to distinguish from that of domestic chicken, but the identification is thought to be secure.[20]

Numbers of wild birds are very varied (Fig. 9.1). Some sites have few or none, but they are plentiful on a few sites, especially by the Late Saxon period (Fig. 9.2). The species found most often at this time are woodcock, lapwing, curlew, partridge, woodpigeon, and the thrush family. Less frequent are wild duck, snipe, golden plover, and godwit. Remains of cranes have been found at several sites.[21] Wild birds were most abundant in Late Saxon Portchester: at this manor within the old Roman shore fort about 230 bones of wild birds were found, from at least fourteen species. These included more than seventy bones of ducks (including mallard, teal, wigeon, and shelduck) and no less than ninety-four of curlew.[22] The curlew must have been an especial target of wildfowling there.

The use of birds for falconry can be inferred from the presence of skeletons and part skeletons of those raptors known to be used in falconry, the short-winged hawks (goshawk and sparrowhawk) and long-winged hawks (peregrine falcon, kestrel, and gyrfalcon).[23] There is archaeological evidence for falconry in Britain from as early as the seventh century, earlier than the first written reference to hawks in Britain.[24] Birds were buried in pagan graves and, after societies converted to Christianity, are found with domestic rubbish.[25] At contemporary sites, the wild birds, such as partridges, pigeons, lapwings, and thrushes, are typical prey of the smaller hawks, but the cranes must have been caught in nets or with the larger gyrfalcon.

Mid-eleventh to mid-fourteenth centuries

The percentage of birds in assemblages of mammal and bird bones is again very variable, depending on context as well as on the nature of the settlement.

In the eleventh and twelfth centuries the number of chicken bones relative to geese and other birds is higher than before the Conquest, and also slightly higher than later (Fig. 9.1). The highest percentages have been found in the towns (Fig. 9.3). This continues to be the case in the thirteenth and fourteenth centuries (Fig. 9.4). The sex ratios again suggest that most of the adult birds were hens (Fig. 9.6), of which a surprising number were in lay, to judge from the incidence

[19] Mead (1967: 88). [20] O'Connor (1991: 261).

[21] See Albarella and Thomas (2002: 23); Bramwell and Wilson (1979: 20, table 45); Dobney and Jaques (2002: 10); Mulville (2003: 356). [22] Eastham (1976: 295).

[23] Cherryson (2002: 308); Prummel (1997: 333, 335).

[24] Dobney and Jaques (2002: 15); Gurney (1921: 29). [25] Cherryson (2002: 309, table I).

Fig. 9.6 Percentage hens of adult fowl, based on the absence of spur or spur scar on the tarsometatarsus. Top: eleventh to mid-fourteenth centuries; bottom: mid-fourteenth to mid-sixteenth centuries

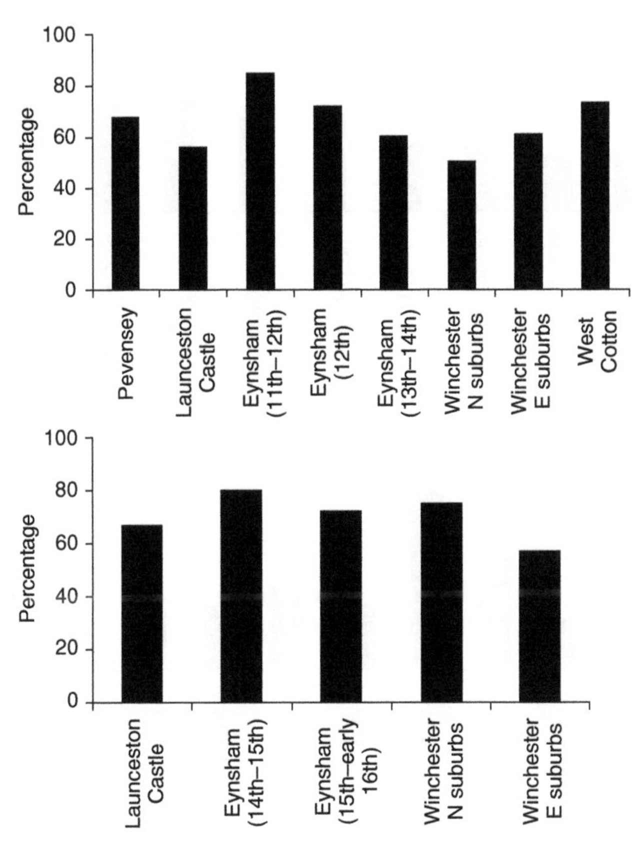

Notes: N, northern: E, eastern.

Sources: Powell and Serjeantson (in press: table 18); Albarella and Davis (1996: 28, fig. 22); Ayres, Locker, and Serjeantson (2003: 383, table 10.37); Serjeantson and Smith (in press); Albarella and Davis (1994: 24).

of medullary bone. At Carisbrooke Castle, for instance, it was seen in four of nine, and at Eynsham in twelve out of thirty femurs. The highest percentage of cocks among the male birds was in the suburbs of Winchester. This probably reflects the fact that the old birds were eaten at home, but it is worth remembering that cockfighting was a medieval pastime and an excess of male birds over the number needed for the domestic flock could suggest that this took place. As many as 40 per cent of the birds eaten at Eynsham Abbey and some urban sites were immature (Fig. 9.7), though the percentage is fewer at the castles, Carisbrooke, Pevensey, and Launceston. The castles may have received renders of laying hens, while the abbeys and towns exercised more choice in the purchase and consumption of birds. While many of the domestic fowl eaten in medieval

Fig. 9.7 Percentage immature chickens of all fowl. Top: late eleventh to mid-fourteenth centuries; bottom: mid-fourteenth to mid-sixteenth centuries

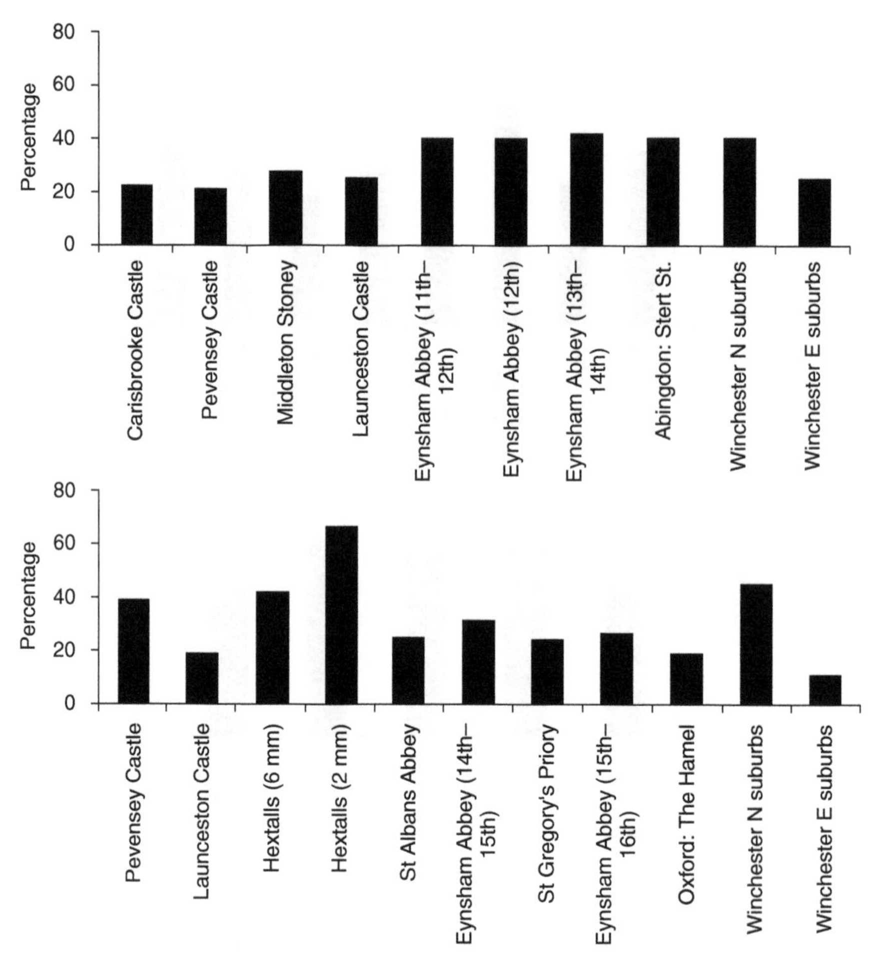

Notes: N, northern; E, eastern.

Sources: Serjeantson (2000); Powell and Serjeantson (in press); Levitan (1984: 121); Albarella and Davis (1996: 38); Ayres, Locker, and Serjeantson (2003: 381, 383, 385); Bramwell and Wilson (1979: 20); Serjeantson and Smith (in press); Bourdillon (1998: 144); author's unpublished data for St Albans Abbey; Powell, Serjeantson, and Smith (2001: 329); Wilson (1981: 189).

England were raised by the peasants (Plate 9.1), bones of very immature chicks, usually an indication that the birds were raised locally, are sometimes also found in towns.

Between the Conquest and the end of the twelfth century, goose bones are fewer than before, rarely more than 20 per cent of all bird bones. Where the percentage is higher (Fig. 9.3) the effect may be a consequence of differential recovery, as the difference between the 12 mm and 2 mm sieved samples from Fishergate (York) indicates. The percentage of geese is higher in the thirteenth and fourteenth centuries, but varies considerably between sites. Very few immature

Plate 9.1 A hen, with white lobes on her neck, believed to be an especially good laying breed, is tethered by her foot, while her chicks feed from a square container on the ground and possibly from seed from the bowl carried by the woman with the distaff. From the Luttrell Psalter, *c.*1320–45, British Library, Add. MS 42130, fo. 59ᵛ. Photograph: © British Library.

goose bones are found at this time. Geese, like chickens, were raised by peasants, but by the end of the Middle Ages some were also raised in large flocks. There is little evidence—in the form of bones of goslings—that they were raised in towns.[26] The skeleton of the goose adapted to its domestic status at this period. The tarsometatarsus developed a thicker shaft, a reaction to the fact that domestic birds fly less and walk more than wild geese, and the wing bones became shorter.[27] Geese were especially useful as a source of feathers, which were both white and grey. Both the down and the flight feathers or quills could be plucked from the live bird. The fact that most geese were fully mature when killed for the pot suggests that they were kept for the feathers as much as for the meat at this time. There is also evidence, in the form of groups of wing bones, for the collection and use of the feathers of slaughtered birds. Collections of the carpometacarpus bone are sometimes found with the radius and ulna, and—when recovery is good—with the distal wing digits, which may indicate that the flight feathers were used. The bones often show cut marks.[28] The whole wing was sometimes used as a brush or a weaving fan.[29] The flight feathers were also used for fletching arrows, but at Winchester and Norwich they seem to have been collected specifically for use as quill pens. At Norwich there were more than twice as many left-hand as right-hand bones: feathers from the left wing are preferred for right-handed scribes. At Winchester a stylus was found close to the pit containing the wing bones, although from a slightly later context.[30]

From the eleventh or twelfth century onwards skeletal remains of pigeons, both adult birds and squabs, as well as buildings confirm the existence of dovecots. The bones of immature birds have been found, sometimes in large numbers, within or near a dovecot: these must be birds which died naturally. Many, almost 30 per cent of which were immature, were found at West Cotton at the base of a

[26] Serjeantson (2002: 51).

[27] Bramwell (1977: 400–1); Reichstein and Pieper (1986). The wider shaft is already evident at Anglian Flaxengate: O'Connor (1982: 42, fig. 54).

[28] Cut marks on goose bones from Winchester are illustrated in Serjeantson (2002: 50, fig. 8).

[29] Scott (1991: 282). [30] Serjeantson (2002: 51).

circular building of the twelfth century, which is thought to have been a dovecot.[31] Other sites where pigeon bones have been found associated with dovecots are Faccombe Netherton;[32] Greyhound Yard, Dorchester;[33] the hospital at St Giles, Brompton Bridge;[34] and the shrunken medieval village at Harry Stoke.[35] In view of their ubiquity in domestic accounts and manorial records outlined in Chapter 10, pigeon bones are surprisingly infrequent in assemblages of food bones. The bones are approximately the same size as those of partridge, woodcock, and teal, but are less frequent in some deposits than these popular wild birds. For instance, they constitute fewer than 2 per cent at Carisbrooke Castle, and hardly more at Eynsham Abbey.

There are few sites which offer hints that ducks were raised domestically. The large number of mallards at West Cotton, however, suggests that ducks may have been raised—and eaten—in that village.

Domestic and wild birds were employed for their impact in terms of display at the meal from at least the twelfth century onwards. Peafowl are said to have been kept on Charlemagne's estates and a bird depicted with William I's palace on the Bayeux Tapestry is also believed to represent a peacock (see also Plate 11.1).[36] Peafowl must always have been kept in captivity. The bones are large enough to be recovered by hand and a dozen or so have been found. The earliest are from high-status sites, but later they have been found in towns and religious houses as well as wealthier sites. The earliest find in Britain of a peacock bone may be a tarsometatarsus at Carisbrooke Castle.[37] It is from a male bird and has cut marks where the toes were detached from the leg (Fig. 9.8). The impulse for the introduction of the peafowl, as the pheasant, must have been the feathers. The peacock looks spectacular compared with native British birds: cooked and served in its skin complete with plumage, it made a grand statement at banquets.

Wild birds were caught for food and show, and some were managed or kept in captivity.[38] The remains are mostly found at the wealthier sites such as the castles, rich manors, and abbeys. The main species correspond well with those referred to in the written sources (Table 9.1), but pheasants—which appear less often in accounts—have not been identified. Swan bones have been found on many sites: many skeletal elements are large, so this is to be expected. The partridge, together with plover, snipe, and woodcock, are the most favoured species, which accords well with historical accounts. In view of their small size, it is surprising that bones of the thrush family have been found quite often, sometimes even in deposits which were not sieved, confirming that thrushes were eaten from an early period.

Bittern, spoonbill, and stork are rarer than might be expected: each has been recorded only once on the sites surveyed. There is a single record of a bustard

[31] Albarella and Davis (1996: 25). [32] Sadler (1990: 505). [33] Maltby (1993: 339).
[34] Stallibrass (1993: table 6).
[35] At Harry Stoke (ASMR 1334) two of the seven abandoned buildings were dovecots, and about sixty pigeon bones were recovered from the base of one of these, associated with thirteenth-century pottery: Serjeantson (1993). [36] Sykes (2001).
[37] Serjeantson (2000: 184). [38] See also Chapter 10.

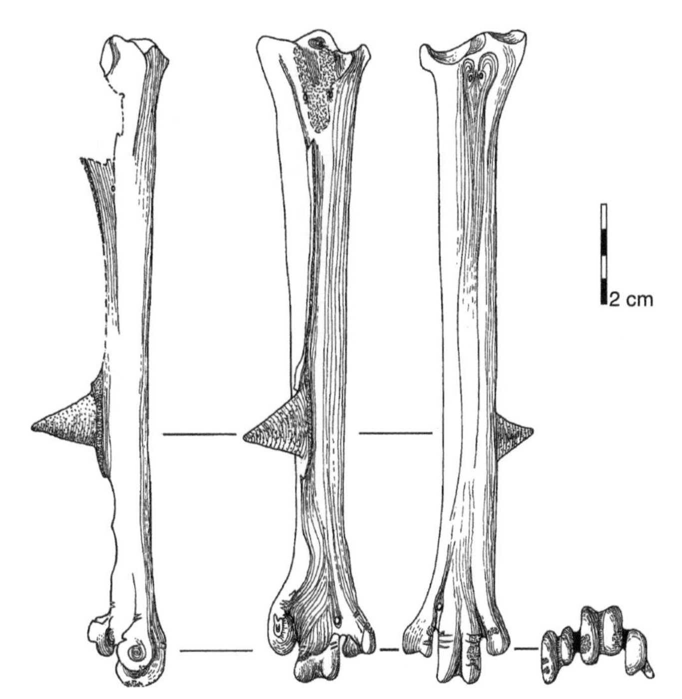

Fig. 9.8 Tarsometatarsus of peacock with cut marks, from Carisbrooke Castle, Isle of Wight, eleventh to twelfth centuries. The cuts on the front and back of the distal articulation show that the foot has been carefully removed. Drawing: D. Webb.

from late medieval deposits in London, but none of the egret. Quail, supplied by the thousand for the Neville feast described in Chapter 10, is found infrequently; lark bones have been found at some sites, but the rarity of these finds is not surprising, since bones would not be seen unless deposits were sieved. The cormorant and coot, recorded on about half a dozen sites, are perhaps more prominent than the records would lead us to expect. Other species recorded in the north of England, but not in the south, include red grouse and capercaillie.[39]

Seabirds were also captured for food, mainly around the western and northern coasts of the British Isles. Birds, such as the manx shearwater, the gannet, and the auks (razorbill, guillemot, puffin), were normally caught at their breeding sites on cliffs and offshore islands. They were a regular if minor element in the diet of the communities of the north and west of the British Isles and could be preserved by salting and drying, in the same manner as fish, and traded.[40] The guillemot and the gannet are not uncommon in the north of England,[41] but are rare in the Midlands and the south. They have been found in early medieval deposits in the Scilly Isles, where the range of bird species is similar to that in the west of Scotland. Gannets and manx shearwater bones are present on sites in the

[39] Dobney and Jaques (2002). [40] Serjeantson (2001*b*: 48).
[41] Albarella and Thomas (2002: 35).

Plate 9.2 A gyrfalcon seizes a duck, probably a teal: a misericord of the late fourteenth century from Winchester College. Bones from both species were found in contemporary archaeological deposits in the western suburbs, the site of the royal mews. Photograph: © Marshall Laird.

west and south-west—at Launceston Castle, Okehampton Castle, and the towns of Exeter and Hereford—but they do not seem to have reached the south-east. The seagull is found more widely and remains of these are found in small numbers, mostly in coastal towns such as Southampton.[42]

Many of the poorer settlements have few or no bones of birds other than chickens, geese, and sometimes ducks and pigeons. It is notable, for instance, that, while several wild birds were found in the wealthier western suburbs, one or two only were found in the poorer northern and eastern suburbs of medieval Winchester.[43]

The sparrowhawk is the most frequent of the falcons that have been found, not only on castles and manors but also in towns and at religious houses.[44] The larger goshawk has been found at Portchester, Faccombe, and Castle Rising, and also at York (Bedern) and Norwich (Castle Mall). The peregrine, today more common than the goshawk, has been found only at Faccombe and Castle Rising. Most unexpected is the gyrfalcon, a hawk which is not native to the British Isles. These birds were brought to Britain from Norway or Iceland, often as gifts to the king, and remains of at least two birds were recovered near the royal mews in the western suburbs of Winchester (Plate 9.2).

[42] Bourdillon and Coy (1980: 118).

[43] Coy (in press: tables 3.36–3.46); Serjeantson and Smith (in press: table 5.28).

[44] Cherryson (2002: table 1).

Later Middle Ages

Many sources suggest that more birds, especially wild birds, were eaten. There are some bone assemblages where this can be shown, but the trend is not seen in every site, perhaps because bones have sometimes come from dumps of secondary waste rather than pits containing primary deposits.

The percentage of chickens, below 70 per cent at several sites, is less than in earlier centuries, but this is relative to an increase in wild birds (Table 9.1 and Fig. 9.5). The percentage of hens among the adult birds is consistently high, between 60 per cent and 80 per cent. In a survey of changes in diet in the later Middle Ages, based mainly on urban sites, Albarella found that immature chickens accounted for between 20 per cent and 40 per cent of finds, and the percentage is much the same in the sites considered here (Fig. 9.7). Hextalls, a wealthy manor with royal connections, is exceptional in this as in other things: more than 60 per cent of chicken bones from the sieved sample were immature. Albarella concluded that more young chickens were eaten in the later Middle Ages, part of the trend he identified towards an emphasis on meat at this time.[45] For abbeys and other wealthy households, however, the percentage of young birds was just as high at an earlier period.

The later Middle Ages saw changes in goose husbandry, and the percentage of goose bones recovered is lower at this time. For the first time a significant percentage of bones are from immature geese: at Winchester the increase is from less than 2 per cent to the 24 per cent found in one sixteenth-century deposit. The same increase was observed at Launceston and Norwich. These immature bones are presumably from the 'green geese' referred to in accounts. Eating young geese further reflects the general selection of animals for meat after the mid-fourteenth century.[46] The quantity of pigeons in food remains—rather than associated with dovecots—continues to be small. At St Gregory's Priory, Canterbury, where deposits were sieved, they make up only between 2 per cent and 6 per cent of all bird bones, and they were fewer than 2 per cent in the hand-collected material and fewer than 3 per cent in the sieved samples.[47] Guinea fowl, introduced to Europe from Africa in the Middle Ages, may have reached England in the thirteenth century,[48] but none has been identified on a medieval site.

The percentage of wild birds is higher in later medieval assemblages. Some, as we see in Chapter 10, came from managed sources, as well as directly from the wild. The very large number from Hextalls, nearly one third of the total number, has distorted the overall percentage, but if the sieved sample from that site is omitted, wild birds still make up approximately 20 per cent of the total. The percentage is high at several sites (Fig. 9.5). At West Cotton the 'other' birds were mostly ducks, which may have been domestic, but elsewhere the increase in wild birds must be genuine. At Hextalls the principal medium-size species were ducks, snipe, lapwing, pigeon, woodcock, and golden plover.

[45] Albarella (1997). [46] Albarella (1997).

[47] Powell, Serjeantson, and Smith (2001); Bourdillon (1998). [48] MacDonald (1992: 303).

The terms 'small birds' and 'stints' were used in written sources for small songbirds and waders respectively, though the terms were sometimes interchangeable. The thrushes were sometimes specified separately but were also sometimes included with the 'small birds'. Larks, which documents suggest were the most popular small bird for the table, seem to be designated separately rather than included with other small birds. The assemblage from St Gregory's Priory, late fourteenth to early sixteenth centuries, provides the best example of the range of small birds eaten.[49] Two samples from the threshold of the refectory and the kitchen floor were sieved using 1 mm mesh and a third sample from the oven floor was carefully recovered. Rather over half of all bird bones from St Gregory's Priory are from the small passerines: those identified were lark, a pipit, a medium-sized tit, a bunting (probably a yellowhammer), a domestic sparrow, the linnet, and possibly other species of finch. At Hextalls they formed one quarter of all wild birds. The consumption of these small birds, especially thrushes and larks, was said to bring health benefits,[50] and they constituted a distinctive part of upper-class diet.[51] They were too small to have had much value as food, however.

Much of the general increase in the consumption of wildfowl at this time may come from the increase in the numbers of small birds eaten. At St Gregory's Priory and Hextalls excellent preservation and recovery has contributed to the high percentage of wild birds, but the large numbers at Portchester Castle, where deposits were not sieved, and Stert Street, Abingdon, confirm that the increase is widespread.

There are no significant species of wild birds found for the first time only after the fourteenth century. Some of the tiny birds are not recorded earlier, but this may be because the bones have not been retrieved or have not been identified to species. Some species, such as partridge, decline in abundance.[52] The crane certainly became rarer, no doubt largely as a result of hunting. Most wealthy late and post-medieval sites, including many in towns, have finds of swans, which correlates well with the establishment of managed swanneries in the later Middle Ages,[53] and at one site at least all the bones are from cygnets.[54] The high percentage of wild birds appears to confirm the increase in the consumption of birds in the later Middle Ages. At some sites, however, very few were recovered, for example, at Launceston Castle (by this time in use as a prison) and in the northern suburbs of Winchester.

Complete and partial skeletons of sparrowhawks, peregrines, and a kestrel have been found in several later medieval deposits, and again they come from towns and religious houses as well as castles.[55] A kestrel was found at St Gregory's Priory: this association between the hawk and the small birds which would have comprised its prey might suggest that the birds there were caught by

[49] The small birds from St Gregory's Priory and Hextalls are discussed by Serjeantson (2001*a*).
[50] According to the later *Via recta ad vitam longam* of 1628, quoted by Simon (1944: 33).
[51] Woolgar (1999: 133–4). [52] Albarella and Thomas (2001: 34). [53] See also Chapter 10.
[54] Stert Street, Abingdon: Bramwell and Wilson (1979: 21).
[55] Cherryson (2002: 309, table I).

falconry, but the quantity of small birds is so great that it is more likely in fact that they were trapped in nets by professional fowlers.

Conclusion

Chickens were eaten by all classes of society throughout the period. They were straightforward for peasants to raise: no specialized skill or knowledge was needed, and they could be fed on household scraps and left to forage. The archaeological evidence from villages is scant, but it is clear that fowls were kept from the beginning of the Middle Ages. From early times, egg production was as important as keeping the birds for meat: there are hints from at least the twelfth century that the laying period was prolonged in hens. From the same time richer households were able to buy and eat immature fowls, pullets, and capons. Geese, also eaten by households of every type from early times, were probably raised in villages and in larger flocks on manors, but do not seem to have been raised in towns. At all periods the feathers were probably as valuable as the carcass of a goose; the more intensive raising of green geese (young geese) for food is seen only in the later Middle Ages. At only one of the sites discussed, West Cotton, were domestic ducks found in numbers great enough to suggest that they were raised locally; they are rarely found in large quantities. There is more archaeological evidence—from dovecots, rather than from food remains—for raising pigeons, but there is surprisingly little physical evidence for their consumption, especially given the numbers in which they were raised on some manors (Fig. 10.1).

The habit of catching and eating wild birds is evident from at least the eighth century, and from the first consumption may have been associated with the sport of falconry. Even when most of the birds eaten were in fact caught by professional fowlers using nets and snares as well as hawking, wild birds continued to be an important food, not only for their nutritional contribution, but also for their symbolic value. The species consumed do not change dramatically over time, but they do vary regionally. To some extent the birds eaten must have borne some relationship to their abundance in the countryside, but a species which is solitary, however abundant, will always be less tempting to the fowler than those that flock together or which breed in colonies, and it is these which feature most often in the written records and excavated finds.

The archaeological evidence provides little support for the view that the peasants were able to eat birds other than chickens, and townsfolk too seem to have eaten few wild birds before the fourteenth century. Eating birds and eggs fitted the ecclesiastical dietary restrictions on the consumption of meat. Like fish they might also be a virtuous and healthy food, and the ecclesiastical sites confirm their importance. Birds, even more than fish and other animals, show how those of higher rank in society could exercise choice over the food they ate—and they chose to eat young poultry and wild birds, including species that were both large and striking in appearance.

10

The Consumption and Supply of Birds in Late Medieval England

D. J. STONE

The consumption of birds in late medieval England is often summed up by reference to one event: the enthronement feast for Archbishop Neville in 1465, which, it was said, included 400 pigeons, 2,000 chickens, 204 cranes, 104 peacocks, 1,200 quails, 400 swans, 400 herons, 100 capons, 400 plovers, 4,000 mallards and teals, 204 bitterns, 200 pheasants, 500 partridges, 400 woodcocks, 100 curlews, and 100 egrets. But while this conveys a sense of the munificence of this occasion and the status and availability of birds in the mid-fifteenth century, it tells us little about their wider role in the medieval world. This chapter explores the documentary evidence for the consumption and supply of birds in England between the twelfth century and the sixteenth, as a complement to the archaeological evidence discussed in Chapter 9. It suggests that birds did play a significant part in lordly diet, albeit usually on a more modest scale, but also that they represent one of the best-documented aspects of peasant economic behaviour and provide tangible evidence of changing patterns of consumption following the Black Death—and in this the historical sources fill one of the most significant gaps in the archaeological record.

Before the Black Death

Birds probably made up almost a tenth of the meat consumed in aristocratic and ecclesiastical households.[1] But whatever the exact proportion, it is clear that a large number and wide range of birds were eaten. In the year from September 1317, while at his manor of Melbourn, Sir John de Argentein, his wife, three young children, and a handful of servants between them consumed seventy-eight geese, sixty-one capons, and a goodly number of pigeons; we do not know precisely how many pigeons, since 'the reeve has no accounts for the dovecot because the cook had the key'.[2] These domestic birds, reared specifically for the

[1] Dyer (1998*a*: 59). [2] Palmer (1927: 29–30, 62–3).

pot, formed the mainstay of the poultry diet in most larger households too; however, on special occasions the range of birds consumed might be widened to include semi-managed or wild species. For example, as well as three geese and forty-seven chickens, the anniversary feast of 20 January 1337 in the household of Dame Katherine de Norwich included one bittern, two herons, four cygnets, four woodcocks, five mallard, thirteen plovers, eighteen partridges, forty-eight larks, and 107 'small birds'.[3]

The King received a certain amount of his fowl through requisition. The orders for provisions sent out to the sheriffs of more than a dozen counties in 1249 give an insight into contemporary perceptions of where certain birds could be acquired. Swans, peacocks, and, at that time, even cranes were considered ubiquitous; herons and bitterns were expected only from the fenland counties of Cambridgeshire and Lincolnshire, the order to these sheriffs being for 'as many as he is able to acquire' (Plate 10.1); all geese, on the other hand, came from London and Middlesex, suggesting the existence already of organized arrangements for marketing these birds.[4]

Other lords obtained their birds in four main ways: by hunting them; by rearing them; by buying them; and by taking them as payment in kind from their tenants. In some regions there was an ample supply of wildfowl. As Thomas of Ely wrote of the Fens in the twelfth century, '[t]here are numberless geese, teals, coots, dabchicks, cormorants, herons and ducks . . . At midwinter, or when the birds moult their quills, I have seen them caught by the hundred, and even by the three hundreds, more or less.'[5] In other landscapes, the hunting of pheasants and partridges was common. In 1347–8, the reeve of Chevington paid 3*s.* 3*d.* to two men for catching twelve pheasants and fifty partridges, which were then sent to London for the Abbot of Bury St Edmunds.[6] Framlingham Castle in fact employed one Edmund *pertricarius*—'the catcher of partridges'—who had falcons and eight hunting dogs in his care in 1286–7 (Plate 10.2).[7] Nor was bird trapping confined to these birds: larks, for example, were taken to Crowland Abbey from its manor of Wellingborough in 1321–2.[8] Of course, birds such as these were not managed and their supply was seasonal and unpredictable; in anticipation of feasts, therefore, lords may well have built up a stock of captured birds that were maintained until required.[9]

A large number of domestic birds were supplied from lords' estates. For instance, in 1329–30 the manor of Cuxham provided the warden and fellows of Merton College, Oxford, with a goose, five ducks, eighteen capons, sixty-three cocks and hens, and 635 pigeons; a further six geese were sent to the warden at Christmas that year, including two females that were too old to lay.[10] Many demesne farms had a scatter of poultry and some also kept a few ducks; at Cuxham, the fishpond served as a duckpond as well, though the ducks not only

[3] Woolgar (1992–3: i. 204). [4] *CCR 1247–51*, 168–9. [5] Darby (1940: 36).
[6] Farmer (1991*a*: 392). [7] Ridgard (1985: 30–1). [8] Page (1936: 128).
[9] Woolgar (1999: 114–15). [10] Harvey (1976: 408–10).

Plate 10.1 Birds, including a bittern and a crane, familiar among fenland species, decorate the lower parts of the north wall and north-west corner of the first-floor chamber in the Longthorpe Tower, near Peterborough, *c*.1320–50. Photograph: C. M. Woolgar.

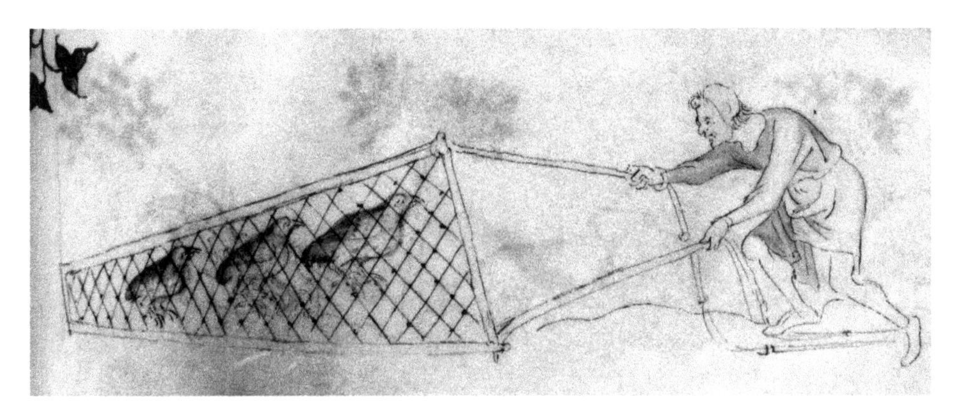

Plate 10.2 Partridges caught in a net, from the early fourteenth-century Queen Mary Psalter, probably made for Queen Isabella, British Library, MS Royal 2 B VII, fo. 112r. Photograph: © British Library.

contributed to human diet, for two were eaten by pike in 1312–13.[11] Peacocks were fully domesticated and were not uncommon on these farms either: no fewer than seventeen of the manors on the Bishop of Winchester's estate kept them in 1301–2.[12] But probably the most numerous of demesne birds were pigeons and geese.

There is no record of dovecots in Domesday Book, but they seem to have appeared in increasing numbers soon after—matching the archaeological evidence discussed in Chapter 9. Integral nests were built into the twelfth-century stone keeps of Rochester Castle and Conisborough Castle,[13] and by the early fourteenth century many manors had at least one free-standing dovecot (Plate 14.1). But not every demesne had one. Indeed, several of the 'home' manors on the Peterborough Abbey estates had no dovecots at all; 57 per cent of the Abbey's requirements in this respect were met by four manors situated 8 to 12 miles from Peterborough.[14] The tender meat of squabs (young pigeons) was especially highly valued, though the supply of them was by no means constant. Squabs were a seasonal product: on the demesne at Hinderclay, for example, where an average of 567 squabs were produced each year in the 1320s, 17 per cent were produced around Michaelmas, 37 per cent around Easter, and 46 per cent over the summer; none, in other words, over the winter months.[15] Furthermore, the production of squabs could change considerably over time. At Hinderclay, the number of pigeons coming from the manorial dovecot can be charted with some consistency over the period 1298–1358 (Fig. 10.1). Numbers fluctuated from year to year, then fell away dramatically in the 1330s, before picking up to some degree in the 1340s and 1350s. A similar pattern is evident elsewhere, which suggests that it was determined either by changing economic fortunes or by climatic swings.

[11] P. D. A. Harvey (1965: 38). [12] Page (1996). [13] Hansell and Hansell (1988: 59).
[14] Biddick (1989: 126–8). [15] Chicago University Library, Bacon 449–56. See also Chapter 14.

Fig. 10.1 Annual issue of the dovecot at Hinderclay, 1298–1358

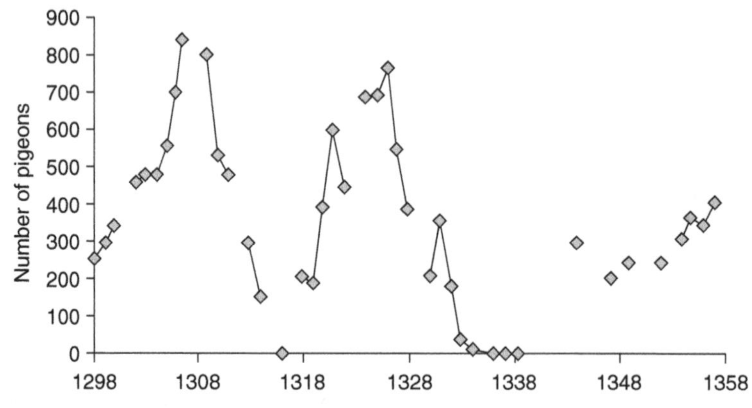

Source: Chicago University Library, Bacon 433–78.

Geese seem to have played an increasingly important part in the diet of many households during the thirteenth and early fourteenth centuries, though, as we saw in Chapter 9, the proportion, relative to chickens, did not augment. The geese, too, were seasonal fare. A few green geese were eaten during the summer, but most were probably consumed in the autumn and winter months: in 1336–7, Katherine de Norwich's household consumed an average of thirteen geese per month between October and January, but very few thereafter.[16] Until the start of the fourteenth century, many of these geese were probably raised on demesne farms. In the 1280s, for example, the monks at Bury St Edmunds annually consumed about forty geese from Hinderclay alone. Yet there are signs that strategies for supplying geese were changing at about this time, with a rising number purchased at market. At Hinderclay, for example, the size of the permanent flock of geese fell considerably at the end of the thirteenth century, from an average of seventy-one in the years 1278–96 to sixteen in the period 1297–1314.[17] Similarly the demesne of Elton, which sent about thirty geese to Ramsey Abbey each year, relied solely on breeding by its own flock to maintain numbers of geese at the beginning of the fourteenth century, but by 1345–6 bought roughly half of its geese at market.[18]

Many of the birds that lords bought probably came from local peasantry. The purchases of Wilton Abbey in 1299 suggest that the nuns were able to buy geese, doves, hens, and eggs from their local market on any day of the week.[19] Indeed, some peasants were clearly involved in commercial goose rearing, judging by the number of geese that they kept: Stephen Clerkson of Brandon illegally kept forty in the lord's meadow; while in 1338 the court at Walsham-le-Willows ordered that fourteen should be retained from Thomas at the Lee and more taken from him if necessary until he answered pleas of trespass.[20] Civic ordinances provide

[16] Woolgar (1992–3: i. 179–227).
[17] Chicago University Library, Bacon 415–45.
[18] Ratcliff (1946: 78, 181, 333).
[19] Farmer (1991*a*: 394–5).
[20] Bailey (1989: 128); Lock (1998: 224).

further evidence of the bird trade. For instance, the 1301 ordinances for York insisted that poultry be sold only on the bridge over the Ouse, presumably to ease the disposal of waste. Geese were to be sold at 3*d*. per bird, mallard at 5*d*. per brace, woodcock or plover at 1*d*. each, while blackbirds and thrushes were literally six-a-penny. Significantly, only one out of thirty-seven York poulterers achieved freeman status at this time, compared to nearly half the butchers: the poultry trade, it seems, was dominated by the poorer sections of society.[21]

In fact, 'manors were generally content to leave the poultry business to the villagers'.[22] Lords certainly received a lot of hens and eggs from peasants in the form of rent. For example, a total of 2,798 hens and 17,095 eggs was collected from the Bishop of Durham's tenants each year in the late twelfth century.[23] Manorial records bring these payments into sharper focus. The 1251 survey of the Bishop of Ely's manor of Tydd records that seventy-nine tenants were each obliged to give a hen and ten eggs as part of their rent and that one tenant held a cottage in return for collecting all these hens and eggs and taking them to Wisbech Castle.[24] Court rolls reveal that birds were also given in payment of fines: for instance, in 1262 Richard Trussert and Roger Minor of Ashton in Wiltshire were fined six fowls for having three cows caught in a meadow before it was mown.[25] Account rolls spell out very clearly the contribution that peasant poultry made. On 29 September 1297 the demesne at Elton had twenty-eight cocks and hens; over the next twelve months the reeve noted that 225 chickens were added to this, sixty of which were the issue of demesne birds, the remaining 165 coming from the tenants.[26]

Bird rearing must have played a significant part in the peasant economy at the start of the fourteenth century. This was a time of immense social and economic stress and the standard of living of many peasants is thought to have fallen to a perilously low level. Yet the consumption or sale of birds is often left out of the equation; while plainly not a cure-all, birds were probably of more importance for peasant welfare than most historians allow. Indeed, assessments of peasant livestock based on records of taxation fail to take account of the fact that poultry was not always assessed. For example, the preamble to the 1283 tax clearly states that geese, capons, and hens were not to be counted as moveable goods; neither was poultry mentioned in the tax assessment for Colchester in 1301, discussed in Chapter 7.[27]

The fact that hens and eggs were the last form of rent taken in kind on many English manors suggests not only that they were useful to lords but also that most peasant households kept a considerable number of chickens. Moreover, in terms of productivity, it has been claimed that domestic fowl production in medieval England reached a remarkably sophisticated level: death rates among fowl were low; caponization, a highly skilled operation, was widespread and an extremely

[21] Farmer (1991*a*: 394). [22] Farmer (1991*a*: 394). [23] Stephenson (1986: 243).
[24] Crosby (1895–7: 65–8). [25] Maitland (1889: 183). [26] Ratcliff (1946: 78).
[27] Powell (1910: pp. x–xi); Carlin (1998: 44).

efficient means of producing meat; and egg production—which the author of the anonymous *Husbandry* indicated should reach 115 eggs per hen each year, although account rolls seldom record more than 100—was not far behind the level attained in the early twentieth century.[28] Judging by the numerous cases of trespass recorded in manorial court rolls, many peasants owned at least a few geese as well. In 1310, for example, five people from Chalgrove were fined because their geese damaged the lord's corn.[29] It was presumably also the peasants who bought the birds sold by demesne farms, and they doubtless hunted a range of fowl, too, legally or otherwise. A degree of snaring in the hedgerows might be tolerated by custom, and at Climsland woodland bird snares were in fact rented for 6*d.* each in the 1330s; but at Weston in 1340 it was ordered that enquiry be made 'Whether there be among you any fowler who with net, trap or other engine without licence taketh crane, heron, wild goose, woodcock, snipe, thrush, lark, pigeon, goshawk or sparrowhawk in [the] park or elsewhere'.[30]

The contribution of birds to peasant welfare was probably of most significance during years of bad grain harvests, not just on account of their own hens and geese, but also because of the availability of birds elsewhere. Some Crowland Abbey manors gave out pigeons to the poor and sick in the worst years,[31] but the initiative often came from the other direction. For example, in 1296 the chaplain of Brightwaltham was convicted of carrying away the lord's fowls and 'caus[ing] them to be taken to his house'.[32] Similarly, in 1318–19 Walter of the Marsh of Westhorpe near Walsham-le-Willows took ten partridges in the lord's warren and a further five next to the wood of John de Walsham.[33] Living in certain environments naturally helped in this respect: the Fens had always provided exceptional numbers of wild waterfowl and been home to professional wildfowlers. On the Cambridgeshire fenland manor of Sutton, for instance, fourteen people were fined in 1317 for catching and taking three bitterns. Nor was it just the bitterns that were stolen, for the following year nine people at Sutton and a further thirteen people at Littleport were fined for stealing bitterns' eggs, 'habitually' in the latter case.[34] This rash of bittern rustling and egg poaching also suggests just how desperate many peasants were during the agrarian crisis of 1315–22; not necessarily on grounds of taste, but because bitterns are secretive and well-camouflaged birds, and coming face to face with one was later thought to be a portent of death.[35] Indeed, such an endeavour could itself prove fatal, for the Cambridge assize rolls of the early fourteenth century record how one boy drowned after going into the fens on stilts to look for ducks' eggs.[36]

[28] Stephenson (1986: 244). Stephenson suggested that the white lobes on the hen illustrated in the Luttrell Psalter (Plate 9.1) were 'the exclusive hallmark of a high egg-yield breed': Stephenson (1977–8: 20).

[29] Dale (1950: 59).

[30] Hatcher (1970: 185); Maitland and Baildon (1891: 95).

[31] For example, Page (1936: 41, 74, 121).

[32] Maitland (1889: 173).

[33] Lock (1998: 77).

[34] CUL, EDC 7/4; Maitland and Baildon (1891: 126). I am grateful to Erin McGibbon for the Sutton references.

[35] Thomas (1984: 76).

[36] Darby (1940: 37).

The later Middle Ages

In the aftermath of the Black Death of 1348–9, living standards generally improved and patterns of demand shifted in a variety of ways: high-quality wheaten bread was more sought after; more ale was consumed; better-quality cloth was bought and worn; even demand for the fur and meat of rabbits increased. The consumption of birds should be seen in the same light, for the number and range of birds regularly eaten rose considerably in the late fourteenth and early fifteenth centuries. By the 1380s, the thirty or so monks of Battle Abbey spent over £12 a year on birds, an amount which could have funded two undergraduates through three years of university.[37] A remarkable list of foods sold by London cookshops suggests that birds were also readily available as high-class fast food at this time, with roast heron for 18*d.*, capon in a pasty for 8*d.*, roast woodcock for 2½*d.*, three thrushes for 2*d.*, five larks for 1½*d.*, and ten finches for a penny.[38]

The relatively high price of geese and capons after the Black Death is another indication of increased demand: while prices of wheat and wool in the last quarter of the fourteenth century were approximately 20 per cent lower than they had been in the first quarter, the price of geese had increased by 2 per cent and the price of capons by 27 per cent.[39] Geese were certainly still popular in many households—for example, the kitchener of Maxstoke Priory (a total household of about thirty people) listed 124 geese in his care in 1449–50.[40] Capons may have become an even more regular feature of aristocratic diet: while Katherine de Norwich appears to have consumed fewer than twenty capons over nine months in 1336–7, the household of the Bishop of Salisbury consumed 338 over eight months in 1406–7.[41] Likewise, pigeons may also have become more popular than they had been before the Black Death: by the early fifteenth century, the households of the Bishop of Salisbury, Robert Waterton of Methley, and Dame Alice de Bryene each consumed an average of more than 100 pigeons a month.[42]

The range of birds consumed on a more regular basis also increased. Swans and herons were seemingly more popular: the household of the Countess of Norfolk, for example, consumed forty-nine swans at Framlingham Castle in 1385–6, while the household of Robert Waterton of Methley worked their way through twenty-five herons in 1416–17.[43] Similarly, the hunting of pheasants at Hinderclay was only mentioned after the Black Death, and even the number of peacocks kept on the Bishop of Exeter's manor of Bishop's Clyst increased at the beginning of the fifteenth century.[44] Of course, there are examples of all of these birds being raised, or hunted and eaten, or displayed before the Black Death, but

[37] Searle and Ross (1967: 75–82); Dyer (1998*a*: 75). [38] Riley (1868: 426).
[39] Farmer (1988: 790–1, 809–10; 1991*b*: 503, 513); Rogers (1866–1902: i. 357–8, 360).
[40] Watkins (1997: 21). [41] Woolgar (1992–3: i. 179–227, 264–399).
[42] Woolgar (1992–3: i. 264–399; ii. 517); Dale and Redstone (1931: 1–102).
[43] Ridgard (1985: 115); Woolgar (1992–3: ii. 517).
[44] Chicago University Library, Bacon 405–510; Alcock (1970: 159).

at that time they were generally birds for special occasions. For example, a heron was bought by the monks of Bolton Priory when Lady Maud de Clere visited in 1309–10, and herons were sometimes among the festive food served in households at Christmas or on anniversaries.[45] By the late fourteenth and fifteenth centuries the consumption of birds of this sort was less of a rarity. With exotic birds seemingly a common gift for the medieval lord, increased consumption brought gift inflation: in the 1310s, Bolton Priory sent swans as a gift to the Earl of Lancaster; but by the 1520s, Thetford Priory was giving cranes, already quite rare, to the King, and bustards to the Duchess of Norfolk.[46]

Household accounts of the later Middle Ages also provide us with a sharper sense of the seasonality of bird consumption. In Alice de Bryene's Suffolk household in the year from 29 September 1412, no birds or eggs were consumed during the season of Lent, in March and April; pigeons were only consumed from late April to mid-November, with a pause in the first part of July; geese were generally eaten between November and February, although a few—presumably green geese—were consumed in July; capons and chickens, on the other hand, were pretty much available throughout the year (Fig. 10.2). Partridges and swans were only eaten in this household between late October and early March, pheasants in a short period at the end of January, and herons in late February and early March and between mid-July and mid-August. The avian menu changed considerably during the course of the year: in August, September, and October there was mainly pigeon, plus a few chickens and the occasional partridge and swan; in December, January, and February there were chiefly geese and capon, plus a few chickens, with some herons, pheasants, partridges, and swans; in May and June pigeons and chicken were the main birds consumed, together with a few capons; and in July, chicken, a few geese, a few capons, and the occasional heron.

The increased consumption of birds in the later Middle Ages was reflected in a variety of ways. In terms of the landscape, for instance, the number of dovecots probably multiplied—there were no fewer than three at Methley by the fifteenth century[47]—and they seem to have become increasingly elaborate in design. In contrast to the predominantly thatched dovecots of the pre-Black Death period, such as those at Wellingborough, Downham, and Elton, the one built at Sawston in 1390 was tiled.[48] The trend towards more elaborate designs is amply illustrated by the remarkable dovecot constructed by Sir John Gostwick at Willington in the early sixteenth century, now in the ownership of the National Trust.

An increasing number of demesnes diversified into the semi-management of wild birds as well. Several, for example, developed swanneries (and employed swanherds) in the late fourteenth and fifteenth centuries. A swannery was started at Wymondley in the 1360s and is first mentioned at Abbotsbury in the 1390s,

[45] Kershaw and Smith (2000: 271); Woolgar (1995: 23).
[46] Kershaw and Smith (2000: 342, 366, 445); Dymond (1995–6: i. 24–5; ii. 427, 513).
[47] Le Patourel (1991: 883).
[48] Page (1936: 125); CUL, EDR D10/2/5; Ratcliff (1946: 213); Teversham (1959: 43).

Fig. 10.2 Main birds and eggs consumed in the household of Dame Alice de Bryene, 1412–13

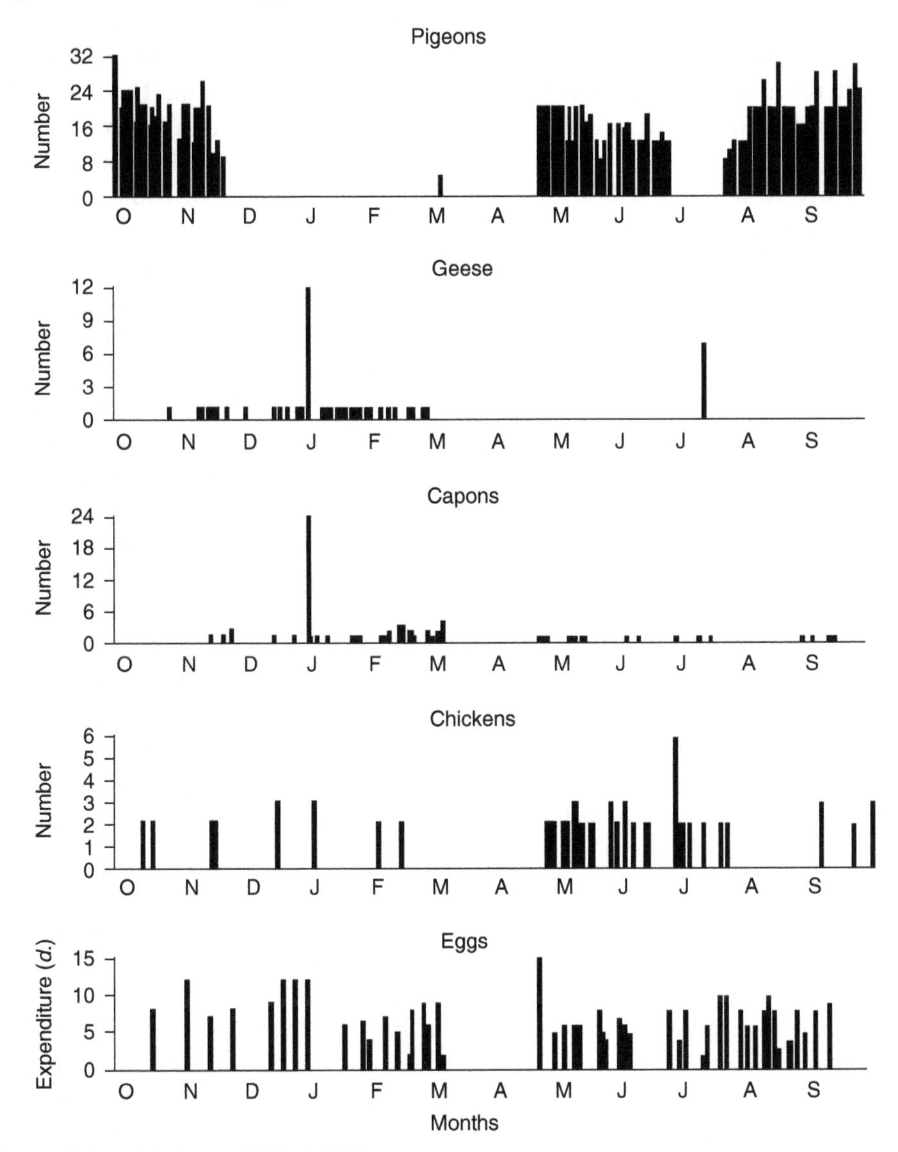

Source: Dale and Redstone (1931: 1–102).

while Richard, Earl of Arundel, had £100 worth of brood swans stolen in 1365.[49] A massive swannery was also developed on the Bishop of Winchester's manor of Downton. There were no swans here at the beginning of the fourteenth century, but numbers grew enormously after the Black Death with fifty or so swans by the late 1360s, about a hundred by the mid-1370s, and 150 to 200 by

[49] Harvey (1991: 267); Fair and Moxom (1993: 29); Stephenson (1986: 244–5).

the 1380s and 1390s.[50] Similarly, the account rolls for Wisbech reveal that the Bishop of Ely kept a hundred or so swans on the pond at nearby Elm in the early fifteenth century.[51] Indeed, the only part of the manor of Buckland still kept in hand in the first half of the fifteenth century was a swannery on the Thames, with swans and cygnets either destined for the lord's table or sold in Oxford.[52] The same situation prevailed at John Hopton's manor of Easton Bavents, where ninety-five swans were fed on oats and rye during the winter of 1464–5, thirty-five of which were delivered to the lord's household.[53]

In the period after the Black Death some demesnes even specialized in raising herons. A heronry was established at Wymondley in the 1350s[54] and at Hinderclay in the late 1360s. This was plainly an opportune moment to exploit heron breeding, for at Hinderclay the price of a heron increased from 6*d.* in 1370 to 8*d.* a decade later, indicating again the rising demand for this sort of bird. Consequently, while some herons were sent to the monks at Bury St Edmunds Abbey and others were given to local dignitaries, many were sold. With sales realizing a total of 26*s.* in 1368 and 18*s.* in 1390, the total size of the heronry at Hinderclay must have been at least 30 to 50 birds.[55] Yet, judging by the distance between the place of rearing and the place of consumption, heronries were not as common a feature of the late medieval landscape as swanneries. For example, in 1406–7, the Bishop of Salisbury's household received eight herons from his manor of Bishop's Caundle, roughly 30 miles from Salisbury, and bought a further twenty herons from Wimborne Minster, more than 20 miles away.[56] Similarly, the Earl of Oxford consumed twenty-three young herons in the year 1431–2, a quarter of which he took from his park at Stansted Mountfitchet, the rest from his manor of Great Bentley, which were respectively 17 and 22 miles from his Essex seat at Castle Hedingham.[57]

Bird rearing seems to have been big business for some peasant farmers too. Indeed, by the late fourteenth and fifteenth centuries the lordly demand for geese and capons was predominantly met from the market. For example, the household of Sir John Dinham of Hartland acquired 134 geese during the year 1372–3: twelve as gifts, thirty-two from his nearby manors of Castel and Butterbury, and ninety from the market.[58] Likewise, the cellarer's accounts for Battle Abbey in 1412–13 record that the value of geese, duck, and chicken brought from manors on the estate was 11*s.* 3*d.*, but that 'bought in the neighbourhood' totalled £7 17*s.* 5*d.*[59] Documents from this period sometimes even name the suppliers. For example, eleven suppliers of birds to Bromholm Priory were identified in 1415–16: the number of poultry they supplied was often quite

[50] Page (1996: 71); Hampshire RO, 11M59/B1/102–50.
[51] Cambridge University Library, EDR D8/3/16–30, D8/4/1–3.
[52] I am grateful to Margaret Yates for this information.
[53] Richmond (1981: 87).
[54] Harvey (1991: 267).
[55] Chicago University Library, Bacon 483–4, 487–503.
[56] Woolgar (1992–3: i. 403, 410).
[57] Woolgar (1992–3: ii. 543).
[58] Woolgar (1992–3: ii. 502).
[59] Searle and Ross (1967: 105).

small, such as seven chickens from Margaret Lessy and six capons from John Prat, but geese came in larger numbers, including fourteen from John Prat, twenty-four from Thomas Harding, and 144 from Nicholas Bere.[60] The named suppliers of birds to Sir William Mountford of Kingshurst in 1433–4 were all female, including Margery Clerke, who supplied twenty chickens, and the wife of John Baggeslow, who supplied sixty geese.[61] The huge quantities of eggs bought by large households may go some way towards explaining why chickens were often sold in smaller numbers, for some farmers clearly kept many hens. In 1499–1500, Thetford Priory purchased eggs worth over £6 from John Bell, which suggests that he must have kept more than 200 hens.[62]

Peasants and townsmen doubtless consumed large numbers of birds in the later Middle Ages as well. Certainly, the *Liber gersumarum* of Ramsey Abbey suggests that many tenants on that estate kept fowls to be fattened as capons: in the years 1421–6, 84 per cent of entry fines on the manors of Houghton, Warboys, Broughton, and Wistow were levied in capons rather than cash.[63] Geese were clearly consumed, as the use of two from Maidwell for a peasant's 'hochepot', or stew, in 1384–5 indicates.[64] Doves probably were too: for example, the dower which Robert David of Tavistock settled on his daughter in 1355 included a dovecot; Alicia Brut built a dovecot on her tenement in Abingdon in 1383–4; and in 1421–2 Thomas Buk took over a croft with a dovecot in Slepe (St Ives).[65] Doves were perhaps even tavern food in the fifteenth century, for John Wymbych the elder of Kentford, the owner of the Stag's Head Inn, also had a dovecot.[66] Even if a villager did not own one, he might still have been able to procure doves from elsewhere, as was the case at Walsham-le-Willows in 1398, when John Manser was amerced 40*d.* because 'he took the lord's doves at High Hall by snares set in the doors [of the dovecot], and killed them'.[67]

As more and more lords either gave up or scaled down their estate swanneries and heronries as the Middle Ages drew to a close, so the raising of these birds by yeomen farmers and the hunting of wildfowl probably became a more prominent means of supply. It was said of the inhabitants of Crowland, in the Fens, for example, that 'the greatest parte of their relyf and lyvyng hath been susteyned in long tyme passed' by the 'great games of swannes', while an entry in a 1522 court roll for Waterbeach ordered that 'none take any fowl, viz. cranes, butters [bitterns], bustards and heronshaws [young herons] within the commons, and sell the same out of the lordship unless he first offer them to the lord of the manor to buy'.[68] Indeed, parliamentary statutes from the beginning of the sixteenth century indicate that managing wild bird resources was too important an issue to be left to local lords. While it is not anachronistic to interpret this as concern about bird numbers, since some species—such as spoonbills and cranes—subsequently

[60] Redstone (1944: 84–5). [61] Woolgar (1992–3: ii. 440, 442).
[62] Dymond (1995–6: i. 128). [63] Dewindt (1976: 160–93). [64] King (1991: 222).
[65] Finberg (1951: 193); Kirk (1892: 47); Dewindt (1976: 163). [66] Northeast (2001: 98).
[67] Lock (2002: 204). [68] Ticehurst (1957: 20); Ravensdale (1974: 51).

became extinct in Britain,[69] the problem was addressed in a typically medieval way. For example, a statute from 1482–3 stated that 'by . . . unlawful means the substance of swans be in the hands and possession of yeomen and husbandmen, and other [persons of little reputation]'; thereafter, it was ordained that keepers of swans must own freehold land worth five marks or more.[70] More explicitly, the 'Acte ayenst destruccyon of wyldfowle' of 1533–4 stated that 'divers persons . . . by certain nets and engines [snares] and policies, yearly taken great number of . . . fowl, in such wise that the brood of wild-fowl is almost thereby wasted and consumed'. It also concluded that gentlemen may hunt them, though only with their spaniels and longbows. Notably, the statute also made it illegal to take the eggs of cranes, bustards, bitterns, herons, spoonbills, mallards, or teal in the period from 1 March to 30 June, though the act did not apply to 'any person or persones that woll distroy any crowes, choughes, ravons and busardes or their egges, or to any other fowle or theire egges not commestyble nor used to be eaten'.[71]

Birds may seldom have been served up in such number as they were for Archbishop Neville in 1465, but they were undoubtedly a conspicuous feature of late medieval England, and not just in terms of diet or environment. Feathers must have been plentiful and some of the uses to which they were put can occasionally be glimpsed in the records. Feather mattresses and pillows are recorded at Thetford Priory in the 1530s, and seem generally to have become more common in great households during the later Middle Ages.[72] Moreover, hunting arrows were finished with peacock feathers in fifteenth-century Farnham, and it has even been suggested that the waves of English arrows in battle resembled a snowstorm since the feathers came from white geese.[73] Quills, usually from geese (as noted in Chapter 9), were in constant demand as writing implements. Birds are also prominent in late medieval literature and art. Birds had long been used as a literary metaphor, but the anatomical precision with which Chaucer described cockfowl and his portrayal of the monk as loving 'a fat swan . . . best of any roast' suggest a relevance beyond mere tradition.[74] Likewise, it is no surprise that at the beginning of the fifteenth century the illuminator of the Sherborne Missal adorned the manuscript with a multitude of birds—including a peacock, crane, pheasants, and herons—and did so in an impressively naturalistic way.[75]

Conclusion

In common with archaeological data, documentary evidence for the study of medieval birds has both strengths and weaknesses. Most notably among the latter, precise identification of birds in the written record is often reliant on contemporary application of avian terminology. Some sources—particularly

[69] Holden and Cleeves (2002: 36, 105). [70] Luders, Tomlins, and Raithby (1810–28: ii. 474).
[71] Luders, Tomlins, and Raithby (1810–28: iii. 445–6). [72] Dymond (1995–6: ii. 597, 608, 640).
[73] Page (1999: 160); Stephenson (1977–8: 22). [74] Benson (1988: 26, l. 206; 254, ll. 2859–64).
[75] Backhouse (2001).

manorial account rolls—are more secure than others in this regard, since they differentiate certain birds by age and sex, specify the origin of capons, and sometimes use English terms interchangeably with Latin ones, but, even so, taxonomic precision remains the preserve of the scientist to a greater extent than the historian. Yet the strengths of documentary evidence far outweigh the weaknesses. The chronology of consumption and supply, for instance, can be sketched with confidence, whether with respect to changing patterns of consumption over the course of the year, adjustments in marketing arrangements, or fluctuations in the number of birds kept on demesne farms from year to year. The evident increase in the number and range of birds eaten in the two centuries after the Black Death is particularly notable, with significant implications not just for diet, but also for our understanding of agricultural diversification and specialization, the development of the landscape, and contemporary perceptions of environmental change. Furthermore, while written (and archaeological) evidence about the agricultural economy of the English peasantry is sparse and in need of careful interpretation, it is clear from a number of documentary sources that bird rearing was a significant activity for many peasants. In terms of their economic welfare, the production and consumption of birds was probably of particular importance in the early fourteenth century, bringing in a small but important market income at this time of acute pressure on resources, as well as contributing to subsistence needs. Indeed, written sources also have the potential to shed a great deal of light on areas which lie outside the scope of this brief survey, not least the geography of medieval birds and the possible impact of bird rearing and consumption on human disease. Late medieval England generated a wealth of documentary evidence for the production and consumption of foodstuffs, yet in terms of the study of birds these riches remain comparatively untapped.

11

The Impact of the Normans on Hunting Practices in England

N. J. SYKES

Hunting might seem an obvious way of obtaining food, but in practice augmenting diet in this way was not necessarily straightforward. This chapter examines the archaeological evidence for consumption of foodstuffs that came from this source, relating it to hunting practices and instructional texts of the later Middle Ages. Chapter 12 considers the historical evidence for the management of hunting preserves and their contribution to diet, particularly in the period from 1250 onwards, which allows us to examine some of the dynamics of supply and demand. Together the chapters provide an overview of the role that hunting wild mammals played in food supply.

Of all the changes attributed to the Norman Conquest, those associated with hunting are probably the best known, with few school-books neglecting to mention the Normans' love of the chase. After 1066 the Norman kings applied forest law to vast areas of land, most famously creating the New Forest. Under forest law, rights to take animals were restricted and unlicensed use of forest resources was punishable by imprisonment or maiming.[1] The conquered population appears to have viewed forest law with contempt; indeed, the authors of the Anglo-Saxon Chronicle devoted approximately one third of William the Conqueror's obituary to complaining about it.[2] Certainly the situation appears to have been fundamentally different from pre-Conquest Britain, where perceptions of game followed the Roman concept that, until caught, wild animals were *res nullius*, nobody's property.[3] The liberal nature of pre-Conquest hunting is indicated by the earliest British texts on venery. For instance, the Old Welsh Nine Huntings (*c.*945) describes the stag as a 'common hunt' and states that 'every person that comes up after he is killed and before the skin is stripped off is entitled to a share of him'.[4] Furthermore, article 80 of Cnut's laws declares, 'It is my will that every man shall be entitled to hunt in the woods and fields of his own

[1] Petit-Dutaillais (1908–29: ii. 173). [2] Garmonsway (1967: 221). [3] Gilbert (1979: 7).
[4] Probert (1823), cited in Clutton-Brock (1984: 168).

property.'[5] The same article continues, though, 'everyone, under pain of incurring the full penalty, shall avoid hunting on my preserves', suggesting that hunting was not completely uncontrolled.

A good understanding of later medieval hunting, as practised by the aristocracy, can be gained from the numerous hunting manuals written between the thirteenth and sixteenth centuries.[6] Documents pre-dating the thirteenth century or concerning peasant hunting are, however, scarce; consequently knowledge of earlier traditions and developments within the sphere of hunting is equally limited. To gain a more balanced view of the origins and evolution of medieval venery and game consumption, the archaeological remains of the wild mammals themselves, synthesized from over two hundred archaeological assemblages dating between the fifth and mid-fourteenth century,[7] are used in this chapter to explore whether the arrival of the Normans changed the way in which game animals were perceived, hunted, and consumed. Central to this investigation is the study of wild mammal frequency (the overall percentage of remains of game animals in zooarchaeological assemblages) and how this changed through the period. Particular attention is paid to inter-site variations in species. Changes over time in the skeletal elements of deer on different sites are also examined. Data for rural settlements are the least abundant, especially for the post-Conquest period, but sample sizes are sufficiently large for most of the other site types to be considered with reasonable confidence (Table 11.1).

Frequencies of wild mammals

Few assemblages dating from the fifth to the mid-ninth century contain high frequencies of wild mammals (Fig. 11.1), and game can have made little contribution to the Early and Mid-Saxon diet. This is not to suggest, however, that

Table 11.1. Numbers of assemblages considered for information about hunting

	Rural	Urban	High-status	Religious house
English sites				
5th–mid-9th century	21	12	6	5
mid-9th–mid-11th century	5	21	9	7
mid-11th–mid-12th century	4	19	18	5
mid-12th–mid-14th century	4	18	13	9
French sites				
9th–11th century	2	18	7	2

Source: Sykes (in press *a*).

[5] Attenborough (1922).

[6] For an up-to-date discussion and syntheses of medieval hunting texts, see Cummins (1988) and Almond (2003).

[7] The data are provided in Sykes (in press *a*) and examples of the principal sites are given in the text.

Fig. 11.1 Variation in representation of wild mammals on English and French sites of different type, fifth to fourteenth centuries, as a percentage of the total bone assemblage, excluding fish

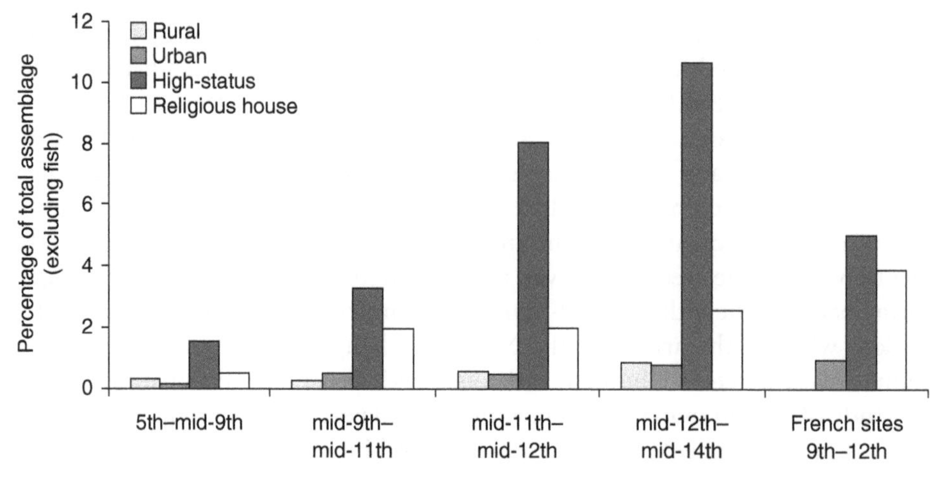

Notes: Based on number of identified specimens (NISP). Numbers of sites given in Table 11.1.
Source: The raw data are presented in Sykes (in press *a*).

the capture and consumption of wild animals lacked social meaning. Early medieval iconography and funerary deposits indicate a clear association between hunting and the social elite: weapons and armour often carry motifs of wild animals[8] and the remains of the animals themselves are frequently represented, along with hunting paraphernalia, in high-status male burials and cremations.[9] The symbolism of hunting, with its references to warfare and power, was more important than the act itself. Likewise, it may be assumed that, in most contexts, the significance of wild animals extended beyond their value as meat. Rather than a strategy for survival, occasional hunting and consumption of game would have been a powerful sign of social identity. This is indicated zooarchaeologically (Fig. 11.1) by the fact that wild animals are far better represented on high-status settlements, such as Ramsbury[10] and Flixborough,[11] than in assemblages from sites of lower status, as at Pennyland[12] and Quarrington.[13]

Inter-site differences in game became more defined during the late ninth to mid-eleventh centuries. All types of site, except rural settlements, demonstrated a rise in wild mammals, but increases were especially marked for religious houses and high-status sites (Fig. 11.1). This shift is symptomatic of the widespread changes in social, economic, and landscape organization that characterize Late Saxon England. With the break-up of the old estates, the newly landed thegnly class was able to establish its own hunting preserves, a bold statement of power and authority, and there is even some evidence that woodland was allowed to

[8] For instance, the boar-crested helmets mentioned by Lucy (2000: 51).
[9] Crabtree (1995); Lucy (2000: 90–4, 112–13).
[10] Coy (1980).
[11] Loveluck (2001: 91–104).
[12] Williams (1993).
[13] Rackham (2003).

regenerate for this purpose.[14] Apportioning ancient commons to specific manors may have curbed peasant hunting rights,[15] perhaps explaining the reduced frequency of game on some rural settlements. Alternatively, urban demand for game may have seen the majority of the wild animals caught by the rural population taken to towns for sale. Such a situation would account for the increased amount of game in Late Saxon urban assemblages.

By the Norman period, the frequency of game on high-status sites had increased further (Fig. 11.1). Wild animals are also slightly better represented in rural assemblages, but those from religious houses and urban sites demonstrate no change from the preceding period. Overall, division between high- and low-status sites became particularly pronounced in this period, hinting at the type of unequal access to wild resources that would have accompanied forest law. Status-based inequality of game even exceeds that demonstrated by French assemblages (Fig. 11.1), supporting the notion that the forest law introduced to England was more severe than it had been in pre-Conquest Normandy.[16]

Game consumption increased at all social levels between the mid-twelfth and mid-fourteenth centuries (Plate 11.1). Patterns of inequality were retained, but

Plate 11.1 The Pilkington charter: Edward I grants Roger de Pilkington free warren in his manor of Pilkington and elsewhere in Lancashire, 1291. The animals include those Roger might have expected to hunt or which were redolent of high status; the birds, including a peacock, encompass many of the species that could be found on an upper-class table. Fitzwilliam Museum, Cambridge, MS BL 51. Photograph: © Fitzwilliam Museum, University of Cambridge.

[14] Hooke (2001: 139, 157).
[15] Biddick (1984: 112); Hagen (2002: 136).
[16] Gilbert (1979: 11).

not accentuated further, confirming the mid-eleventh to mid-twelfth centuries as the period during which the most dramatic change occurred. It is encouraging that the zooarchaeological data are consistent with historical perceptions of post-Conquest forest law, supporting the belief that the Normans both delighted in hunting and restricted rights to it. To advance our understanding of Norman influence on hunting practices, the evidence must be examined at the more detailed level of frequency of species.

Changing patterns in wild mammal representation

Examination of species reveals a discrepancy between the zooarchaeological and historical record (Fig. 11.2–11.5). Whilst documents regularly cite wild boar (*Sus scrofa*) as one of the more common game animals,[17] faunal assemblages suggest that this species was rarely hunted or consumed in medieval England. The zooarchaeological scarcity of wild boar could be the result of problems of identification, as it can be difficult to separate the remains of wild and domestic pig, especially where the two have interbred. Even where researchers have studied tooth size, a reliable criterion for distinguishing species, however, evidence for wild boar has remained slight. It must be assumed that this species was exceptionally uncommon in England by the thirteenth century and that the wild boar recorded in many medieval documents, for example, as at the Christmas feast held by Henry III in 1251,[18] were in fact domestic pigs. Some, at least, might be specially fattened in preparation for feasts, as noted in Chapter 7. Emphasis on

Fig. 11.2 Relative percentage of different wild mammals on rural sites

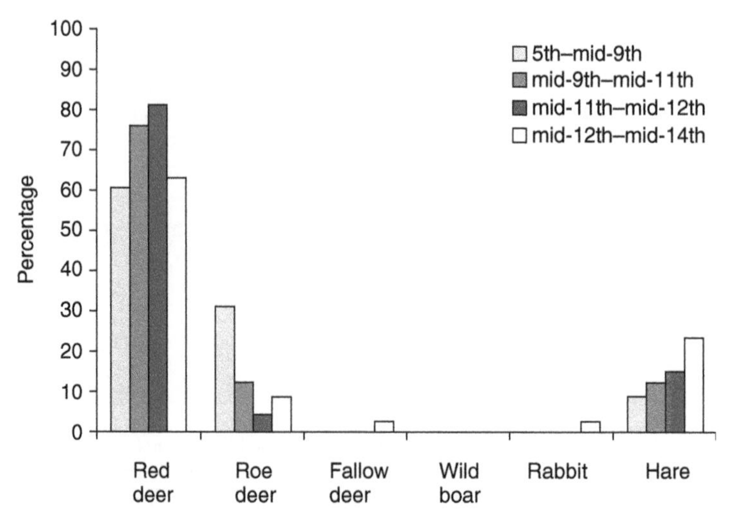

Notes: Based on number of identified specimens (NISP). Numbers of sites given in Table 11.1.
Source: The raw data are presented in Sykes (in press *a*).

[17] Rooney (1993: 3). [18] Rackham (1997: 36, 119).

Fig. 11.3 Relative percentage of different wild mammals on urban sites

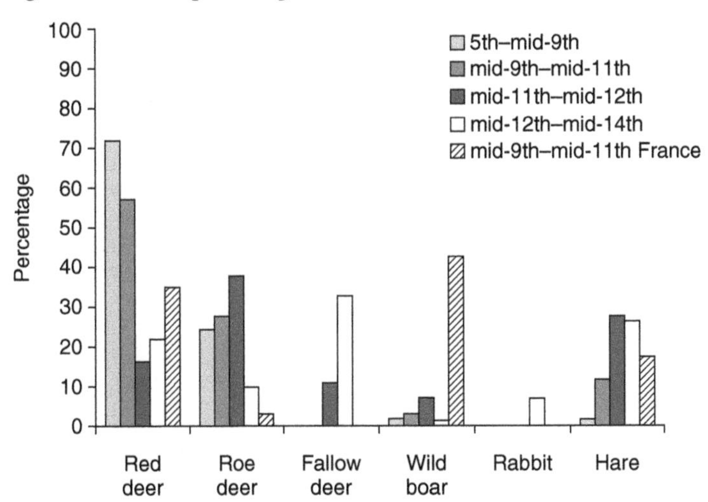

Notes: Based on number of identified specimens (NISP). Numbers of sites given in Table 11.1.
Source: The raw data are presented in Sykes (in press *a*).

Fig. 11.4 Relative percentage of different wild mammals on high-status sites

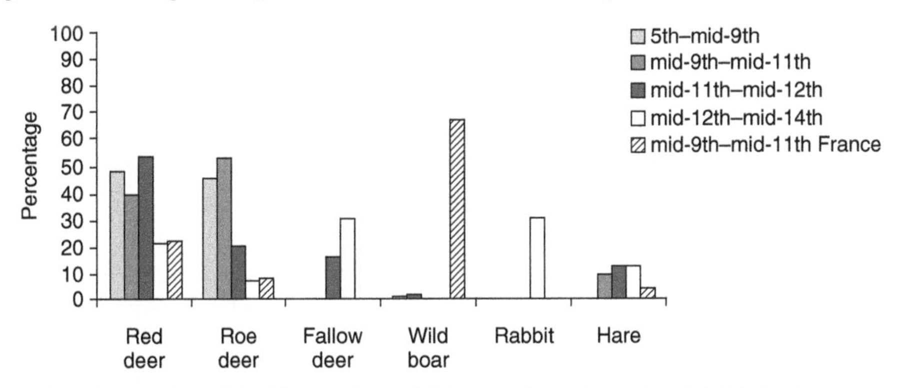

Notes: Based on number of identified specimens (NISP). Numbers of sites given in Table 11.1.
Source: The raw data are presented in Sykes (in press *a*).

wild boar in English hunting manuals is attributable to the fact that most of these texts were based on continental originals,[19] in which the centrality of wild boar is unsurprising: zooarchaeological evidence confirms this species as the animal most frequently hunted by the elite of France (Fig. 11.4) and Germany.[20]

Perhaps the most obvious temporal and inter-site variations in species are shown by the three cervid species: the native red deer (*Cervus elaphus*) and roe deer (*Capreolus capreolus*) and the imported fallow deer (*Dama dama*). Characteristic

[19] For instance, the *Master of Game* of Edward, Duke of York, is a translation of Gaston Phebus's *Livre de chasse*.

[20] Sykes (in preparation).

Fig. 11.5 Relative percentage of different wild mammals from religious houses

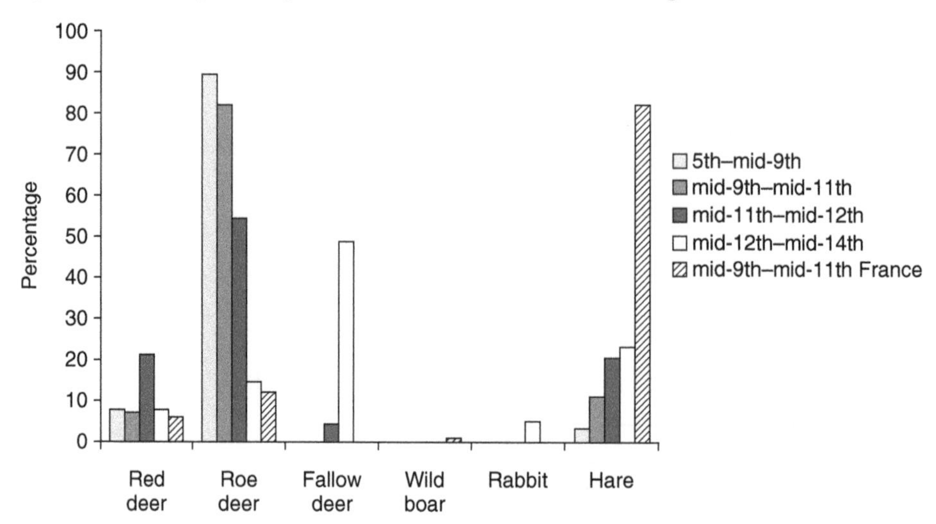

Notes: Based on number of identified specimens (NISP). Numbers of sites given in Table 11.1.
Source: The raw data are presented in Sykes (in press *a*).

patterns for deer are exhibited by each settlement type. Rural assemblages are, in all periods, typified by a high percentage of red deer (Fig. 11.2). By contrast, assemblages from religious houses of the seventh to the mid-twelfth century, including Eynsham Abbey,[21] Flixborough[22] and North Elmham,[23] show an overwhelming preponderance of roe deer (Fig. 11.5). Why ecclesiastics should have consumed the venison of roe deer in preference to that from red deer is unclear but it is interesting to note that the same is true on the Continent (Fig. 11.5).[24] It may be significant that in both Britain and France ecclesiastics were often granted rights of chase, in effect the freedom to hunt lesser quarry such as roe deer and hare. Another explanation for the preponderance of roe deer at religious houses could lie in the symbolism of roe deer: in later medieval art and literature roe deer are depicted as faithful, chaste, and abstemious.[25] In a period when it was believed that the character of an animal could be acquired through the consumption of its flesh,[26] it seems possible that venison from roe deer was deemed a suitable meat for men of the cloth. The abundance of hare remains in assemblages from French religious houses can, likewise, be linked to the perceived properties of hare flesh: a number of medieval penitentials authorize consumption by ecclesiastics of hare flesh on the basis of its medicinal value.[27]

Ecclesiastics were not the only section of society that consumed roe deer: it is also well represented on high-status sites of the fifth to mid-eleventh centuries (Fig. 11.4). By the mid-eleventh and twelfth centuries, however, high-status

[21] Ayres, Locker, and Serjeantson (2003). [22] Loveluck (2001: 91–104).
[23] Noddle (1980). [24] Yvinec (1993: 497–8). [25] Cummins (1988: 89).
[26] Scully (1995: 42); Salisbury (1994: 44). [27] Laurioux (1988); Hagen (2002: 132).

assemblages demonstrate a shift away from roe deer towards red deer, a trend apparent even for religious houses (Fig. 11.5). This move has been noted at many multi-period sites, including Faccombe Netherton[28] and Eynsham Abbey.[29] It could represent a decline in the roe deer population, but more probably the shift from roe to red deer reflects changing preferences and hunting strategies. In Saxon England aristocratic hunting appears to have centred on enclosed wooded parks,[30] environments for which roe deer are adapted and within which they will form groups.[31] In the Norman forest, however, the hunting landscape changed from these enclosed parks to unbounded tracts of moor, heath, and agricultural land. It is in these open environments that roe deer disperse, whereas red deer form herds.[32] It must have been this shift from woodland to forest hunting that promoted red deer to the premier beast of venery: certainly by the fourteenth century roe deer had lost both their status as beasts of the chase and their protection under forest law.[33]

The Anglo-Saxon Chronicle mentions William the Conqueror's love of red deer and the controls he placed on hunting them. Imposition of restrictions may explain the significant post-Conquest decline in red deer on urban sites (Fig. 11.3). No similar decrease is apparent for rural assemblages of the mid-eleventh to mid-twelfth centuries; indeed, the frequency of red deer increases slightly (Fig. 11.2). This may reflect peasant defiance of the new game laws, that they continued to hunt red deer regardless of the restriction, if not more so. Strong evidence for poaching was recovered from the deserted village of Lyveden where the remains of a heavily butchered red deer skeleton, which had been stripped rapidly of its flesh, were found hidden down a well shaft.[34]

Changes in the ratio of red and roe deer are not the only shifts in cervid representation that occurred between the pre- and post-Conquest periods; fallow deer appear for the first time in deposits of the mid-eleventh and twelfth centuries (Figs. 11.2–11.5). Recent research confirms that the Normans introduced this species but suggests that, rather than being brought from Normandy, fallow deer were imported from Sicily along with other aspects of hunting practice.[35] As exotic animals, they would have been kept in enclosures and the post-Conquest fashion for emparkment can be attributed to the burgeoning numbers of fallow deer.[36] The difficulty of poaching these jealously guarded animals may account for their absence from rural assemblages; peasants would have found it easier to take red and roe deer from the forests that often surrounded their own agricultural land. Similar limiting factors ought to have prohibited the urban population from consuming venison from fallow deer, but quantities did percolate into post-Conquest towns (Fig. 11.3). Some may have been purchased from professional hunters or parkers, but documents record that venison also arrived through a

[28] Sadler (1990). [29] Ayres, Locker, and Serjeantson (2003).
[30] Hooke (2001: 154–60); Liddiard (2003: 4–23). [31] Darling (1937).
[32] Clutton-Brock, Guinness, and Albon (1982). [33] Whitehead (1972: 210).
[34] Grant (1971). [35] Sykes (2004). See also Chapter 12.
[36] Rackham (1997: 123). See also Chapter 12.

system of organized trafficking: court rolls from the thirteenth century highlight the illicit trade running between the Midland forests and Bristol.[37] The roll for Cannock mentions one man who had hidden a deer carcass in a cartload of timber.[38] Problems concealing carcasses may explain why, by comparison with other types of site, post-Conquest urban assemblages demonstrate high frequencies of hares and, after their introduction in the late twelfth century, rabbits, both animals that could be smuggled with less risk of discovery.

The Conquest and changes in hunting techniques

With a change in hunting landscapes, together with the introduction of fallow deer, it might be expected that post-Conquest hunters would not only target different species of deer but also use different techniques to catch them. Based on the documentary evidence it has been suggested that the 'drive' (also known as 'the bow and stable') was the most common pre-Conquest hunting method—deer were chased into enclosures, known variously as *haga* or *haia*, and were then killed by archers.[39] Bow and stable would have been an effective method for obtaining large quantities of venison in a single event, although to be successful it would have required the participation of many people. Domesday Book mentions that citizens of Hereford, Shrewsbury, and Berkshire were obliged legally to act as drivers[40] and it seems likely that the ability to muster manpower was used as a conspicuous display of royal or thegnly resources. The drive may have continued as the preferred hunting method into the post-Conquest period, but historical sources suggest that a new hunting style, the chase *par force*, also emerged. Later medieval hunting manuals describe the chase *par force* as a wide-ranging hunt of day-long duration in which a single deer was stalked, killed, and excoriated (skinned, disemboweled, and butchered) in a ritualized and formulaic manner.[41] By comparison with the drive, the *par force* technique was not an efficient means of obtaining venison and it must be assumed that sport and, more particularly, social display were the main functions of this style of hunting.

The origins of *par force* hunting and the date at which it was introduced to Britain are unknown, although the Normans have been proposed as candidates for its introduction.[42] At face value it would seem difficult to prove whether or not this was the case, but the documentary evidence supplies enough detail to allow excoriation or 'unmaking' to be detected in the zooarchaeological record. Excoriation of the deer was the culmination of the hunt, where the huntsmen could demonstrate their 'nobility' through their knowledge of the procedure and the French terminology surrounding it.[43] Various texts outline the unmaking procedure: the earliest is in Gottfried von Strassburg's *Tristan* (written in 1210),[44] but the fullest description is provided by the late fifteenth-century *Boke*

[37] Birrell (1982). [38] Birrell (1982: 18, 20). [39] Gilbert (1979: 54); Cummins (1988: 51).
[40] Loyn (1970: 366). [41] See Cummins (1988) and Almond (2003). [42] Gilbert (1979: 58).
[43] In medieval England, knowledge of hunting language marked a person as noble: Rooney (1993: 12–13). [44] Hatto (1967).

of St Albans.[45] The texts suggest that, in the case of a stag, its testicles and penis were first removed. These were hung on a stick (the *forchée*) which was used to collect various organs and titbits: the *forchée* would later be carried at the front of the homeward bound procession. Skinning was then undertaken, and the animal was split from the chin down to the genitals and out to each leg before being flayed down to the spine. At this point the feet were removed from the carcass but were often left attached to the skin. The skin was then spread out to protect the venison from the ground, but also to collect the blood, which was later mixed with the intestines and bread, and fed to the dogs. After skinning, the shoulders and the haunches were removed and the rest of the carcass was disembowelled, butchered, and the meat and antlers carried home in the skin, presumably using the feet as handles. Certain parts of the carcass were given to particular people: for instance, the '*corbyn* bone' (the pelvis) was cast away at the kill site as an offering to the *corbyn* (raven), and the left shoulder was presented to the forester or parker as his fee.[46]

If carcasses were treated in this way, patterns of skeletal elements ought to be apparent in the zooarchaeological record and, indeed, this is the case (Fig. 11.6). Deer assemblages from later medieval manor houses, castles, and religious houses are typified by an abundance of foot bones, especially those of the hindlimb. There is an almost complete absence of the body parts, in particular the pelvis and elements of the forelimb, that would have been given away at the kill site. That shoulder joints did not arrive back at high-status settlements is reinforced by the evidence from late medieval household accounts, which commonly record haunches of venison but seldom mention the presence of the forequarters.[47] The final destination of the shoulder joints is difficult to establish from the archaeology: few assemblages of deer bones show an over-representation of forelimbs. It may be that venison from the forequarters was distributed more widely than that of the haunches. In such circumstances a range of contexts may contain shoulder elements in frequencies too low to attract attention from zooarchaeological researchers. Another possibility is that the type of sites where shoulders of venison were consumed have been little investigated by archaeologists. Credence is perhaps added to the latter suggestion by the evidence from the late medieval hunting lodge at Donnington Park.[48] Partial excavation of the lodge and its enclosure ditch yielded an animal bone assemblage with a preponderance of fallow deer remains. Interestingly, the patterns of body parts are the reverse of those found on most high-status sites, showing instead a large number of forelimb bones, with the humerus by far the best represented element. Nearly all the specimens in the Donnington Park assemblage derived from the left side of the carcass, evidence that the parker who inhabited the lodge regularly received his allotted shoulders of venison.

[45] See Brewer (1992).

[46] Detailed descriptions of unmaking procedures and the variations between different manuals are provided by Cummins (1988: 32–46), and Almond (2003: 73–83). See also Chapter 12.

[47] See Chapter 12. [48] Bent (1977–8: 14–15).

Parkers, foresters, and hunters could not have lived on venison alone and it seems highly probable that renders of venison were converted into cash by selling some of the meat to the urban population: this would explain the representation of fallow deer in town assemblages. Unfortunately, because deer remains form only a small component in most urban assemblages, their skeletal representation is seldom examined in detail and it cannot be stated with certainty whether or not the remains found on these types of site are generally those of the forelimb. Recent excavations in Reading,[49] however, produced a large medieval assemblage with deer bones present in quantities sufficient to allow patterns of body parts to be constructed. On this site deer were represented predominantly by bones of the forequarters, confirming that venison shoulders were brought to the town.

Deer bone assemblages from rural settlements tend not to show skeletal patterning akin to those from either high-status or urban sites. Instead all parts of the body are usually present, indicating that rural sites were not supplied with butchered joints of venison and that, when hunting, peasants did not excoriate the deer following the unmaking rituals.

If it is accepted that excoriation was an integral part of the *par force* hunt, the distinctive suite of body parts resulting from it should provide the key to ascertaining the date for the introduction of this hunting style. Studies of multi-phase assemblages, including Eynsham Abbey,[50] Faccombe Netherton,[51] Goltho,[52] Portchester Castle, and the Cheddar palaces,[53] indicate that patterns, in which bones of the hindlimb are the most frequent (Fig. 11.6), appear for the first time in the late eleventh century: where dating permits, they first become apparent shortly after 1066. Prior to this deer are represented by all parts of the skeleton, with bones of the upper forelimb among the most numerous elements.

Before it is concluded absolutely that *par force* hunting was part of Norman hunting practice imported into post-Conquest England, it is worth examining the evidence from France. Interestingly, studies of deer assemblages have demonstrated that patterns of body parts akin to those of post-Conquest England do not appear in northern France until the late twelfth and thirteenth centuries.[54] The only other location where they are found is Sicily, notably at Brucato.[55] At this site the assemblage dominated by hindlimbs is specific to fallow deer; the red and roe deer are represented by all parts of the skeleton. While data from a single site need not be representative of the wider situation, they may suggest that the unmaking rituals were of southern, rather than northern, European origin. This does not exclude Norman involvement. Sicily was conquered at Norman hands by the mid-eleventh century and remained under their rule for over a hundred years. Strong connections between the Normans of England and Sicily provided

[49] Sykes (n.d.). [50] Ayres, Locker, and Serjeantson (2003). [51] Sadler (1990).
[52] Jones and Ruben (1987).
[53] The data for Portchester Castle and the Cheddar palaces are presented in Sykes (in press *a*).
[54] Sykes (in press *a*). [55] Bossard-Beck (1984: 615–71).

Fig. 11.6 The relative frequency of body parts of red deer recovered from later medieval manor houses, castles, and religious houses

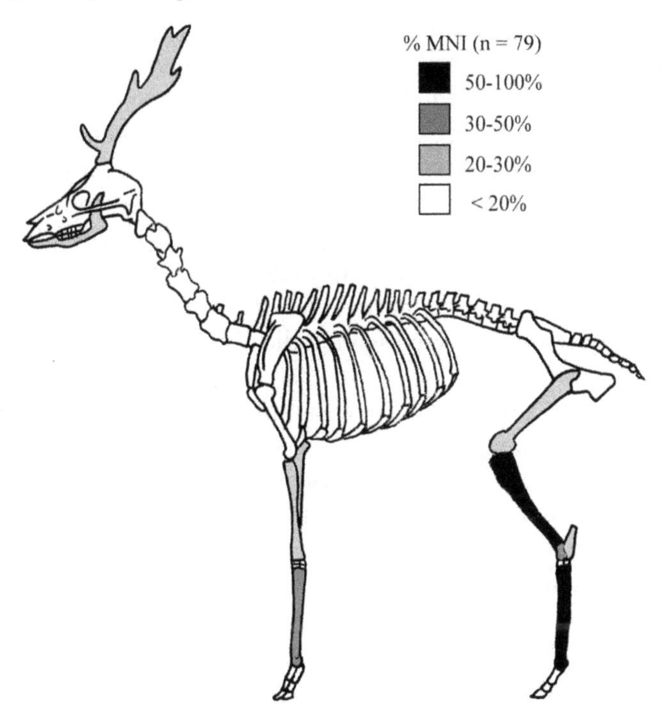

Note: Calculated as a percentage of the minimum number of individuals (MNI).
Source: The raw data are presented in Sykes (in press *a*).

ample opportunity for the exchange of ideas. Hunting practice no doubt formed part of this: Henry II's park at Woodstock is thought to have been based on the model *parco* at Palermo of King Roger II of Sicily.[56]

Associations between the excoriation rituals and fallow deer are indicated by the fact that, in Sicily, England, and northern France, both appear in the archaeological record at approximately the same time. If the unmaking traditions were originally specific to fallow deer, this has implications for the connection of excoriation with *par force* hunting. The practice of maintaining fallow deer in parks would have precluded this form of wide-ranging hunt, a method better suited to the open forest. Furthermore, several hunting manuals suggest that fallow deer lacked the stamina to make them a worthy *par force* quarry, instead recommending that they be taken through the bow and stable method:[57] archaeological finds of fallow deer bones displaying arrow wounds confirm that they were hunted in this way.[58] It must, therefore, be assumed that not only were the excoriation rituals employed regardless of hunting method but that, after their introduction, they were rapidly applied to all cervid species, not just fallow deer.

[56] Rowley (1999: 181). [57] Cummins (1988: 87). [58] Sadler (1990: 487).

That it was the unmaking rituals, rather than the total *par force* hunting package, that were imported to post-Conquest England appears to find literary expression in Gottfried von Strassburg's *Tristan*.[59] In the hunting scene, which is set in Cornwall, the King's party had already caught their stag using the *par force* technique when Tristan arrived. Tristan did not object to the way in which the animal had been hunted but took issue when the English huntsmen attempted to butcher the deer following the customs of the country, the method being to split the animal into four equal quarters. At this point Tristan demonstrated how to excoriate the hart according, he says, to the traditions of his homeland (Brittany). It is interesting to note that, although much of the terminology he uses is French, the English hunting party refer to some of it as Arabic:[60] Arab influence was very strong in Sicily, again indicating this island as the source for the methods of excoriation.

While it cannot be proved that new hunting techniques were imported to post-Conquest England, it is clear from the skeletal representation of deer that novel butchery practices associated with unmaking rituals were introduced shortly after 1066. By ritualizing what must previously have been a relatively straightforward *par force* hunt, the post-Conquest elite were responsible for the creation of new traditions. Customs are seldom fabricated without reason and one motive for their creation is to legitimize change,[61] such as that which would have been brought about by the Conquest. The archaeological record is sufficiently consistent to suggest that the customary distribution of body parts was maintained even when deer were hunted by professionals to supply aristocratic households, as discussed in Chapter 12.

Conclusion

Within most pastoral societies, hunting is a social action undertaken as an expression of power and authority.[62] It was clearly used in this way during the Saxon period, but the socially divisive function of hunting became particularly marked in post-Conquest England. Restriction of wild resources to the elite and the creation of new hunting traditions embellished with French terminology would have been powerful devices through which the post-Conquest aristocracy could display their control over the subjugated population. These expressions were also embedded in the consumption of the game itself. 'Venison' is derived from the Anglo-Norman *venesoun*, literally 'the product of hunting'. The meat was distributed on the basis of rank: while the lords consumed the prized portions, such as the liver and testicles, and the parkers received the shoulders, persons of lower standing were offered the remaining offal or umbles. Indeed,

[59] Hatto (1967: 78–86). [60] Hatto (1967: 80). [61] Eisenstadt (1969).
[62] Kent (1989: 132); Hamilakis (2003: 239).

the saying 'to eat humble [umble] pie' is derived from the association of the consumption of the poorer cuts with low status.[63]

The decades following 1066 were undoubtedly a watershed in the development of English hunting. Creation of forests not only restricted hunting rights but also influenced methods of venery and, thus, the species hunted. The import of fallow deer and new hunting rituals further divorced the post-Conquest hunt from the traditions of Saxon England. Zooarchaeological data record the unparalleled increase in aristocratic hunting following the Conquest: it is no surprise, therefore, that love of hunting should have become so closely identified with the Normans.

[63] Goody (1982: 142).

12

Procuring, Preparing, and Serving Venison in Late Medieval England

J. BIRRELL

The aristocracy of medieval England liked to have venison on their tables, especially at celebratory meals and when entertaining guests. The venison could be enjoyed for its own sake and it also added to the variety of meats presented. But just as important was the special status that venison derived from the fact that, as we have seen in Chapter 11, it was not accessible to everybody. It was neither bought nor sold in the normal way; it had to be hunted, and over much of the country it could only be hunted in one of the many long- or newly established game preserves. Venison might be eaten only on occasion and in smaller quantities than other meats, but this is no measure of its true importance. It was highly prized, and the desire to serve it had many consequences. This chapter examines some of these. Chapter 11 considered the introduction of new rituals of hunting to England and here, too, the processes by which this precious meat was procured and served at aristocratic tables are discussed, with particular reference to the historical evidence. It therefore concentrates on the century and a half beginning in the mid-thirteenth century.

Hunting preserves and the consumption of venison

By the middle of the thirteenth century, England was covered with hunting preserves. They were of three types: some seventy royal forests survived, which between them included many of the stretches of countryside most prolific in deer; an elite among the aristocracy had acquired one or more chases, that is, in effect, private forests; lesser lords made do with deer parks. These last were especially numerous; by the end of the thirteenth century, royal licences to impark had been acquired even by men of relatively lowly knightly status, and there were probably more than 2,000 parks in England as a whole.[1] All three types of deer in

[1] For the royal forest, see in particular Bazeley (1921); Young (1979). For chases, see Cantor (1982); Birrell (1990–1). For parks, of a vast literature, see the useful summary by Stamper (1988).

medieval England—red, fallow, and roe—were widely found, though the herds in parks, especially small and medium-sized parks, were predominantly fallow. The latter, introduced into England after the Conquest, had spread rapidly, and in many forests, especially those in the lowlands and south, they had become more numerous than red deer by the end of the thirteenth century.[2] Hunting preserves served many purposes. They gave status, but also material benefits in the form of increased control of valuable resources, such as pasture for livestock and timber and wood, and their value cannot be assessed solely in terms of their deer populations. But they were primarily created to provide hunting opportunities and a supply of venison, and this probably remained, in most cases, their prime purpose.

The venison from these hunting grounds was not consumed only by their owners. First, however jealously they were guarded, poaching was rife at all social levels. At one end of the scale, powerful barons were tempted by the unrivalled hunting opportunities offered by the royal forests; for them, the sport was probably as much of an attraction as the bag. At the other, peasants laid snares or traps for deer and snapped up animals abandoned by aristocratic hunters, perhaps disposing of the venison on the black market as frequently as eating it themselves.[3] Second, a steady trickle of venison from hunting preserves reached local people legitimately. As we have seen in Chapter 11, for many forest workers—keepers, foresters, huntsmen—joints of venison were a perquisite of office. Others—estate officials, local people of high status—could expect occasional gifts of venison. This was a way in which owners of hunting preserves demonstrated largess, encouraging thereby the loyalty and respect of those they rewarded, and reinforcing, as they did so, the social hierarchy. Thus, it was to 'the best people of the county' that the Black Prince sent gifts of deer from his Cornish parks and it was to 'divers knights, ladies, squires and other persons of gentle estate' living nearby that John of Gaunt distributed deer from Needwood Chase in the run-up to Christmas 1372.[4] The biggest source of gifts of venison, however, was the King. In the mid-thirteenth century, Henry III regularly gave away over 200 deer a year. Many of his gifts were to mark some special occasion in the recipient's life or career—a wedding, a funeral, a knighting, an inception—an association that can only have helped to reinforce the special status of this meat.[5] Lastly, many an abbey in or near a royal forest or chase received a regular supply of venison, as tithe was paid on this as on any other produce. A memorandum in the Tutbury Priory Cartulary notes that it could count on twelve, fourteen, or sixteen deer a year from Needwood in the mid-fifteenth century.[6]

[2] There is a useful discussion of deer populations in Rackham (1980: 181–4, 193–5).

[3] For peasant poaching, see the discussion in Chapter 11 and Birrell (1996).

[4] Dawes and Johnson (1930–3: ii. 15) (Aug. 1351); Armitage-Smith, Lodge, and Somerville (1911–37: ii. 102) (Nov. 1372).

[5] For the gifts, see *CCR*, *passim*; see also Rackham (1980: 181).

[6] Saltman (1962: 257).

Most of the venison from hunting grounds, however, was destined to be eaten in the household of the owner. But hunting grounds and households were often far apart, which meant that it was necessary for venison to be transported long distances. Deer abounded, for example, in the Cumberland forest of Inglewood, and it supplied huge quantities of venison to the royal household in the thirteenth century, most of which had to be carried to royal larders in the south of England.[7] There were several deer parks on the Cornish estates of the Black Prince, who rarely visited the county; but venison from these parks was shipped to London for his household at Kennington, to Southampton for his household at Sonning, and even to Bordeaux when he was campaigning there in the summers of 1363 and 1364.[8]

Hunting deer

Deer were best hunted on a seasonal basis. Males were at their best in summer when they were 'in grease', that is, had built up fat in preparation for the rut. The fattest harts and bucks were to be caught in the relatively brief period between mid-June and early September, though they were often hunted earlier. By Michaelmas, the season was over.[9] Hinds and does, conversely, were best hunted in autumn and winter, their season lasting until February or Lent. Fresh venison could be obtained for much of the year, but it was at its best for only a limited period. Medieval aristocrats may have loved the chase, but hunting for sport did not fit easily with hunting for meat. It was simply impracticable for the owners of deer preserves to supply their own tables: the greatest lords had too many hunting grounds, while even the least among them might be occupied elsewhere at the crucial times of year. Much, perhaps most, of the venison consumed in aristocratic households was hunted by servants, and hunting was a job that employed many people and required equipment and dogs.

The King, with his large household and many hunting preserves, employed several teams of huntsmen, whose working lives were spent travelling from one royal forest to another, hunting to order. It was usually they who were dispatched to take 'the King's venison this present season' in specified forests, like the trio sent to Wychwood in July 1312 with a retinue of berners and fewterers, keepers of the more than fifty hounds that accompanied them.[10] But it needed more than a handful of hunting teams to satisfy the royal demand for venison,

[7] For example, 200 stags in July 1234, 100 in 1246, 100 in 1247, 200 in 1251, 160 in 1255: *CCR 1231–4*, 487; *CCR 1242–7*, 321, 455; *CLR 1251–60*, 8, 235.

[8] Dawes and Johnson (1930–3: i. 92; ii. 2, 68, 204; iv. 533). A further 100 harts and 100 bucks from the Forest of Wirral were to be sent to him at Bordeaux in 1363: Green (1979: 186).

[9] The royal huntsmen were often instructed not to hunt after Michaelmas: see for example *CCR 1237–42*, 102.

[10] The huntsmen were William Balliol, John Lovel, and Robert le Squier: *CCR 1307–13*, 464; see also 284, 324, 465, 494–5, 532. For the royal huntsmen, see Cummins (1988: 183–4, 266).

and all sorts of other persons were regularly called on to hunt for the King. It was one of the regular duties of the forest keepers; other forest officials, sheriffs, justices of the forest, and household members were also employed. Even relatives and favoured barons who were keen hunters contracted with the King to take venison for his household in return for a share in the bag.[11]

The richer aristocrats, too, employed specialized hunting servants, though never on anything approaching the same scale as the King. The Bishop of Winchester kept at least two huntsmen and also fewterers and berners on his Hampshire estates in the thirteenth century, but he seems to have been exceptional. Specialized hunting servants, who might be underemployed for much of the year, were an expensive luxury even in the large households of the very great. It cost the Bishop £26 6*s.* 8*d.* to keep a huntsman, three grooms, and a pack of thirty hounds at Bishops Waltham for 316 days in 1332–3. This was far more than most lords could afford at a time when many a member of the lesser gentry had an annual income not far in excess of £40.[12] Many a baron made do with a single huntsman like that keen hunter John Giffard of Brimpsfield, and even much richer men might manage with part-time hunting servants, like Bishop Mitford of Salisbury, who took on a huntsman, a page, and a fewterer on a temporary basis for the winter hunting season early in the fifteenth century.[13]

In practice, all sorts of people from high officials to the most menial servants were drawn into hunting to supply aristocratic tables. The Black Prince instructed his constable and his parker 'to take this season's grease' in Berkhamsted Park in August 1347; the Bishop of Winchester used his household knights as well as his huntsmen; it was Sir Henry Percy's parker who supervised the taking of deer at Petworth in the mid-fourteenth century.[14] No doubt lesser servants, too, were involved, though they are less likely to be named in the documents. It is striking how many servants appear among the poachers presented at forest eyres, the commissions of itinerant justices sent to determine all pleas relating to the forest. Cooks, packmen, palfreymen, woodwards, carters, shepherds, ploughmen, porters, and 'grooms' are more prominent than specialized hunting servants; they are often poaching without their master but with his connivance and using his dogs, which it seems likely they had learned to use in his service.[15] The recipients of gifts of venison were often allowed to hunt their allotted deer themselves, under the eye of a local keeper. This gave enthusiasts a

[11] One such was Richard of Cornwall: *CLR 1240–5*, 314; *CCR 1247–51*, 322; *CCR 1259–61*, 421.

[12] Roberts (1988: 71–2, 79). For knightly incomes, Dyer (1998*a*: 30–2). Of the mass of information about the cost of royal hunting teams, I will give just two examples: a three-week expedition to Devizes by one huntsman and team in 1285–6 cost £6 2*s.*; in July 1312, the wages of one huntsmen with two berners, two fewterers and thirty-six dogs came to 3*s.* 2*d.* per day: Byerly and Byerly (1977: 208, nos. 2037–43); *CCR 1307–13*, 465.

[13] Birrell (1994: 56–8); Woolgar (1992–3: i. 416–17). For the relative rarity of hunting servants in aristocratic households, see Mertes (1988: 49).

[14] Dawes and Johnson (1930–3: i. 117; see also iv. 533); Roberts (1988: 71); Salzman (1955: 37, 51).

[15] For example, Birrell (1982: 12, 13).

precious opportunity to hunt, while saving the donor expense, as the Black Prince was well aware: those to whom he granted deer in August 1351 were, he instructed, 'to come, if they wish, on certain days to certain appointed places, there to take at their own costs the does so given'.[16]

Preparing and preserving venison

The circumstances in which deer were hunted inevitably influenced the way the meat was prepared and presented. Fresh venison was appreciated, and action often taken to ensure it was served. When Bishop Swinfield of Hereford's journeys round his diocese took him anywhere near one of his hunting grounds, one or two beasts were typically taken for his table, often even before the season was properly under way, as if the desire for the fresh meat was irresistible.[17] When a celebratory occasion loomed, deer were often hunted or even poached in preparation. Thus, John fitz Reginald hired two notorious poachers to take a deer on his behalf before dining at the house of Philip Matson of Matson, near the Forest of Dean, in October 1276. More conventionally, when, on 10 August 1353, the Black Prince invited guests for Thursday, 15 August 1353 (the Assumption), he instructed his servants to take six roe deer in Macclesfield Forest 'quickly' and have them sent to Chester 'by next Wednesday'.[18] Henry III made sure there would be fresh venison at his daughter's wedding at York at Christmas 1251 by ordering a dozen does and six roebucks to be taken in the nearby forest of Galtres and delivered alive to his cook 'by the third day before the feast'.[19] The only venison eaten by Bishop Hales of Coventry and Lichfield during the summer of 1461 was fresh, but he was then staying on his Staffordshire manors, within easy reach of Blore and Beaudesert Parks and Cannock Chase.[20]

It was far from easy to ensure a regular supply of fresh venison. Like all meats, it did not keep well, especially in summer. Henry III often gave orders for a few deer to be caught in a local forest or park only a week or two before a specific feast, and it seems likely that it was intended that the venison be eaten fresh. In an order dated 28 March 1249, for example, ten beasts were to be taken in Havering Park, some 14 miles east of London, and sent to Westminster in time for Easter, which fell that year on 4 April. Other orders issued in similar circumstances seem to reflect a fear that the venison might go off even in such a short period. In an order issued on 12 April 1237, for example, it was specified that the

[16] Dawes and Johnson (1930–3: ii. 15).

[17] Webb (1854–5: i. 70, 83) at Colwell in early April and at Bishops Castle in May; see also Webb (1854–5: ii. p. clxviii).

[18] Birrell (2001: 149–50); Dawes and Johnson (1930–3: iii. 112).

[19] *CCR 1251–3*, 23.

[20] He ate his first buck rather early, on 25 May, the second on 6 June, the rest between 21 June and 7 September. The Bishop's accounts are printed in Woolgar (1992–3: ii. 451–86).

venison of twenty does to be caught in the same park for the Easter festivities at Westminster (19 April) was to be salted 'if necessary'.[21]

In practice, most great households ate both fresh and salt venison. The Bishop of Winchester was well provided with parks on his Hampshire manors and able to indulge a taste for fresh venison, particularly on special occasions, but he also regularly served it salted.[22] In any case, we should perhaps not see salt venison as too inferior an alternative. Much would depend on the skill of the salter and on the condition of the meat when it reached him. The quality would also depend on how heavily the flesh was salted and how long it was stored. In general, venison lends itself well to this method of preserving, with the possible exception of the small roe deer, which carries little meat. No less an authority than the *Master of Game* observes that 'the venison of [bucks] is right good if kept and salted as that of the hart'; the *Ménagier de Paris* believed that some cuts—he specified breast—were at their best when salted.[23] Medieval cooks were experienced in dealing with salt meat and skilled in preparing and presenting it. Like other salt meats, venison was generally parboiled before being baked, like ham today, which would help get rid of the salt. In any case, it is likely that much salt venison was eaten fairly quickly after it was made, and only lightly salted with this in mind. It was common for deer to be hunted on royal orders for specific feasts just a few weeks ahead, as we have seen; every year, for example, there was a flurry of orders for venison for Christmas in late November or early December, for St Edward's Day (13 October) in September, and for Easter in March. Only light salting would have been necessary in these cases. It may be relevant here that Bishop Swinfield of Hereford, when progressing round his estate, often ate up any salt deer remaining in the larder before moving on, as if recognizing that it would deteriorate if left for a later visit.[24]

Nevertheless, it was a routine procedure on big estates for deer to be hunted according to season, when the meat was at its best, and the venison prepared and stored in larders till needed, and in this case heavier salting would be necessary. Large quantities of venison might remain in larders from one year to the next: there were fifty-one carcasses in the four Lancastrian larders of Amounderness, Leicester, Tutbury, and Belper in October 1313, and seventy-three, a year later.[25] Account rolls offer glimpses of the various processes involved. At Tutbury, a larderer was taken on for this purpose for five weeks during the accounting year 1313–14, and it was presumably he who prepared the seventy-eight bucks and dozen does taken that year in Needwood Chase. At Framlingham in 1286–7, the job was done by Walter the Cook; making the venison in the larder took six days, and three casks (*doliis*) were prepared and cleaned to receive it.[26] Two different sorts of salt were used, an unspecified quantity of 'gross' salt, presumably the coarser and less pure 'bay' salt produced by the evaporation of sea water, and a

[21] *CCR 1247–51*, 150; *CCR 1234–7*, 430–1. [22] Roberts (1988: 73).
[23] Baillie-Grohman and Baillie-Grohman (1909: 39); quoted in Crossley-Holland (1996: 122).
[24] Webb (1854–5: i. 46, 91). [25] NA DL 29/1/3. [26] Ridgard (1985: 29).

quarter of 'white', a more refined salt, perhaps even from brine springs. It is possible that Walter was deliberately mixing the two to get the advantages and minimize the disadvantages of each; the gross salt was cheaper but slower to penetrate the meat and likely to give it a coarser flavour.[27] Such use of high-quality salt was not unique to Framlingham. In the knightly household of Sir Henry Percy at Petworth, the very expensive 'small' (*minut'*) salt was used for preserving fallow deer in the mid-fourteenth century.[28]

A fairly lavish use of salt is often suggested, though the calculations can only be very approximate. Quite apart from uncertainties regarding the quality and purity of the salt used, we do not know exactly how much meat was being salted. Red deer are much the larger, roughly comparable in size to cattle; though significantly smaller, fallow deer carry more meat in proportion to their size. It must be emphasized that the size of deer varies greatly not only with the season but according to age, condition, and diet. The live weight of a stag on Scottish hill-land today may be in the region of 277 lb, a hind of 194 lb, but Scottish red deer nowadays are particularly small and were so when the great nineteenth-century expert on deer parks, E. P. Shirley, spoke of 'a good fat buck' weighing between 100 and 120 lb; a modern adult buck may weigh anything between 100 and 210 lb, a doe between 75 and 120 lb. In any case, the larder weight of the beast would be considerably less than the live weight.[29] Not all the meat on the carcass was preserved: the process was expensive, and only the better joints might be deemed 'worth their salt'. Where the rates can be calculated, they are generally between half and a whole bushel of salt per fallow deer and two bushels per hart. Average costs are difficult to calculate not only because of the paucity of precise information but because of the variations in the price of salt; from a few shillings a quarter in the first half of the fourteenth century, it rose to an average of 6*s.* 4*d.* per quarter in the period 1351–1400. It is at least clear that it was by no means cheap to preserve venison by this method. It cost 16*s.* 8*d.*, for example, to salt one hart and thirty-one fallow deer from Duffield Frith in 1313–14 (using 4 quarters of salt) and 6*s.* to salt ten harts at Pickering in 1325–6 (using 20 bushels of salt). At Petworth, the 15 bushels of 'small' salt used to salt twenty-one fallow deer in 1349 cost 18*s.* 9*d.*, and 4 bushels used to salt six deer the year after cost 5*s.* 8*d.* On this manor, where the expenses of the servants who caught the deer are also noted (13*s.* 3½*d.* in 1348–9, 4*s.* 6*d.* the year after), we see how insignificant they are compared with the cost of the salt.[30]

[27] There is a useful discussion of salting in Horandner (1986: 53–5). See also Hagen (1998: 40–1).

[28] Salzman (1955: 37, 51).

[29] Anon. (1981: 24); Chapman and Chapman (1982: 5). A much wider possible weight range—between 165 and 750 lb for red, between 75 and 440 lb for fallow—is quoted for France by Benoist (1984).

[30] For salt prices, Bridbury (1955: 3, 176–7). One bushel (56 lb) of salt was used per doe at Belper (and probably at Tutbury) and 0.8 of a bushel per buck at Leicester in 1313–14, 1 bushel per fallow deer at Tutbury in 1370–1: NA DL 29/1/3 and SC 6/988/14; approximately 0.7 of a bushel (of 'small' salt) per fallow deer at Petworth in 1348–50: Salzman (1955: 37, 51); 0.5 of a bushel per buck for the King in

The operation was on a much larger scale in the case of the royal forests, where it was common for scores or even hundreds of deer at a time to be hunted in season for the royal larder. This venison was generally salted and packed in barrels before being put to store locally or, more often, transported at once to a royal larder. No single system developed for organizing this formidable task, one consequence of which, for the historian, is that the cost, which must have been significant, is difficult to analyse. It was sometimes the huntsmen themselves who were made responsible, but more often it was the keeper of the forest or the sheriff. In August 1243, for example, the constable of St Briavels was to find salt for forty bucks that were to be taken by royal huntsmen in the Forest of Dean, and have them carried 'by cartloads' to Windsor. Similarly, in August 1246, the keeper of the forests of Somerset was instructed to salt 'as it was taken' the venison of forty stags to be caught by the King's huntsmen and, when the operation was complete, deliver it to the keeper of the royal larder at Westminster.[31] We are not told what sort or quantity of local labour was employed, but the work would have had to be done quickly, and at short notice, as it was important to get the carcasses out of the forest in good condition and salted as soon as possible. It is interesting in this context to remember the temporary *bucheria* to be constructed in Weardale for the Bishop of Durham's annual 'great hunt'.[32]

By the early fourteenth century it was customary for one or more larderers to accompany the royal hunting teams, and they presumably superseded the local men previously hired for the task.[33] Perhaps this is a sign of greater concern for quality. At all events, this concern regularly surfaces in the otherwise formalized records, in connection with both the hunting and the preparation of the venison. The royal huntsmen sent to Staffordshire in August 1228, for example, were instructed to take harts that were 'fat' (*crassos*); bucks were to be taken in the New Forest in September 1247 'where they are fatter and better' (*ubi pinguiniores et meliores*); some venison was judged not 'good and fat enough to send overseas' in the summer of 1242, when Henry III was in France (it was palmed off on the household of the young Prince Edward).[34]

All three species of deer were eaten and generally meticulously differentiated in hunting and larder accounts. It is difficult to tell the extent to which the proportions reflected availability or preference. If the latter, it is rarely explicitly

1288–9: Byerly and Byerly (1986: 334, no. 2869); 2 bushels per hart at Pickering in 1325–6: Turton (1894–7: iv. 231). Two bushels of salt per hart was also the rate at the French court in the 1390s: Cummins (1988: 255).

[31] *CLR 1240–5*, 190; *CLR 1245–51*, 75. See also, for example, *CLR 1245–51*, 164 (the sheriff of Northamptonshire claims 35*s*. 4*d*. for salting 100 does and carrying them from the Forest of Cliffe to Westminster for the feast of St Edward, 1247) and *CLR 1251–60*, 17 (the sheriff of Essex claims 28*s*. for the carriage of 94 bucks from Havering to Canterbury for Christmas 1254).

[32] Austin (1982: 37).

[33] *CCR 1307–13*, 324; *CCR 1313–18*, 239–40.The larderers were paid 1½*d*. or 2*d*. per day, the same rate as the men who handled the dogs, considerably less than the huntsmen themselves.

[34] *CCR 1227–31*, 77; *CCR 1242–7*, 535: it was to be salted and received at Westminster in time for the next feast of St Edward (13 Oct.); *CCR 1240–5*, 144.

stated, though the *Master of Game* seems to favour the taste of the fallow buck.[35] There are hints that roe enjoyed a special status, but as easily digestible rather than better tasting. The flesh of the roebuck is described by the *Master of Game* as 'most wholesome'; roe was often sent by the King as a gift to invalids;[36] and, as we saw in Chapter 11, roe featured more often on the table at religious houses. Medieval recipes seem not to differentiate between types of deer, speaking simply of 'venison'; nor did the *Ménagier de Paris* when discussing deer meat. A preference for certain parts of the animal, on the other hand, emerges clearly, the same joints recurring in the aristocratic contexts of recipe books and household accounts, that is, the hindquarters, sides, and ribs. Sides (*latera*), haunches (*hanchias*), and *cauda*, perhaps rump or hindquarters, appeared on Bishop Swinfield's table. In 1296–7, Joan de Valence ate haunches (*hanchias* and *quissa*), *cauda*, and ribs (*costa*). Henry III presented his sister Eleanor with a side of venison in July 1231.[37] The *Ménagier de Paris* said that the three joints fit for a feast for the lord were saddle, loin chops, and breast, the latter best salted.[38] The meat of the less prized chines, together with loin, was to be put into pasties for Henry III in 1239.[39] The joints of venison specified in two mid-fifteenth-century cookbooks are ribs (which were boiled), haunches (baked), fillets (roasted), and hocks (baked in pastry). There is also a recipe for a 'good potage' made from umbles, a soup that also appears in Richard II's recipe book.[40] The offal was apparently sorted and certain pieces separated as more or less desirable delicacies, the rest fed to the hounds. According to Twiti's *Art of Hunting*, for example, the hounds were given the neck, liver, and entrails.[41] But umbles (or 'barbilles') were often among the perquisites claimed by foresters, and the umbles (*escaetis*) of bucks taken in the New Forest were among the venison to be delivered to the New Forest bailiffs by the royal huntsmen.[42] As discussed in Chapter 11, forest officials were often also given the less desirable forequarters; the shoulders, said Twiti, went to the man who broke the stag.

Quantifying the supply of venison

How much venison could owners of hunting preserves expect to have at their disposal? How realistic were the expectations so precisely quantified in the late

[35] Baillie-Grohman and Baillie-Grohman (1909: 29). According to Cox (1905: 50), summer harts and bucks are 'more of a delicacy' than females.

[36] Baillie–Grohman and Baillie–Grohman (1909: 43–4); two roes were sent to the Bishop of Carlisle when he was ill at Reading in 1244: *CLR 1240–5*, 236. See also *CCR 1234–7*, 253. The belief that the roe was more wholesome can be traced through Gaston Phebus to Modus, who is quoting Avicenna, according to Cummins (1988: 88).

[37] Webb (1854–5: i. 15, 20, 23, 40, 41, 44, 46, 49–50, 83, 91, 105); NA E 101/505/26, mm. 11, 12; *CCR 1227–31*, 533.

[38] Quoted in Crossley–Holland (1996: 122).

[39] *CLR 1226–40*, 390.

[40] Austin (1888: 10, 61, 70, 73, 81); Sass (1975: 58).

[41] Danielsson (1977: 17, 51). See also Baillie-Grohman and Baillie-Grohman (1909: 198; umbles are discussed on 244).

[42] Cox (1905: 119, 133–4); *CLR 1240–5*, 142.

fifteenth-century Black Book of the Household of Edward IV: 300 deer annually for a duke, 120 for an earl, eighty for a lesser earl, and twelve for a knight?[43] The nature of the evidence and its chronological dispersal makes a clear answer to such questions impossible, but some useful points can be made. The subject can be approached in a number of ways. Manorial accounts may record the number of deer hunted and stored, or the costs of the hunting, preparation, and transport of venison. The Earls of Lancaster were among the wealthiest in the country and owned a number of chases. In the accounting year 1313–14, ninety deer were hunted for the Earl in Needwood Chase, eighty-seven in Duffield Frith, and seventy-one in Leicester Frith. This was a year when Earl Thomas spent some time in the Midlands, and 175 deer were consumed by the households in his residences at Tutbury, Melbourne, Donnington, and Kenilworth. A further twenty-three deer were hunted in another, more remote Lancastrian chase, Amounderness, but none of them appear to have been needed for any of the Earl's households. Even though a number of deer were given away to local people (sixty-three from Needwood and Duffield, fifty-one from Leicester, and eleven from Amounderness), many deer remained in the larders at the end of the year. In 1313–14 at least, the Earl had more venison than he needed.[44]

The household of Dame Margaret Brotherton at Framlingham, which could draw on the venison from a dozen fairly local parks and from Yardley Chase, consumed a total of ninety deer in 1385–6.[45] Lower down the social scale, the Petworth household of Sir Henry Percy, who had two parks on the manor, consumed seventeen and a half deer in 1347–8. Deer were hunted for Sir Henry in the two following years, but most seem to have remained in the larder, only two being dispatched to his London household, and one given away.[46] Like the Earls of Lancaster, Sir Henry (who had other estates as well as Petworth) was clearly not short of venison.

Another approach is through household accounts, though we again face the problem of the figures being for isolated years, which may or may not have been typical. Bishop Swinfield of Hereford consumed about fifty deer in the months October 1289 to July 1290, a period when he was partly travelling, partly resident on some favoured manors, and briefly in London, and often within reach of his hunting grounds. If his consumption in these ten months was typical, his household was getting through about sixty-seven deer a year. This was considerably more than two other households of roughly equivalent status at the beginning of the fifteenth century: the Bishop of Salisbury consumed only

[43] Myers (1959: 96, 100, 103, 109).

[44] NA DL 29/1/3. Needwood and Duffield continued to supply large quantities of deer: forty deer were hunted there in 1370–1: NA SC 6/988/14. Tutbury Priory received twenty-four deer in tithe from Needwood in 1434, and twenty from Duffield, though the cartulary notes that twelve, fourteen, or sixteen were more normal: Saltman (1962: 257).

[45] Ridgard (1985: 86).

[46] Twenty-one deer were taken in 1348–9, six in 1349–50; twenty-four remained in the larder at Michaelmas 1350: Salzman (1955: 31, 43, 51, 58).

twenty-one deer in a slightly shorter period (October 1406–June 1407), equivalent to about thirty beasts a year; the household of John de Vere, Earl of Oxford, consumed thirty-six deer from his parks in the accounting year 1431–2. Bishop John Hales of Lichfield ate far more venison during his stay on his Staffordshire manors in the summer of 1461, twenty-three deer in four months, equivalent to nearly seventy a year.[47] Dame Alice de Bryene, in contrast, served no venison at all in the year 1412–13.[48]

Royal gifts, in particular those made on a regular basis to family members, may give some useful pointers that complement these widely differing figures from household and ministers' accounts. Henry III, for example, made regular grants to both his sisters. Between 1231 and 1235, when Eleanor was a widow, he sent her between thirty and forty-six deer a year (181 in all), and though the number dropped slightly over the next couple of years, she still received an average of about thirty deer a year from her brother over the six-year period. When Henry's other sister Isabella was living in Marlborough Castle in 1231, he sent her two bucks a week throughout the season, and made sure she was adequately supplied at three major feasts by dispatching additional animals at Christmas (five), Easter (six), and St John the Baptist (six). Similarly, Edward I saw that his mother had plenty of venison at Christmas 1272, with a gift of twenty deer.[49]

Yet another approach is through hunting preserves and in particular the deer parks that proliferated during the thirteenth and later centuries, and which might help suggest the potential consumption of the lesser lords for whom manorial or household accounts have rarely survived. How many deer could a park support, and how much venison could a park supply? It depended, obviously, on the size of the park and also on the extent to which it was dedicated to deer. Reliable estimates of the size of park herds at this early period are, unfortunately, few. The large park of Havering, adjoining the forest of Essex, supported about 500 deer on its 1,000 acres in the mid-thirteenth century.[50] Counting was becoming more common in the fourteenth century, though it was still by no means the norm. The number of deer in Blansby Park, in the Forest of Pickering, were estimated for accounting purposes in the 1320s, and put at 1,300 in 1326 and 1,500 the year after.[51] In 1337, the deer in the Duchy of Cornwall's rather smaller parks in Cornwall were counted with at least a show of great precision: the number ranged from fifteen and forty-two in the two smallest parks to 200 and 300 in the largest, with a total of 887 deer in the seven parks.[52] Contemporary estimates are much more common for the fifteenth century: the Bishop of Durham had 540 deer in his four main parks, each of between

[47] The accounts are printed in Woolgar (1992–3: i. 261–430; ii. 522–48, 451–86); see also Dyer (1998*a*: 59–61).

[48] Dale and Redstone (1931); see also Dyer (1998*a*: 59).

[49] The grants to Isabella and Eleanor are in *CCR 1227–31*, *CCR 1231–4*, and *CCR 1234–7*; both sisters may have had other sources of supply. Edward I's grant is *CCR 1272–9*, 5.

[50] McIntosh (1986: 18).

[51] Turton (1894–7: iv. 227, 252).

[52] Hull (1971: 2, 24); see also Hatcher (1970: 179).

100 and 200 acres, in 1457; the Duke of Buckingham had 300 deer on about 800 acres at Madeley in 1521; and in a much larger park of about 1,000 acres at Thornbury, he had in 1507 about 550 deer.[53] The stocking rates suggested by these few figures are low by modern standards, though we should remember that E. P. Shirley estimated that a minimum of one acre was needed for every fallow deer. But medieval deer farmers were probably wise to avoid overstocking their parks; overcrowding was bad for the deer, encouraging disease and the risk of malnutrition.[54] At all events, it seems reasonable to assume that the average deer park of some 100 to 200 acres could maintain at most a herd of between fifty and 100 deer.

How many of these animals could be hunted without damage to the deer population? The King was taking about forty-four beasts a year from Havering Park in the mid-thirteenth century, that is, about 9 per cent of the estimated total herd of 500.[55] In Cornwall, where the Duchy's parks had been said to contain 887 deer in 1337, 'this season's grease', that is, the number of bucks taken that summer, was forty in 1351; in the summer of 1363, the equivalent figure was sixty 'if so many can be taken'.[56] We have no record, sadly, of the number of does taken in these years; nor can we necessarily assume that the herds had remained at their 1337 size, but the figures would suggest a cull rate slightly lower than that of Havering Park.

If we take these few figures as a guide, we may perhaps extrapolate that a landowner like the Bishop of Durham, with four parks containing about 540 deer, could expect to get at most fifty deer annually, and the Duke of Buckingham at most thirty deer from Madeley Park, with its 800 acres and herd of about 300. A modest knight with a single park of about 100 acres, which might have a herd of about fifty deer, would be able to take only a handful of deer each year for his own use.

These figures are extremely tentative, for all the reasons already quoted, but also because they do not allow for factors such as disease or poaching, both of which could, in bad years, decimate herds. The lord of Okeover might have been exaggerating when he claimed, in 1441, that 100 of his 125 deer had been stolen from his park, but the figures remind us of a constant problem owners of hunting preserves had to face.[57] Deer populations, especially in parks but also, if to a lesser extent, in chases and forests, did not necessarily flourish of their own accord and needed regular care and attention. Deer suffered especially if they were undernourished or overcrowded. In bad years, even in the royal forests, populations might suffer badly from disease, and deaths from murrain were counted in the hundreds.[58]

[53] Drury (1978: 97); Cantor and Moore (1963: 37); Franklin (1982: 156). [54] Birrell (1992).
[55] Rackham (1980: 191–2). [56] Dawes and Johnson (1930–3: ii. 15, 204).
[57] Birrell (1992: 115; see also the figures on 124).
[58] Turton (1894–7: iv. 139); see also Cox (1907: 514–15).

Conclusion

It is clear that the quantities of venison consumed in aristocratic households varied greatly. Personal taste and individual circumstances no doubt help to explain the discrepancies, as well as the availability of venison. That said, what is perhaps most striking is the enormous gulf revealed between the great lords with their large and numerous hunting reserves and the humbler knights with only one or two, often quite small, parks. For the latter, the occasional bounty in the form of a gift or a successful poaching expedition must have been very welcome, whereas in some years at least, great lords had access to as much venison as they wanted. In whatever quantities it was eaten, venison did not come cheap. It has been possible to give only the briefest indication here of the various processes and costs involved, both directly in the hunting, salting, and transport of venison, and indirectly in the creation and maintenance of hunting preserves, with all this implies in terms of resources. All meats were valuable in the Middle Ages, and to serve meat in abundance was a way of demonstrating wealth and status. But venison was a special case, a meat which only great landowners with vast estates could afford to serve frequently and freely, and a meat which elsewhere must have remained only an occasional luxury.

PART II

STUDIES IN DIET AND NUTRITION

13

Group Diets in Late Medieval England

C. M. WOOLGAR

In late medieval England, there were two principal reasons for group diets, that is the fashioning or restriction of consumption for a common purpose. Foremost was the link between diet, religion, and virtue with the state of the body beyond the physical. The moral impact of this belief was wide; the dietary consequences were in practice restricted to that part of the population, a minority, who had regular access to meat and who could afford substitutes, often expensive, such as fish. The second principal motive for group diets—identity or social competition—was similarly restricted to those with the resources to choose their food. Two further grounds had a narrower impact. Diet was an essential component of valetudinarianism and medicine. This led to consistent patterns of consumption among some groups of the young, the old, and for some individuals and groups, male and female. There were, finally, uniform patterns that resulted from communal living or common purpose, for example, the result of the provisioning of armies, or food in an institutional setting. In each case the group might live physically together, but one need not conceive group diets solely in this way. A pattern of diet might identify an individual as part of a wider or dispersed grouping. Equally, one might find in a group living together, such as an aristocratic household, perhaps as many as half a dozen distinct dietary regimes for different parts of the establishment.

Diet, religion, and virtue

Christian belief and the pattern of consumption that it established were the most influential determinants of group diet in the late medieval period: they provided a standard against which everyone might be measured. In 1307, the miraculous resuscitation of Gilbert, the son of a London goldsmith, was examined by papal commissaries investigating the life and miracles attributed to Thomas Cantilupe, the late Bishop of Hereford. In answer to one of their questions, two witnesses,

I am grateful to Barbara Harvey, Professor Paul Freedman, and Professor John Walter for their comments on a draft of this chapter.

one Gilbert's sister, the other his nurse, held that Gilbert's parents were of good standing and devout: his mother regularly abstained from flesh on Saturday, his father on both Friday and Saturday.[1] Examined for heresy in October 1514, Thomas Watts of Dogmersfield confessed his belief that, apart from Lent and the Ember Days, there were no days on which fish and meat might not be eaten; but that in order to avoid attracting the suspicion of heresy, he ate bread and cheese on many days traditionally regarded as fast days—that is, on days when he would normally have been expected not to eat meat, but fish, he chose to blur his rejection of the orthodox dietary pattern for one that was both in line with his beliefs and ambiguous enough to spare him persecution.[2]

Throughout the Middle Ages, Christianity precipitated the rise and decline of patterns of consumption that avoided eating flesh and the vices of carnality associated with it, particularly lechery and gluttony. Commonly held beliefs were that when the outer man fasted, the inner man prayed; that the prayers of those practising abstinence were more likely to be effective spiritually; and that fasting, one might be more likely to merit spiritual revelation.[3] Fasting was seen as preparation for spiritual acts, such as communion.[4] Although the practice of fasting was widespread in England before the Conquest, the dietary consequences—as we have seen in Chapter 8—are not now absolutely clear to us. By about 1300, however, fasting had produced a pattern of consumption among the laity that frequently led to abstinence from meat, often substituting fish, on Wednesdays, Fridays, and Saturdays; throughout the season of Lent; the eves of the great Marian feasts; the eves of the feasts of the apostles; and the three days before the Ascension.[5] This regime was scrutinized at confession,[6] but how far it might be adhered to varied from individual to individual. Personal devotion might add further days of abstinence, for example, the Nativity of St John the Baptist (24 June), the feasts of All Saints and sometimes of All Souls (1 and 2 November), the feast of St Katherine (25 November), and, in the later Middle Ages, the feast of Corpus Christi. Double fasts—when one of the additional fasts fell on a day of abstinence, or a Friday in Lent—could produce a more austere regime. On these occasions, individuals might abstain from dairy products as well as flesh, or consume solely bread and ale.

The basic pattern was common expectation, but accepted as exempt from it were those below the age of 21, pregnant women, the elderly, the sick, and the poor, as well as those whose work involved hard, manual labour; but it was agreed that omission rested on the individual conscience.[7] There might be

[1] BAV MS Lat. Vat. 4015, fos. 51^{v}, 52^{v}, 53^{v}, 54^{v}. [2] Hampshire RO 21M65/A/1/19, fo. 73^{r}.

[3] Cazier (1998: 198–90); for dietary practices generally, Woolgar (2000: 36).

[4] Powicke and Cheney (1964: i. 32).

[5] Although preachers advocated fasting on the Ember Days (the Wednesday, Friday, and Saturday of the four, seasonal Ember Weeks) in preparation for the day of ordination, the following Sunday, I have not traced this as a frequent occurrence: Erbe (1905: 253–4). [6] e.g. Goering and Payer (1993: 37–8).

[7] Erbe (1905: 82).

further dispensations from the regime, for example, granted by the papacy for some members of the aristocracy; or it might be mitigated through personal lapses of determination or devotion. There is some evidence that suggests that keeping the Friday fast was regarded as sufficient—or even exemplary—in some circles.[8] In secular households it was unusual to have a complete fast, although on one day of the year, Good Friday, it was not uncommon to go as far as restricting consumption to bread and ale alone.[9]

These diets had for their model the examples of saints and ascetics, whose practices were generally marked by extreme denial and abstinence. They can be characterized not only by their deliberate abstinence from some forms of food, but also from the enjoyment of food in general. There have been good studies of this phenomenon, for example, in connection with female saints and religious;[10] and it was widely considered a typical manifestation of sanctity and essential to devotion. St Hugh of Lincoln employed a rigorous diet to keep his body in subjection, first, as a priest at Chartreuse, living sometimes on dry bread and water, especially in the season of Lent when he fasted in this way on three days of the week, and for four in the last week. After his elevation to the see of Lincoln, when he was required to eat in public, in company, he abstained entirely from meat, frequently eating fish. Although he drank wine from time to time, he did so in moderation, partly as an example to others.[11] The 7-year-old St William of Norwich, unlike his elder brothers, practised abstinence on Mondays, Wednesdays, and Fridays, as well as on the vigils of the feasts of the apostles and other saints' days, prescient of his future sanctity.[12]

The 1307 inquiry into the life and miracles of Thomas Cantilupe provides a detailed account of the saint's eating habits. The witnesses, from his household, frequent guests, and friends, answered a series of interrogatories: tellingly, the one that provides information about diet was 'What qualities of excellence or virtues had you noted in his life ?'[13] Hugh le Barber reported that Thomas fasted on Wednesdays, Fridays, and Saturdays, that he fasted on bread and water on the eves of the Marian feasts. Following the custom of the exceptionally pious, he kept by fasting two or three Advents in the year, in anticipation of Christmas, Pentecost, and the Nativity of St John the Baptist—although many of his household ate meat at these times. He rarely ate supper (*cena*) and did not drink after lunch (*prandium*), that is, at the afternoon drinking, unless there were visitors.[14] When he did drink, his wine was considerably diluted with water. He greatly liked lampreys from the Severn in pasties and he regularly had them prepared

[8] Robertson (1875–85: i. 147); BAV MS Lat. Vat. 4015, fos. 66^{r}, 68^{r}.

[9] e.g. Longleat MS Misc. IX, fo. 77^{v}.

[10] e.g. Walker Bynum (1999: 186–94).

[11] Douie and Farmer (1985: i. 37 n. 4; i. 76, 125).

[12] Jessopp and James (1896: 13–14). See also Erbe (1905: 12): St Nicholas as a baby suckling from his mother only once a day on Wednesdays and Fridays.

[13] 'Interrogatus quas excellencias vite vel virtutum notaverat in ipso', abbreviated from the introductory questions in BAV MS Lat. Vat. 4015, fo. 4^{r}.

[14] fos. 18^{r}, 19^{v}.

and brought to him. He would then smell them and ask bystanders whether they thought they were well prepared—and when they answered positively, that he could happily eat them, he would decline to do so. Hugh had seen him and heard reported that he would order partridges and other birds to be prepared; but once they had been prepared, he would abstain from them totally or take a single bowl.[15] These details were substantiated by other witnesses.[16] Richard of Kimberley noted that his Christmas fast, celebrating Advent, extended for six weeks before the feast (rather than the more usual four);[17] and Nicholas of Warwick, a judge and royal councillor, recorded how Thomas, in the thirty years preceding his death, had never eaten to satiety.[18] William Cantilupe, who had carved for Thomas, reported that when he had various pottages or sauces he sometimes made a mixture of them, as William believed, to lessen the flavour.[19]

Cantilupe is not untypical of a group of saints or holy men whose rejection of carnality through dietary practice formed an exemplar for lay and religious alike. Some ascetics were more extreme. St Godric (d. *c*.1172) reputedly lived from the roots of grasses or the fronds and leaves of trees, or flowers, rejecting gifts of food from admirers (although forced to accept some, so as not to hurt feelings). He cooked up as his food a mixture of grasses, plants, and vegetables, which he kept for long periods of time. He also avoided eating during the day.[20] Such austerity in diet was a significant influence in some religious circles, but was deliberately eschewed in others.

The different monastic Orders had varying models for diet, based on what, in origin in the Rule of St Benedict, had been a vegetarian model. They were not, on the whole, productive of individuality in dietary practice, but sought a corporate pattern of regulation. The Rule set out the dietary norm, forbidding, for example, consumption of the flesh meat of quadrupeds except by the sick. By the fourteenth century, within the Benedictine Order, significant changes to this pattern had been negotiated and defined, to add meat and meaty dishes, as well as locations for consumption other than the refectory envisaged by Benedict. A compromise position was confirmed by Pope Benedict XII in 1336.[21] The thirteenth-century customary of the Benedictine abbey of Eynsham advised novices that they were not to eat because they enjoyed it, but as a necessity.[22] In fact, as we can see in Chapters 8 and 9, the necessities of the community at Eynsham embraced a varied meat diet, with an apparent abundance of fish and fowls.

Other Orders had different expectations. The spiritual renewal of the Cistercians of the twelfth century seems to have been marked by rigour in diet, but the pattern may have been mitigated in various ways. A mirror of instruction for the Cistercian novice, by Stephen of Sawley (d. 1252), forbade abstinence, but demonstrated how consumption of food should be moderated.[23] At Beaulieu

[15] fos. 23^{v}, 24^{r}. [16] fos. 30^{v}, 31^{r}. [17] fo. 34^{r}. [18] fos. 15^{v}, 16^{r}. [19] fos. 56^{r}–57^{v}.
[20] Stevenson (1847: 71, 73, 80–4). [21] Harvey (1993: 34–71, especially 38–41); see also Chapter 15.
[22] Gransden (1963: 49–50, 100–1). [23] Mikkers (1946: 57).

Abbey, in 1269–70, financial records confirm that it was still exceptional for Cistercian monks to eat meat, although those in the infirmary and lay servants might do so: there were thus several different diets in regular operation within the monastery.[24] It was also generally required by the Order that the Cistercian travelling outside the cloister maintain the same dietary pattern and manners that he would have sustained within.[25]

The Augustinian canons were openly much more carnivorous than the Cistercians and much more akin in their diet to the upper levels of secular society. The summary of meat consumption at Bolton Priory in 1377–8 shows the larder stocked with the carcasses of 3 bulls, 32 cows, 16 heifers, and 38 oxen; 88 pigs; and 220 sheep, besides further unspecified quantities of meat, as well as the usual fish. In 1402–3, the Augustinian canons of Southwick Priory followed the pattern of abstinence in Advent before Christmas (with, in 1402, the exception of marrow bones). Lenten abstinence started on the Monday before Ash Wednesday and the cellarer's account noted nothing consumed on Good Friday, and fish alone on the Monday, Tuesday, and Wednesday before Ascension Day.[26]

Where there was variety between individual monasteries of the same Order, it was probably more the product of local circumstances than of spiritual aspiration. Regimes of abstinence could, however, become ends in themselves. St Hugh of Lincoln, almost certainly addressing himself to heads of houses outside the Benedictine Order, rebuked them for over-zealous adherence, forcing their monks to abstain from all sorts of meat, when this was not part of their rule.[27]

Stricter regimes of abstinence might also feature within secular households. Where friars and chaplains might be resident, individuals might chose to ape the more extreme patterns of abstinence of the clergy. Aymer de Valence, the future Earl of Pembroke, stayed with his mother over Christmas 1296. The friars in this household fasted throughout Advent in 1296, up to and including 9 January 1297, with the exception of 27 December. Aymer joined their fast on Christmas Day itself and on 8 January 1297. In Aymer's case, this was not an isolated incident. Another Valence household account, almost certainly for Aymer, for Lent 1300, contains purchases, alongside ordinary bread, of *peyn cendre*, bread made with ash.[28] There are other instances of loaves made with ash: Reginald of Durham describes the bread made by St Godric as containing one third ash, and the practice of mixing bread with ash may be linked with penance.[29]

There is good evidence linking other dietary practices to penance. Margery Kempe often did penance by fasting on bread and water;[30] and fasts, in various

[24] Hockey (1975: 178–9, 183, 187, 308–13).
[25] Griesser (1956: 251).
[26] Kershaw and Smith (2000: 232, 558); Hampshire RO 5M50/69.
[27] Douie and Farmer (1985: ii. 196–7).
[28] NA E 101/505/26, mm. 8r–12r, cited in Woolgar (1999: 37 n. 50; 91), attributing Woolgar (1992–3: i. 170–3) to Aymer de Valence. I tentatively identified this as bread made with sanders, a red colourant (Woolgar 1992–3: ii. 758), but I am now convinced that it is ash.
[29] Stevenson (1847: 79–80); Erbe (1905: 254), for the association of bread baked in ashes with the Ember Days.
[30] Meech and Allen (1940: 7).

degrees, were imposed as a penance by ecclesiastical courts and at visitations.[31] Abstinence was also associated, in the twelfth century, with mourning.[32] Its rejection was indicative of heresy. Lollards objected to fasting on the grounds that the practice lacked biblical authority, holding that all foods could be eaten at all times and that abstinence enfeebled the body.[33] Some elements of fasting, particularly eating fish, were costly and abstinence therefore had a differential impact on society.[34]

Social competition and diet

In diet, as in so much in medieval society, social competition was a key element in delimiting different groups. A share of the best or most precious meats or fish, of the roast meats, an entitlement to additional courses, to wine as opposed to ale, all were keenly sought as marks of distinction in the competitive atmosphere of the households of the upper classes. Even the most humble position in the household gave access to food unimaginable in other contexts. Thomas Cantilupe's inverted sense of social competition, which led him to send away delicate foods to the poor and the sick, was crowned for one observer by his habit of having the pottage that had been made for the grooms, the lowest in rank in his household.[35] Royal household ordinances illustrate this well. Those of Edward II, for 1318, set out a series of daily allowances, giving individuals such as the clerk of the spicery a portion of bread, half a pitcher of wine, half a gallon of ale, a portion of the ordinary meat of the kitchen, with a portion of roast. His assistant, the second clerk, had the same livery of bread, a gallon of ale, but no wine, and one portion of ordinary meat, but no roast (Plate 13.1). The clerk of the spicery was a middle-level official: those of the rank of esquire and above, in general, had access to the roasts; valets and those who were graded lower did not. Others of higher rank had more courses and larger rations, which they may have used to feed others, their assistants, or to give in alms.[36] These distinctions remained throughout the Middle Ages and into the early modern period.[37]

In the mid-fourteenth century, there were two statutory attempts to limit consumption of food. In 1336, the focus was on the efforts of the lesser gentry and others to follow the expensive dietary habits of the aristocracy.[38] The sumptuary legislation of 1363 went further, concentrating as far as food was concerned on those ranking as grooms or below. It restricted those of this rank to one meal a day of meat or fish, with other items, such as milk, cheese, and butter, according to their estate. All those occupied in husbandry, tending animals, threshing corn, dairy workers, and carters—perhaps aimed at upstart manorial workers, after the

[31] Poos (2001: 137, 175, 220); Thompson (1914–29: i. 9).

[32] Foreville and Keir (1987: 305–6).

[33] Thomson (1965: 41, 95, 105, 128); Tanner (1977: 15–16); Hudson (1988: 141–2, 147–8). This is the implication of heresy trials, but Lollard sermons are more ambiguous: Cigman (1989: 121–4).

[34] See Chapter 8.

[35] BAV MS Lat. Vat. 4015, fo. 121^{r}.

[36] Tout (1936: 241–84, at 247–55).

[37] Myers (1959: 145); Anon. (1790: 162–94).

[38] Luders, Tomlins, and Raithby (1810–28: i. 278–9).

Plate 13.1 The medieval kitchen at the Bishop of Winchester's palace at Wolvesey. The large opening is all that remains of one of the kitchen's massive fireplaces, where roasting would have taken place. Photograph: C. M. Woolgar.

Black Death, demanding greater allowances in the changed labour market—and those worth less than 40*s*. in goods and chattels, were to eat in an appropriate way and not excessively.[39] Excess was reserved for the aristocracy.

Domestic regulation and practice used diet to mark status in myriad ways. In the households of Thomas Arundel, as Bishop of Ely in the 1380s, and of Joan Holland, Duchess of Brittany, in 1377–8, only freshly baked bread was eaten; whereas in other households, such as that of Bishop Mitford of Salisbury, in 1406–7, it might last for four or five days.[40] The entitlement to and quality of pittances in monastic houses marked similar boundaries.[41] The household book of the Earl of Northumberland, in 1511, indicated at considerable length who might be served with particular birds, both wild and domestic.[42] The range of spices employed in the greatest households far exceeded those in common use. The two priests of Munden's Chantry at Bridport, in the 1450s, aspiring to the habits of their betters, had pepper, cloves, ginger, cinnamon, vinegar, saffron, and mustard, although in small quantities and at the great feasts.[43]

One further group distinguished by diet were the poor supported by great households. They might be maintained physically within the household, or fed at the gates, frequently in multiples of a dozen or thirteen. Those fed at the gate

[39] Luders, Tomlins, and Raithby (1810–28: i. 380–2).
[40] Woolgar (1999: 125); AD Loire-Atlantique E206/1.
[41] See Chapter 15.
[42] Percy (1905: 102–7).
[43] Woolgar (1992–3: ii. 765–6); Wood-Legh (1956: pp. xxviii, 23, 25).

might achieve something from the choicest foods on the table, but the poor might also have a special dietary regime. In the household of Dame Katherine de Norwich, in 1336–7, they were fed a diet of herring and loaves of both a lesser quality and smaller size than the household loaf.[44] Although this diet might at first appear one of perpetual abstinence, use was made of a special dispensation that allowed the poor to eat meat when others ate fish—or to receive other food of higher status. Thus, in the household of Joan de Valence, the poor were given mutton on 22 December 1296 when the rest of the household had fish;[45] and in the household of Dame Katherine de Norwich, on Good Friday 1337, although the poor did not get their regular herring, they did receive wastel bread.[46] In these examples, the diet of the poor was less substantial than that of other household members, but the regular provision of food created a standard of living much greater than that enjoyed by the poor at large.

Diet and health

Medieval dietetics, based on the analysis of Galen and its subsequent elaboration, gave individuals—like everything else—their own make-up in terms of the four humours. There were differences between the old and the young, men and women. The humours were in turn affected by external influences, known as the six non-naturals: air; food and drink; sleep and waking; movement and rest; consumption and excretion; and the emotions.[47] The prose Salernitan questions of *c.*1200, which circulated in the schools of northern Europe, illustrate this system in practice. They held that it was easier for the old to fast than the young, as the old were cold and humid, which made up for lost sustenance, but the young were soft and their bodies easily lost sustenance, which could not be so readily replenished.[48] Foodstuffs equally had their own humoral composition, which might be altered by cooking: roasting would warm a food, but dry it; boiling would warm it, but make it moist.[49] To find a mix that was the most beneficial was a complex process that exercised physicians. Broadly speaking, one might seek to even out the imbalances in one's complexion by consuming food that moderated it. The addition of meat to the diet of the sick, or to those who were convalescent, was for this reason a standard of many monastic regimes, a 'group' application of the theory.[50] These practices can be difficult to identify, partly because much humoral theory appears in contradictory forms and partly because, at its most refined, it was designed for the individual rather than group practice.

Other group diets

If religious groups and enthusiasts or even the virtue of common men might be defined by diet, other groupings might also be shaped in this way. Countries

[44] Woolgar (1992–3: i. 177–227); see also Chapter 2. [45] NA E 101/505/26, m. 10^{r}.
[46] Woolgar (1992–3: i. 223). [47] Albala (2002: 48–162). [48] Lawn (1979: 101).
[49] Scully (1995: 42–9). [50] Rawcliffe (2002: 58).

might have sufficient distinctiveness in their foodstuffs for differences to be noticed. To a foreigner, England was marked by its food, the unfamiliarity of which was identified as one of the factors behind the brevity of the stay of the first Prior of Witham (founded largely with French Carthusian monks).[51] There may have been sufficient variation in the nature of staple ingredients to create a sense of difference, although in the later Middle Ages it was possible in Europe to distinguish cuisine on a regional basis. Both the structure of the aristocratic meal and its constituents varied between countries.[52] Whether ethnic groups might be identified on the basis of diet, as opposed to customs associated with food, is a more open question. The results of the excavations of the Jewish cemetery at York are ambiguous in the differences they record between those buried there and the skeletal remains of others in York. Although there were some general differences in stature, particularly of males, it is not clear what might be attributed to genetics and what to nutrition.[53] Good, contemporary evidence for the dietary practices of the Jewish community in England is lacking. While it differed from the Christian community, it was not allowed to do so in some obvious ways. In 1253, for example, Jews were forbidden by statute to buy or eat meat during Lent.[54] Many of the basic foodstuffs would have been the same, even if customs of preparation and consumption followed a different course. There is a strong possibility that the new Anglo-Norman aristocracy imported at the Conquest some specifically Norman (and perhaps Sicilian) practices. Here one might look especially at some aspects of hunting, discussed in Chapter 11, and hawking for food, although the growth in fish consumption, both of marine and freshwater fish, which appears at about the same time, is probably part of a wider European phenomenon.[55]

There were other uniform patterns of diet. Some, such as provisions for the military, or arrangements for workers with food as a part of the recompense for their labour, diverge from normal expectations although following the general outlines of diet expected in terms of abstinence. The military placed greater reliance on preserved foods, especially on meat. The calculations that were made for English troops in Scotland, around 1300, were based on 244 days of meat to 122 days of fish—whereas in an aristocratic household of this period the division was almost equal. The calorific intake suggested for the soldiers is substantial, around 5,500 kcal per person, although one cannot be sure that all grooms and attendants have been taken into account; but it is still lower than some of the allowances in great households in the fifteenth century noted in Chapter 7.[56] With another group, harvest workers, the food supplied by demesne lords gave them a better diet than at other points of the year, although it may have been a

[51] Douie and Farmer (1985: i. 47). [52] Hieatt (2002); Flandrin (1992).
[53] Lilley, Stroud, Brothwell, and Williamson (1994: 450, 455, 525–6).
[54] Powicke and Cheney (1964: i. 473). [55] See Chapter 8.
[56] Prestwich (1967); Woolgar (1999: 90); Davies (1963).

way of using up stocks of food that was reaching the end of its life.[57] Both groups required a high calorific content to their diet in order to support heavy manual work.

Conclusion

In terms of our understanding of nutrition, group diets would have had little effect on the many who might seldom taste meat and whose budgets did not extend to choice; but they were especially important to contemporaries as a reflection of the inner virtue of the consumer. Dietary patterns expected by religious practice might place one among the saints, heretics, or at one of very many points in between. Luxury, in terms of consumption of meat and fish, was wholly compatible with abstinence and religious aspirations, provided the foodstuffs did not fall into categories proscribed for the particular day or season, or did not do so without licence. Social competition, however, brought very different consequences in terms of nutrition. On flesh days, the aristocracy focused their consumption on lighter meats, particularly birds and wildfowl, or meats that were otherwise restricted, such as game; their servants had available astonishingly large quantities of meat, particularly of beef and mutton, but with quantities of pork diminishing in the later Middle Ages. On fish days, the highest echelons had fine seafish and freshwater fish; the rest of the household, staples such as stockfish and salt fish. These features of diet then became the aspirations of those at lower levels.[58] The advantages or disadvantages that flowed from this were in turn mitigated by medical advice and dietetics, again, on a selective basis.

These dietary patterns were not unique to England. Elements are found across medieval Europe, with variations dependent on local produce, piety and persuasion, or medical influence. The Cathar *perfecti*, for example, were marked out by abstinence from meat, cheese, and eggs, with consumption centred on bread, fish, costly spices, and vegetables; the population in general around them followed a regimen with a mixture of flesh and fast days.[59] Subtleties and nuances of diet told much about consumers, which is why medieval inquisitors recorded these details. Our present understanding of the link between good diet and virtue may have a different axis, but the connotation has endured.

[57] Chapters 7 and 8; Dyer (1994*b*: 92–5).

[58] Woolgar (1999: 132–5; 2001); Dyer (1994*c*).

[59] Pegg (2001: 80–1); Weiss (2001: 76, 105, 113–14, 135, etc.); Riera i Melis (2000).

14

Seasonal Patterns in Food Consumption in the Later Middle Ages

C. C. DYER

Historians have rightly focused their attention on the fluctuations in supplies of food, and especially grain, from year to year. They have also highlighted the long-term shifts, such as the rise of meat consumption in the fourteenth and fifteenth centuries, which has been discussed in earlier chapters. This chapter, while taking note of movements over long periods of time, is mainly concerned with seasonal changes, because these were an important dimension of people's experience of eating and drinking. The questions to be addressed here are how and why did medieval diet vary through the year? In order to proceed from the known to the less well documented, the questions will first be answered in relation to aristocratic households, and then for the remainder of society. Throughout society explanations will be partly functional and economic, as food supplies were affected by the weather, the cycle of growth (for crops) and reproduction (for animals and birds), and storage and distribution of commodities. We must also take into account cultural factors, as religious and family calendars had their impact on diet, and consumers' preferences.

Upper-class and institutional households

The accounts kept by household officials, and especially their records of daily expenditure and consumption, demonstrate clear patterns in eating and drinking through the seasons. These were sometimes characteristic of all households, but also resulted from individual choices and local supplies. To take two examples which show distinctive patterns, most households had particular seasons at which veal was eaten, but William Mountford's household, which spent most of the accounting year 1433–4 at Kingshurst in north Warwickshire, ate some veal in almost every week. This may have reflected the abundance of calves in a region which specialized in cattle farming, but perhaps Mountford or some other influential member of his household had a strong preference for this meat.[1] Hamon Le Strange's household, which lived through the fateful year 1348–9 at Hunstanton on the Norfolk coast, had a predilection for plaice, which must have been locally plentiful.[2] These fish appeared intermittently on most aristocratic menus, but Le Strange and his companions ate them constantly, except in May and June of 1349, which probably marks the temporary disruption of the fishery by the Black Death.

Such was the wealth of the aristocracy, whether in terms of acreages of grain on their demesnes, or of rents in the form of cereals from their tenants, or of cash rents, that they were assured of steady supplies of bread and ale. Exceptions include the serving of cider as a substitute for part of the ale drunk in the household of the Countess of Pembroke when she stayed at Inkberrow in Worcestershire in late September and October 1296.[3] The consumption of other food and drink—meat, fish, game, dairy products, fruit, vegetables, and wine—was all subject to marked seasonal fluctuations.

The most obvious influences on consumption came from the fasts and feasts imposed by the Church—discussed in Chapters 13 and 15—which created a weekly pattern as well as seasonal changes reflecting the liturgical year. In Lent not just meat and poultry, but usually also animal products such as cheese and eggs, were completely removed from the menu for six weeks. The aristocratic consumers were still served with a variety of high-protein food. In addition to the preserved fish, particularly herring which were often bought just before Lent, the fishermen and the network of traders in fish could supply many species freshly caught in the sea, rivers, or ponds. When the Bishop of Bath and Wells was staying in London on 25 February 1338 he and his companions ate preserved fish: red herring, stockfish, haberdines, and salt salmon; and among fresh fish, lamperns, *merling* (whiting), oysters, plaice, and smelts. A large piece of porpoise was also on offer.[4] Households felt deprivation in Lent, not just from the absence of meat, but also from the reduction in the numbers of meals just before Easter.[5]

Every household celebrated the festive seasons of Christmas, Easter, and Whitsun, though in different ways and not always on the same day. The largest and most elaborate meals in the Christmas period might be served on 25 December, or on 1 January, or on 6 January, or on some other day. Households recognized a variety of other festivals, giving priority to particular saints days depending on local cults and personal preferences. John Hales, Bishop of Coventry and Lichfield, arranged an elaborate dinner with a number of varieties of freshwater fish at Lichfield for the Assumption of the Blessed Virgin Mary (15 August) in 1461, while Prior William More of Worcester in 1519 marked the Nativity of the Virgin (8 September) with a meal in which venison made a prominent

[1] Woolgar (1992–3: ii. 436–43). [2] Norfolk RO, NH 8. [3] Woolgar (1999: 128).
[4] Thompson (1924: 145). [5] Woolgar (1999: 90–2). See above, Chapter 13.

appearance. The fellows of Merton College, Oxford, enjoyed veal, piglets, and frumenty on Corpus Christi Day (17 June) 1484.[6] The vigils of such special days would be marked by fasts. Households would also have their own events to commemorate, such as weddings and funerals. Dame Katherine de Norwich organized an elaborate feast with game, poultry, and a boar's head at Norwich to mark the anniversary of her husband's death on 20 January 1337.[7]

If the year was punctuated by episodes of fast and feast resulting from religious observance, other less sudden or dramatic changes reflected the availability of food, together with complex combinations of opportunity, customs, and preferences. Beef and mutton were consumed more frequently than any other meat. They were eaten in most households throughout the year, in both fresh and salted forms (Fig. 14.1). For beef the most noticeable peak came in December and January, with a lull in some households in November.[8] Mutton was consumed in increased quantities in June, July, and August, and again towards Christmas. The consumption of younger animals was strongly influenced by seasonal availability. Calves were slaughtered and eaten between April and July, after the cows had given birth in the spring and early summer, and veal returned to the tables of the wealthy in December, January, and February (Fig. 14.1). Lambs were also eaten at a young age from April until July; the tradition of the Easter lamb (together with kids, veal, and piglets) was practised in households such as that of Richard Mitford, Bishop of Salisbury, in 1407. The much less wealthy chantry priests of Bridport in Dorset marked Easter with lamb and veal in 1454.[9] In some households older lambs were consumed in the autumn and as late as December.

Patterns of pork consumption diverged between households. Some observed a tradition (which still existed in living memory) of avoiding fresh pig meat in the summer, presumably because it was regarded as risky to health. So the Mountfords in 1434, and the Le Stranges in 1349, did not serve fresh pork in the first case between June and November, and in the second between April and August.[10] Alice de Bryene's Suffolk household by contrast in 1413 increased its eating of both bacon and fresh pork between May and August (Fig. 14.1).[11] Most households consumed a great deal of the meat, along with others, at midwinter. Piglets were a festal food, reserved to celebrate Christmas and its aftermath, between December and February (Fig. 14.1). Bishop Richard Mitford, who seems to have had a partiality for piglets, had them served regularly through the year, but observed Christmas by increasing their number, culminating in a grand meal containing eighteen piglets (and much else) on 2 January 1407.[12]

[6] Woolgar (1992–3: ii. 474); Fegan (1914: 90–1); Fletcher and Upton (1996: 73).
[7] Woolgar (1992–3: i. 204–5).
[8] e.g. in the household of Sir Hugh Luttrell, 1405–6, Somerset RO, DD/L P37/7.
[9] Woolgar (1992–3: i. 360); Wood-Legh (1956: 4).
[10] Woolgar (1992–3: ii. 440–3); Norfolk RO, NH 8.
[11] Dale and Redstone (1931).
[12] Woolgar (1992–3: i. 317).

Fig. 14.1 The fluctuations in consumption of beef, mutton, bacon, pork, piglets, and veal in the household of Dame Alice de Bryene at Acton in Suffolk in 1412–13. In the case of salted meat the number of carcasses or joints consumed each month is indicated. For fresh meat, the graph is based on the amount of money spent on each variety. The calculation sometimes involves estimation when the accounts give a figure for the purchase of 'beef and pork' together. Fig. 10.2 gives the consumption of birds in this household.

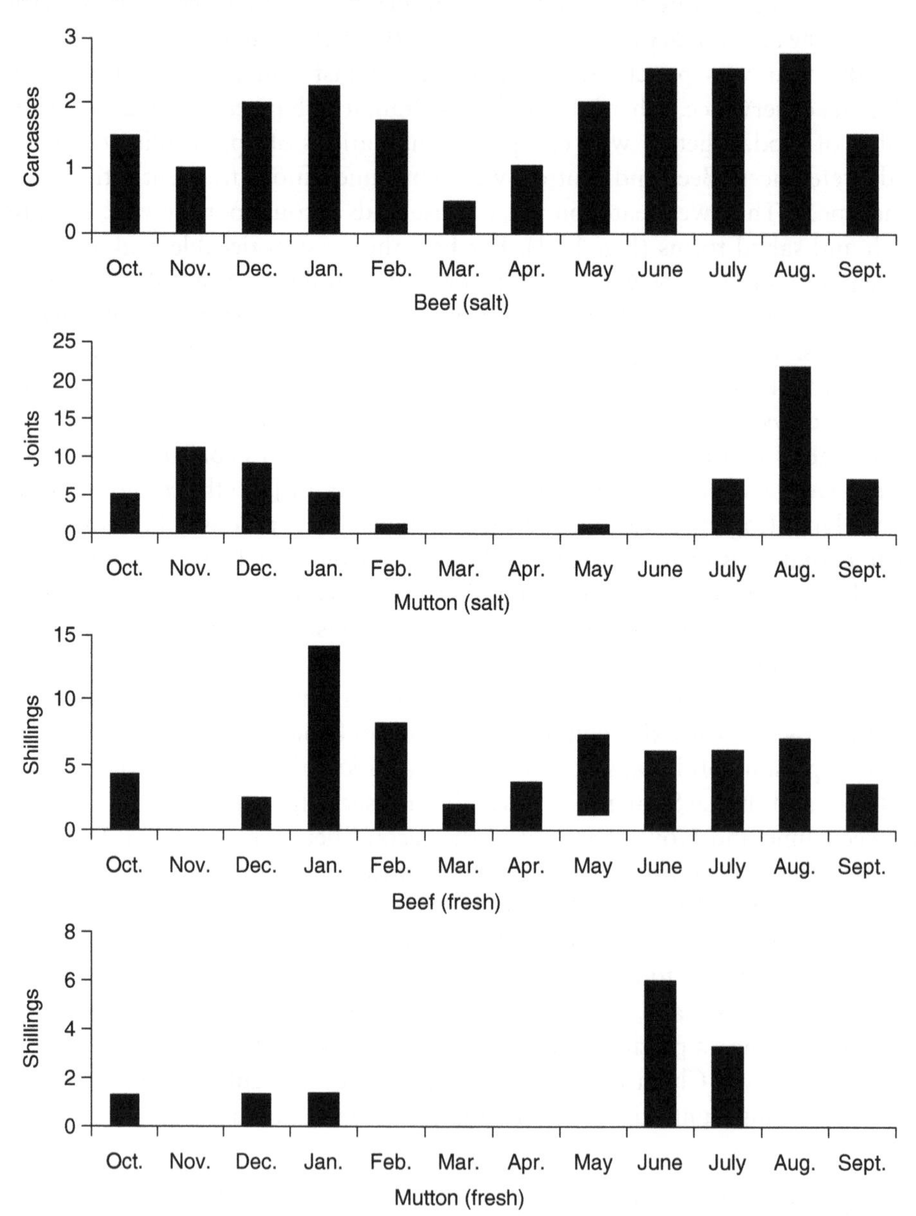

Source: Dale and Redstone (1931: 1–102).

Fig. 14.1 *Continued*

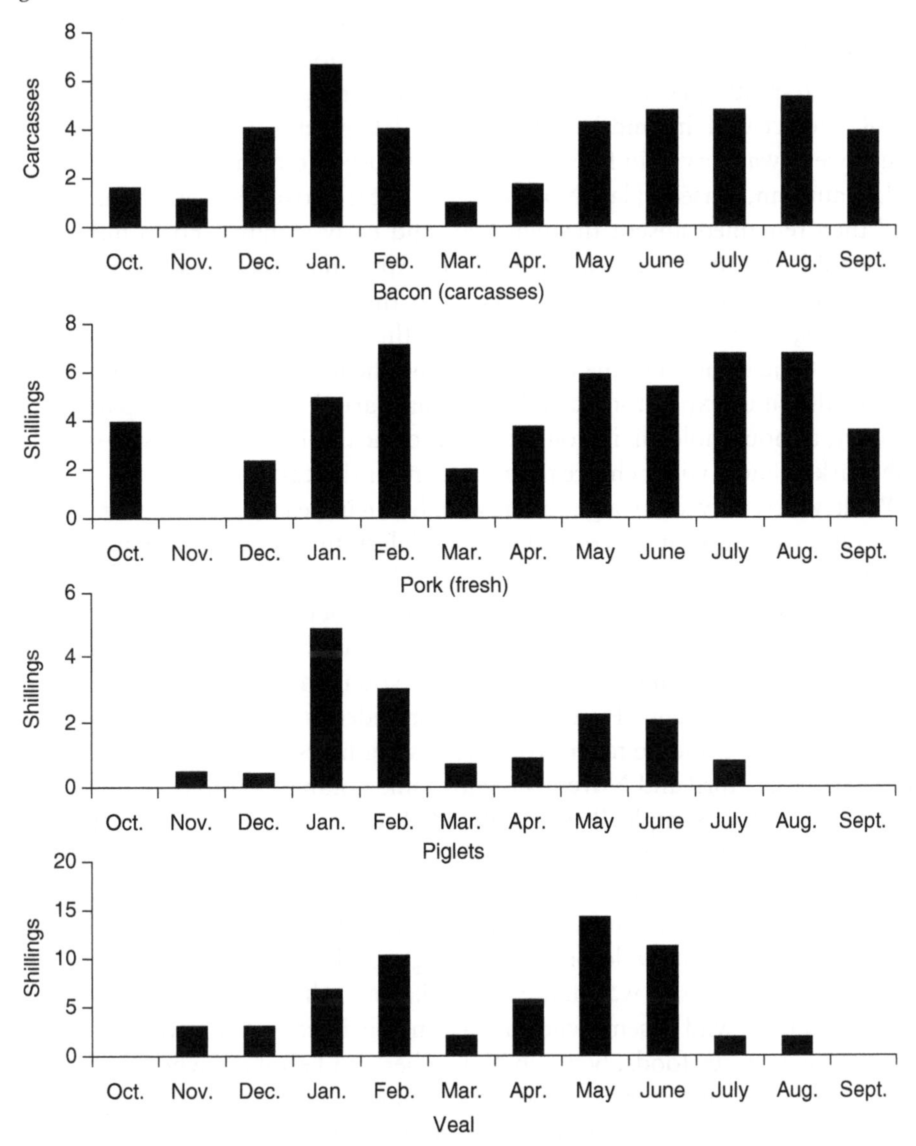

The decision to serve fish at particular times was influenced by the rules of the Church, but the varieties of fish available depended on the season. In Lent in the wealthiest households, the suppliers sought to enable an array of different dishes to be offered at each meal. The freshwater species, such as bream, pike, and tench, and the less common seafish, such as gurnard, rays, and turbot, were served. Sometimes the range of shellfish was increased, with the appearance of mussels or cockles. Throughout the rest of the year households depended on some basic staples among the preserved fish, notably red and white herring, dried cod, various salted white fish, and salt salmon. The fresh fish that were added to these to provide variety, especially for the high-status members of the household, depended to some extent on the migration of the fish and the fishing seasons. So sprats appeared on the table of the Le Strange family in mid-November 1348, and ceased in mid-February 1349.[13] Conger eel, a speciality of the south-west, was served in Bishop Mitford's household on almost every fish day in late autumn, ceased in late November, reappeared over the Christmas period, and then remained absent through Lent and early summer. For a number of households the mackerel season began in April and usually ended in June.[14] Some households were supplied with oysters throughout the winter and spring, but others, such as Mountford's, confined their consumption of this shellfish to Lent.[15] The location of the households and the limitations of the distribution system also must explain some of the seasonal variations in different parts of the country, as households living on the coast, or near a large town's well-supplied fish market, had a wider choice than those in rural locations inland.

Poultry were served throughout the year, but in increased quantities at times of celebration as the meat was regarded as of high status. Geese appear at all times, but in the summer younger, higher-quality birds were available, and often a number, fattened on the stubbles, were eaten at the end of the harvest season.[16] Doves or pigeons, contrary to the recently invented legend that they provided a substitute for fresh meat in midwinter, were eaten young in the summer and autumn (Fig. 10.2). The dovecots (Plate 14.1) were provided with internal ladders so that the young birds could be taken from the nesting holes, and they were served at meals between April and November, with some notable examples of dove feasts in May when, for example, Bishop Mitford's household disposed of ninety-six of them at once.[17] Demand for rabbits and their young, which must be regarded as domesticated animals in view of the careful management of their warrens, tended to reach its peak in midwinter and especially over the Christmas period.[18] Some game, mainly deer and wild birds, was caught and eaten in the summer, and at that time much venison was hunted in parks and salted for later consumption. Most meat from wild or semi-wild animals and birds was consumed, as in the case of other high-status foods, between November and February, when there was a second hunting season for deer, and active wildfowling.[19]

We would expect that dairy produce would be especially subject to seasonal fluctuations, as milk was abundant in the summer. Most hens' eggs were also laid

[13] Norfolk RO, NH8.

[14] Woolgar (2000: 42), and accounts cited above, including Richard Mitford and the Bridport chantry priests.

[15] Woolgar (2000: 43).

[16] e.g. 7 Sept. 1225 in Eleanor of Brittany's household: Woolgar (1992–3: i. 135).

[17] Woolgar (1992–3: i. 380–1).

[18] e.g. in the Alice de Bryene household.

[19] See Chapter 12. Birrell (1992: 122–3). A remarkable record of game consumption is given in the household of John de Vere, Earl of Oxford, in 1507 (Essex Record Office, TB 124); in January, as well as deer and rabbits, the household obtained blackbirds, cranes, curlews, herons, 'oxbirds', partridges, pheasants, plovers, redshank, swans, teal, woodcock, and smaller birds.

Plate 14.1 The fourteenth-century dovecot at Kinwarton in Warwickshire, built for Evesham Abbey. It contains 600 nesting holes. Inside the wooden device for supporting a rotating ladder survives. This gave ready access to the nests, enabling young birds to be taken for the table. Photograph: © William Muirhead and the National Trust.

in late spring and summer. As we have seen in Chapter 7, dairy products account for a relatively small proportion of expenditure in large households. Milk in liquid form was bought, partly for young children if the household contained them, but mainly as a cooking ingredient for such dishes as frumenty, and eggs were used in abundance in the kitchen. Fresh milk, butter, and eggs could be obtained throughout the year, even in December and January. The management of dairies depended on making cheese in the summer and autumn, which could then be sold and consumed through the months when the ewes were unproductive and the milk supply from cows and goats (though the last produced only a small proportion of the total) was greatly reduced. Cheese was eaten through the winter, but not in very large quantities, and fresh cheese was enjoyed as a delicacy in the milking season. Just as the monks of Westminster were served with cheese flans in early summer, secular households like William Mountford's and Hamon Le Strange's were eating cheese, presumably new, in April and into the months of May and June. Similarly the chantry priests of Bridport bought some cheese and butter in the winter, but indulged in a great deal of eggs, butter, milk, and cheese between Easter and midsummer.[20]

Preserved fruits, mainly dried dates, figs, currants, and raisins, together with nuts such as almonds, were often bought in preparation for the Christmas season, but also before Lent, as these luxuries apparently relieved the monotony of the constant succession of fish dishes. Fresh fruit and vegetables were eaten according to the season, but also in relation to the cycle of fast and feast. So most households bought increased quantities of vegetables, such as leeks and onions, and even cabbages, in Lent, as these evidently formed an appropriate accompaniment to fish, and no doubt formed part of the austere regime of the fast. Onions would have been kept in store since the autumn, and leeks can be grown through the winter, but producing quantities of green vegetables in February and March must have put some pressure on the gardeners. Apples and pears were consumed when they ripened at the end of the summer, but they were a traditional Christmas delicacy, and the varieties must have been chosen for their keeping qualities, as the great majority of purchases of these fruits were made in December and January, and they were even obtained in the early summer. Some fruits and vegetables had to be consumed immediately, and households like that of the Duchess of Buckingham in 1466 would buy and serve strawberries and peascods in June.[21]

Most wine was imported and drunk within fifteen months of the grape harvest, so the consumers were to some extent influenced by the rhythms of the agricultural year on the Continent, and the trade across the Channel and Bay of Biscay. The wine fleets arrived from Gascony in the late autumn, and households would obtain their barrels typically in November. A second period for imports came in the spring. Bishop Mitford's household drank wine at a remarkably steady rate of around

[20] Harvey (1993: 61–2); Wood-Legh (1956: pp. xxv–xxvi). [21] BL Add. MS 34213.

4 gallons per day, rising to more than double that figure on special occasions, such as All Saints (1 November); but from Christmas Eve until Epiphany (6 January) daily consumption only once fell below 8 gallons, and touched daily highs of 12, 16, 20, and 33 gallons. The Bishop of Hereford's household in 1289–90 drank more wine, perhaps because a substantial proportion came from the lord's own vineyard. Normal daily consumption moved between 4 gallons and 14 gallons, with lower daily allowances in October, November, and February, and rather higher levels of consumption between late March (including Lent) and June. On special days, such as Christmas and Easter, the daily total could exceed 40 gallons.[22]

Lower ranks of society

Although there is less evidence for the diet of the mass of the population, enough is known to show that they were even more strongly influenced by seasonal fluctuations. In some respects they followed similar practices to those of the upper class, but often their diet reflected scarcities imposed by nature and the limited resources of the consumers.

Lent was observed by almost everyone, and the rejection of dietary rules by the religious dissident minority, the Lollards, made them stand out from the orthodox norms. John Reve, a glover of Beccles, in 1430 is reported to have said that fasts, such as that at Lent, were not necessary, and that it was lawful for Christians to eat meat at any time: 'I have eaten flesh on Fridays and other days' (Lent, Ember Days, and vigils of saints).[23] The observance of fasts by wage earners is apparent when meals formed part of their pay. So a group of building workers, employed by the fraternity of Stratford-upon-Avon for eight weeks early in 1431, ate no meat until the last two weeks which fell after Easter.[24] The general lack of demand for meat in Lent meant that the shambles, the stalls of the butchers, closed down completely in towns such as Leicester, and the butchers presumably took a rest or found employment elsewhere.[25] The surge in demand for fish gave commercial justification for fairs held in the early spring, like the Ash Wednesday fair at Lichfield.[26] Such occasions are known to have attracted aristocratic consumers who bought in bulk, but the sale of fish can also be observed in small and local markets. At the market court held at Hingham in Norfolk on 8 March 1466, in the third week of Lent, the servant of a Norwich merchant was accused by the 'searchers of the fish market' of selling large numbers of rotten herring, 'unhealthy for the king's subjects'. The case reveals the existence of a busy trade in Lent in herring which, as long as the fish were wholesome, would not appear in the court records.[27] The millions of herring

[22] Woolgar (1992–3: i. 264–430); Webb (1854–5: i.).
[23] Tanner (1977: 111). See also Chapter 13.
[24] Shakespeare Birthplace Trust RO, BRT 1/3/40.
[25] Laughton and Dyer (2002: 164).
[26] *VCH Staffordshire*, 16, 118. For Exeter's successful Lenten Fair, see Kowaleski (2000: 447).
[27] Norfolk RO, MCR/B/26.

which were caught in the North Sea were sufficiently cheap to be bought by a wide range of consumers, including the peasants, artisans, and wage earners who acquired their food at such markets as Hingham.

The rise in meat consumption at the time of the major festivals, and especially in midwinter, had a general impact on the levels of trade, which must reflect changes in diet beyond the small minority of very wealthy households. At the Leicester butchers' shambles in 1377–8, the tolls reached high levels in October and November, and again at Easter when meat eating was resumed.[28] At Melton Mowbray in the fifteenth century market tolls show high cattle and pig sales in December. Many sheep were sold in the months of October, November, and December.[29] Home-killed meat would also have been most readily available early in the winter. As in later periods, the occasion of the killing of a pig or other animal, which often happened in November, would make fresh meat, and especially the perishable offal, temporarily available to households unable to afford meat as a regular part of their meals.[30]

More specific evidence for the celebration of Christmas throughout society comes from numerous incidental references. The brewers of Tutbury in Staffordshire in 1406 claimed that they were exempted from amercements (fines in cash) for offences against the assize of ale at Christmas, which suggests that they expected this to be a busy and profitable season.[31] Occasionally custumals show that lords of manors gave Christmas meals to their tenants, to which the peasants themselves were expected to contribute. At North Curry in Somerset in 1314 the meal included poultry, beef, and bacon, with plenty of ale.[32] These meals lapsed with the end of labour services in the fourteenth century, though some lords and ladies kept up the custom, presumably as a goodwill gesture rather than in fulfilment of a formal obligation. On 1 January 1413 Alice de Bryene at Acton in Suffolk gave a meal for '300 tenants and other strangers', in which beef, pork, and goose figured prominently.[33] The majority of ordinary people were not invited to a great house at Christmas, but still managed some indulgent eating and drinking. When a young heir to a peasant holding, John Ansty of West Monkton, in 1333–4 was guaranteed annual supplies of food and clothing by his guardian, as well as the usual allowance of grain, he was also promised a three-day feast (*gestum*) at Christmas.[34] Even paupers in an almshouse, like those at Sherborne in the early fifteenth century, were provided with a special allowance of bread, ale, and meat at Christmas, Easter, and Whitsun.[35] Community feasts were held at various times of year. Fraternities in both towns and villages would hold feasts to mark the days of the saints to which they were dedicated. For example, Corpus Christi guilds would hold their celebration at the end of

[28] Laughton and Dyer (2002: 176–7). [29] Laughton and Dyer (2002: 173–4).
[30] See also Chapter 7. [31] NA DL 30/110/1650. [32] Homans (1942: 358).
[33] Dale and Redstone (1931: 28).
[34] Longleat MS 6369. I owe the transcript of this document to Dr J. Z. Titow.
[35] Dorset RO, D204/A5-A10.

May or in June (the date varied as it was related to Easter). The churchwardens often held their church ales, at which food was served as well as drink, at Whitsun, again in the early summer.

The customs which influenced consumption in wealthy households were also observed more generally. Apples and pears were sold in towns in midwinter. For example in London on 13 January 1301 an apple seller (*costardius*) was crying his wares after curfew near Gracechurch Street, when two men took five apples, causing a disturbance which led to the violent death of one of the thieves.[36] Geese were fed in the fields after the corn had been cut and carried, and the feast of the 'reap-goose' was a high point among the meals served to the harvest workers and those who directed them, as well as the lords of the manors to whom the fields and their crops belonged.[37] A common manorial custom provided the haymakers, usually in June or July, with a sheep with which to prepare a meal at the conclusion of their work in the lord's meadows.[38] This small example of summer mutton eating coincides with the serving of that meat in wealthy households, and the increased trade in sheep in the early summer, recorded in the market tolls at Melton Mowbray.

Finally, peasant families held feasts at the various landmarks of their lives. We know most about funeral meals as they were recorded in the earliest, rare, probate accounts. In 1329 we are told of the funeral expenses of William Lene of Walsham-le-Willows. He was an exceptionally affluent peasant for his time, and the guests at his obsequies (which probably included a gathering a week after the burial as well as the funeral feast) were served with a variety of meats—beef, mutton, pork, goose, and poultry, with spices—at a cost of 47*s*. This was equivalent in value to thirty of his flock of a hundred sheep. In Yorkshire in the late fifteenth century similar peasant feasts are recorded costing as much as 40*s*. and 62*s*., and one included wine. A wealthy Warwickshire peasant, Thomas Chattock of Castle Bromwich, was able to serve venison, a gift from his lord, at a wedding celebration in 1497.[39]

We can see that the better-off townsmen and villagers could hold feasts, change from meat to fish at the fasting time, and observe various seasonal dietary customs, in parallel with the aristocracy. There was much poverty, however, and changes in the availability of food had a very strong influence on the less privileged sections of society. We have already noticed in this chapter and elsewhere that the aristocracy consumed dairy products on a limited scale. The peasants and poorer townspeople in contrast regarded 'white meat' as one of their principal sources of *companagium*, 'that which is eaten with bread', or as we would say, animal protein. In Blackbourne Hundred in 1283, 1,393 households contributed to the royal taxes, and the collectors assessed their contribution on the basis of their grain and animals.[40] Among the latter, 2,181 cows were

[36] Sharpe (1913: 14–15). [37] Dyer (1994*b*: 89). [38] Jones (1977: 98–107).

[39] Lock (1998: 135); Borthwick Institute of Historical Research, York, D. and C. wills, 1464, Jakson; 1468, Hall; 1482, Kyrkeby; 1494, Gaythird; Watkins (1993: 23).

[40] Powell (1910: pp. xxx–xxxi).

counted and 7, 213 ewes, and no doubt other milk-yielding animals were exempted, concealed, or missed. A minority of these animals belonged to the lords (who were liable to tax along with the peasants) and the produce went to their households. Some milk and cheese was sold in nearby towns such as Thetford and Bury St Edmunds, but most was consumed locally. The majority of tax-paying households owned one or two cows, and a few ewes, and we can assume that their diet was greatly improved in quality in the months between May and October when milk was in steady supply. Through the winter they would have eaten the cheese that had been made during the summer.

Peasants and townsmen were also regular consumers of the vegetables and fruit which appear intermittently in the records of the households of the wealthy. We should remember that the aristocracy's purchases came mainly from the surplus produce of peasant gardens. Vegetables and fruit would have been most readily available in the summer and autumn, with some harvesting of vegetables through the winter, apparent from the leeks which the aristocracy bought in Lent. Stored apples, onions, and cider helped to keep those with gardens with at least small supplies until the following summer. The less affluent consumers would be able to obtain the relatively cheap preserved fish, such as dried cod, salted white fish, and herrings. Unlike the lords they would not lay in large stocks, but would buy them as they could afford them through the year. We presume that everyone took advantage of the arrival of seasonally plentiful fish, like the mackerel in the early summer, or the mullet which appeared off the south Devon coast in February.[41] The opportunistic nature of eating among the relatively poor would mean that, unlike their social superiors, their seasonal diets varied with their location, with shellfish figuring much more prominently in the diets of coastal villages in Lent, and more wild fruits, nuts, and presumably small birds in the woodland communities, although archaeological finds (Chapter 9) provide little confirmation of the last.

The high point in the diet of most ordinary people was the grain harvest, a period almost unnoticed in most elite households, yet occupying a considerable span of time, from late July well into September. We know that the harvest workers employed on lords' demesnes ate well, with plenty of bread, and regular supplies of ale, meat, and fish, even in the period of relatively low living standards between 1280 and 1320.[42] These wage earners were attracted into employment by lords offering high remuneration in cash and food, and this suggests some competition with peasant employers offering comparable rewards. Urban workers were often tempted away from their normal tasks to work in the fields, so the whole of society enjoyed some weeks of good living when grain was plentiful and other types of food available. The landless labourers would be eating better than they did at any other time, and even the disabled and elderly of the village would not go hungry as they could glean ears of corn left after the harvest.[43]

[41] Fox (2001: 123). [42] Dyer (1994*b*). [43] Ault (1972: 27–32).

The annual variations in the quantities of cereals harvested must be regarded as the cause of the most important fluctuations in diet, as grain and pulses were the basic foodstuffs which provided the lesser ranks with as much as 80 per cent of their calories. The long-term changes in the frequency of good and bad harvests can be briefly summarized. Chronicles report seriously deficient harvests in the early eleventh century and at the end of the twelfth.[44] Using the more objective evidence of yields and prices we can identify a concentration of bad years in the period 1290–1320, including the Great Famine of 1315–17. Episodes of poor harvests persisted into the mid-fourteenth century, but after 1375 grain was relatively cheap, and after a few years of bad weather and poor yields in the late 1430s, good harvests (that is, viewed from the consumers' point of view, when basic foods were plentiful) predominated until *c.*1520.[45]

These harvest fluctuations were experienced variably through the year. The contrasts between seasons were experienced in a more extreme fashion in the bad harvest years. Even in a relatively normal year grain prices fluctuated: between October and December a large quantity of grain was sold as all producers needed money to pay off debts and to make purchases. The magnates were buying, for example, wine and spices; the peasants were replacing worn-out boots and shoes, and renewing agricultural implements. The peasants sold a higher proportion of their grain at this time than did the lords, because they tended to lack high-quality storage for unthreshed grain, and they were expected to pay much of their rent in cash at Michaelmas (29 September) and Christmas. Low prices reflected both the abundance of grain and this spate of selling. After Christmas prices tended to rise until midsummer. In the Bristol area in an average year in the late fourteenth century, for example, the July price of wheat could be 20 per cent higher than in the previous January.[46] Those with remaining stocks (usually the larger producers) were influenced by predictions of the future harvest. If it promised to be good, they would sell quickly, leading to a fall in price well before the new corn came onto the market. In a bad year, after Christmas the already high price would rise steeply through the spring and early summer as supplies became scarce; sales might be delayed if the next harvest promised to be another poor one. Cornmongers would influence prices, by building up stocks when prices were low just after the harvest, and releasing grain onto the market when they judged that they would obtain the best rewards.

If we examine how these fluctuations affected food consumption in the period when bad harvests were most frequent and severe, at the end of the thirteenth and in the early fourteenth century, the peasant producer with about 15 acres of arable, who normally harvested enough grain to feed his household and pay the rent, might in a bad year be running out of grain by May or June, and would be

[44] Britton (1937: 41, 73–7). [45] Dyer (1998*a*: 258–73). [46] Dyer (1980: 133–4).
[47] Farmer (1988: 753–4).

joining the ranks of the purchasers of grain. Even in the countryside a substantial minority of households, those with 8 acres of land, or even less, would normally buy grain throughout the spring and summer. To these should be added the town-dwellers, then accounting for near to a fifth of the population. There were thus many buyers and few sellers, which ensured high grain prices.

The impact of bad harvests on diet can be reconstructed in this way. Poorer households would maximize their use of seasonal foods, which they obtained from their own resources, such as garden and dairy produce, but they would be forced to curtail their purchases of inessential foods such as meat. Those able to produce cheese would sell it in order to buy grain. This supposition is supported by the tendency in years of high grain prices for the price of cheese to fall. In the same years pigs, animals kept solely for their meat, declined in price, as those rearing them sold their stock.[47] More of these relatively expensive foodstuffs were evidently coming on to the market, and few could afford to buy them. The poorer households would make the most efficient use of the grain that they had in store, or which they purchased at great expense. They would use inferior grains, such as barley, to make bread. They would boil grain in pottages, which involved no wastage of calories, and they would reduce the amount and quality of grain used for brewing. Their diet would as a result be reduced not just in quantity, but would also be imbalanced between cereal foods and those containing animal protein. The diet would also be considerably less pleasurable, adding to the general sense of deprivation. We can appreciate why during the 'hungry time', in May and June, the village poor expected to be able to pick green peas in the open fields, and why the more substantial villagers, while willing to concede this custom to help the poor, put strict restrictions on the practice—only the plants at the ends of the rows could be picked, and the peas should be for home consumption only, not for sale.[48]

From the late fourteenth century the trend towards better harvests meant not just that bread and ale were more easily affordable, but the cheapness of bread left cash to spare for other foods: most people could eat a higher proportion of meat and fish. They would appreciate this improved diet partly through their changing experiences of different seasons. Meat eating would be spread more evenly through the year, and early summer in most years would be regarded as a hungry time by only a small minority.

[48] Ault (1972: 38–40).

15

Monastic Pittances in the Middle Ages

B. F. HARVEY

Monastic diet and the place of pittances

St Benedict of Nursia (*c.*480–*c.*550), whose Rule enshrined the principles of monastic observance accepted in the Western Church throughout the Middle Ages, was at once precise and a little ambiguous in his treatment of food and drink.[1] At dinner, the daily meal taken at a time varying with the liturgical season, there were to be two cooked dishes, described as *pulmentaria*, and a third consisting of fresh fruit or vegetables when in season. This, St Benedict believed, would be a sufficient provision, although he did not explicitly exclude a third such dish. We should envisage the *pulmentum* from which the cooked dishes derived their name as pottage, consisting mainly of cereals and vegetables but flavoured or diversified from time to time with small quantities of eggs, cheese, and other foods, including fish. It was perhaps to obtain one or more of these other foods, which may have been brought separately to the table, that a monk might find it necessary to use 'an audible sign', as permitted in the Rule and there distinguished from speech. The two dishes of *pulmentum* were, it appears, to be treated as alternatives, two being provided so that a monk unable to eat the one might make his meal out of the other; and the fresh fruit and vegetables were also to be alternative dishes. In addition, each monk was to have a loaf of bread weighing a 'full pound', and a measure of wine which, however, it would be meritorious to forgo.

On days when supper was also eaten—that is, daily from Easter to mid-September, but after Pentecost with the exception of Wednesdays and Fridays—a third part of the loaf was to be saved for the occasion, but no other indication is given of the food at the second meal. Except when ill, no monk was to consume the flesh of quadrupeds, and whatever the meal, frugality was enjoined. Yet there might be additional measures of food and drink if circumstances so demanded, and the abbot so decided; and he might also invite monks to his table, where—as we should understand—they would partake of superior

[1] Fry (1980: caps. 39–41); Knowles (1963: 456–65).

dishes not provided at other tables, as guests did when they were present.[2] Dishes were to be brought to the tables by a rota of monks, each of whom served for a week at a time, and except for the voice of the monk chosen to read aloud from an edifying work for the duration of the meal, the latter normally proceeded in silence.

In the Middle Ages—a period extending for the purposes of this chapter from *c.*900 to *c.*1500—the dietary provisions of the Rule, together with the complicated liturgical cycle to which they are related, tended to distance monastic diet from secular diets outside the cloister at the higher levels of society to which monasteries as institutions, though by no means all their members on entry, now belonged. Cloister walls, however, were permeable, and nothing seeped in more easily than secular views on eating and drinking.[3] But how could the changes these seemed to recommend be accommodated in two dishes which were not even complementary but alternatives, and, moreover, vegetarian? When Peter the Venerable, Abbot of Cluny (1122–56), was challenged to justify the three or four *pulmentaria* which his monks now enjoyed at dinner, he pointed out in a spirited reply that St Benedict had not commanded that there should be only two.[4] The ambiguity of St Benedict's words in the passage in question and his emphasis on the abbot's duty to give his monks extra food and drink in special circumstances were the seeds from which grew a variety of occasions for eating and drinking, in addition to those envisaged in the Rule, and extra dishes for consumption at dinner and supper, the meals that are prescribed there.

This chapter is devoted to the extra dishes—including extra allowances of drink—at dinner and supper. On occasion, these constituted a so-called full service which, for the monks themselves, may have superseded every item normally provided at the meal in question: superior bread, wine or other beverage if appropriate, and cooked food were all provided.[5] If, as seems likely, the common dishes of an ordinary meal were also put on the table on these occasions, a full service helped the poor, too, for they received the leftovers at the end of the meal. More frequently, we should probably envisage an extra dish or two of cooked food, but nothing more. Until the late twelfth century, a dish of this kind was referred to in a number of ways, but thereafter it was normally a 'pittance' (*pietantia*), a term already much used, but previously without the settled meaning it now acquired.[6] Nevertheless, in the history of diet, the dish is more important than the name attached to it, and even in the period of settled vocabulary, it will sometimes be necessary for us to take notice of extra dishes known by other names.

[2] Fry (1980: cap. 56). [3] Cf. Montanari (1988: 65).

[4] Constable (1967: i. 52–101, especially 67).

[5] Slade and Lambrick (1990–1: i, nos. L188 and L340). Watzl (1978: 56, no. 5; 66, no. 17; 119, no. 103; etc.); Chadd (1999–2002: ii. 695). In a pioneering discussion of the sociological significance of monastic pittances, Dave Postles has defined the latter as, typically, extra meals, though small ones, and not as extra dishes at the normal meals; see Postles (2003: 179–80). This, however, is incompatible with the witness of monastic customaries and other monastic sources; see, e.g., Zimmermann (1971: sections I/49–59 *passim*, I/205); and for the thirteenth century, see e.g. below, 219.

[6] Cf. Zimmermann (1971: introduction, 47–8).

First, however, we must consider the two dishes of *pulmentum* prescribed in the Rule, for these became defining features of dinner in a medieval refectory, without which it could not claim to be 'regular'—that is, in accordance with the Rule. In the Middle Ages, these dishes were served one at a time, and very likely separated by one or more pittances. This arrangement is one of many indications we have that they were no longer alternatives but served to everyone present.[7] By a convention already old in the tenth century, their contents might be described as *pulmentum* however little cereal they contained, provided they were cooked. The dominant ingredient of the one dish was often beans and that of the other, vegetables of a different kind.[8] Frequently, however, they are referred to in our sources as 'common dishes' (*generalia*), a term signifying that the contents were identical for every monk present and the individual portion appropriate for a main dish. These conventions were evidently in use at Christ Church, Canterbury, *c.*1179, when Gerald of Wales, a guest at dinner on Trinity Sunday, noticed that after sixteen other dishes, if not more, no one had any appetite for the vegetables served at the end of the meal as a common dish.[9] In this chapter, the term 'common dishes' signifies the two dishes of *pulmentum* in the strict or loose sense of the word, as appropriate.

Pittances and the different Orders

When tracing the history of pittances, we must distinguish the practice of Benedictine and Cluniac monks on the one hand from that of the new monastic Orders founded in the twelfth century, among which that of the Cistercians was the most numerous, on the other. The Benedictines, who were known as black monks from the colour of their habits, and the Cluniacs followed the Rule of St Benedict according to the elaborately liturgical tradition which developed after St Benedict's own time; this was renewed in the tenth and eleventh centuries by reforms spreading from a number of different centres, but principally from Cluny itself.[10] Benedictines and Cluniacs accepted the propriety of pittances from an early date. Already in the tenth century, wealthy communities, which tended also to be large ones, not only consumed such dishes on the major feast days and anniversaries in their calendars, but were also provided with a third dish of *pulmentum* on many ferial, or ordinary days, if not in all cases normally on these.[11] The less

[7] Below, 224–5.

[8] Spannagel and Engelbert (1974: 255); Davril and Donnat (1984: 59–60); Migne (1853: cols. 703, 728). Cf. Maunde Thompson (1902–4: ii. 126), where vegetables are apparently distinguished from the *fercula*, by which we are to understand the dishes of *pulmentum*; and for a similar practice at Beaulieu Abbey, see below, 218 n. 16.

[9] Brewer, Dimock, and Warner (1861–91: i. 51; iv. 40–1). For *generalia*, see Zimmermann (1971: introduction, 48–9).

[10] Dubois (1974: 7–22) and Morris (1989: 63–8).

[11] Davril and Donnat (1984: 59–60). In tenth-century Cluny, the extra dish was explicitly permitted on four days in each week during Lent and when the Office of the Dead was celebrated: Dinter (1980: 57, 281). Its use in other seasons is likely.

wealthy and smaller houses have left all too few records, but here, too, though later, the practice has left its mark.[12] This dish, however, was *pulmentum* in the loose sense referred to above, and at this time the contents might actually be fish, cheese, or eggs, or, in the case of cheese and eggs, both together. It was referred to in various ways, according to the day of the week and the size of the individual portion. Thus at Cluny, in the late eleventh century, where any dish served to everyone present in an identical form and in the proportions of a main dish was apparently a 'common dish', the third dish was so described on Sundays, Tuesdays, Thursdays, and Saturdays, but as a pittance on Mondays, Wednesdays, and Fridays, each of which was a fast. On the fasts, individual portions were relatively small, and each monk might receive, for example, eggs or uncooked cheese but not—as on other days—eggs and cheese, the cheese being, moreover, cooked.[13] Whatever the name used, the third dish at dinner represents a quantum leap in the use of extra dishes and suggests that from a relatively early date these tended to become routine features of meals in Benedictine and Cluniac houses. In the early thirteenth century, in a work written, very likely, by Robert Grosseteste, murmuring against a prior or cellarer who withheld a pittance is treated as a sin that a monk in such a house might need to confess.[14]

The Cistercians, or white monks, were at first more literal than the older established Orders in their interpretation of the Rule and inclined to think that what St Benedict had not explicitly sanctioned he had forbidden. To the end of the twelfth century, the General Chapter of the Order clung to the belief that a pittance should never become a matter of routine and was for this reason hostile towards daily pittances—and to the General Chapter an extra dish of *pulmentum* at dinner was a pittance whatever the day of the week.[15] But this attitude could not actually satisfy appetites, and the third dish crept into Cistercian dinners, as did other pittances at the will of the abbot. In 1269–70, a monk's daily pittance at Beaulieu Abbey, a moderately wealthy Cistercian monastery, consisted of fish, and so apparently did the two common dishes served with it at dinner.[16] Even the Carthusians, the strictest among all the new Orders, followed the prevailing trend. From 1259, the monks of this Order were allowed a pittance, though of a very modest kind, on two of the three days in the week that had previously been days of abstinence from all food except bread and water.[17]

The purpose of pittances

Pittances served several purposes, but from the point of view of the monks themselves, principally two. First, they emphasized the distinction between ordinary

[12] See, e.g., Franklin (1988: no. 329), where the 'first' pittance on the feast of the Transfiguration, which is referred to by implication, was, almost certainly, the daily one.

[13] Herrgott (1726: 147); Migne (1853: cols. 728, 761). See also Scott (1981: 162); and for *generale* as the third dish of cooked food at dinner, Chadd (1999–2002: ii. 693).

[14] Goering and Mantello (1986: 152–3). The reference to a prior and not an abbot suggests that the work may have been addressed to a Cluniac monk.

[15] Canivez (1933–41: i. 66 (*s.a.* 1157), 83 (*s.a.* 1175), 178 (*s.a.* 1194), 186 (*s.a.* 1195)).

[16] Hockey (1975: 306, 308–12). Pottage is mentioned separately.

[17] Hogg (1980: 134).

days and feast days, which pervaded the calendar of every community following the Rule of St Benedict. The foundation of anniversaries, on which a benefactor's death was commemorated with a requiem mass or in some other way, might enlarge the number of special days or, if the anniversary in question coincided with a feast already in the calendar, strengthen an existing distinction. In monastic calendars most days in the year became liturgical feasts of one kind or another, but in the long term rising costs of living and declining real income tended to deprive lesser feasts of extra pittances and concentrate this form of observance on the major ones. But how many of these should we envisage? Major feasts were often graded in the calendars as feasts in copes—so-called because the monks wore copes at the principal mass of the day—and second only to these were feasts in albs, when albs were worn without copes. By the end of the thirteenth century a typical black monk or Cluniac monastery probably celebrated between forty and fifty such feasts in the course of a year, and a Cistercian monastery a number that was perhaps smaller, though still significant.[18] With the addition of vigils and the days immediately following the feast in question, which might also be marked by extra dishes, and of anniversaries too, the relevant number in each case might well be significantly larger. At the end of the thirteenth century, the abbot and obedientiaries of the abbey of Saint-Magloire, in Paris, were expected to provide pittances for the monks on forty-three days, including vigils and the days following some of the feasts that are mentioned; and, in addition, on twenty-three anniversaries the convent received cash payments from the abbot which it may have spent on pittances.[19] In the early fifteenth century, the Cistercians at Heiligenkreuz, in Lower Austria, were entitled to pittances on about 205 such days in the year, including anniversaries, and at least forty-five of these were already days for pittances by 1300.[20]

Yet the pittances associated with anniversaries were to some extent malleable features of the refectory life, since select ones might be served on a day other than the actual date of death of the benefactor who endowed it—a date perhaps chosen to suit the needs of the monks for an extra pittance on the day in question as well as the religious sensibilities of the benefactors consenting to it.[21]

Arrangements of this kind, and the very existence of daily pittances, direct our attention to the second purpose of the pittance system as it developed in the Middle Ages. This was to provide food, including, on occasion, drink, of a superior quality to that of the ordinary daily provision, and as far as possible to do so whatever the liturgical status of the days on which they were served. From an early date, it had been a common practice in monasteries to serve some food of this kind, and notably so-called 'spices', in the form of dried fruit, during the long Lenten fast, when monks might have only one meal a day; and where wine

[18] Wormald (1939–46: *passim*). See also Chadd (1999–2002: ii. 694, 703–4).

[19] Terroine and Fossier (1966–98: ii, nos. 348–9). On two other anniversaries the abbot provided a full service (*plena procuracio*). See also Terroine and Fossier (1966–98: ii, no. 346).

[20] Watzl (1978: 48–9 and *passim*). Daily pittances during Advent and Lent are excluded from these totals; see Watzl (1978: 80, no. 36, and 137, no. 132).

[21] Watzl (1978: 57, no. 6; 73, no. 26; and 114, no. 94); and see Postles (2003: 179).

was the normal drink, it might be of a superior quality during this season and during Advent.[22] But pittances were the means of multiplying the occasions on which such food and drink were served and greatly increasing the variety. If, for example, bread was included, white bread might be specified if the community did not normally enjoy the relatively fine wheaten bread to which this description applied, or the much finer kinds known respectively as wastel and simnel if it did; or the bread might be of the normal kind but freshly baked for the occasion.[23] Or was fish an item? In well-favoured houses, the fish would certainly be fresh, and might be of a relatively expensive kind. When the monks of Westminster consumed their so-called Oakham pittance, they were entitled to the best fish that could be purchased on the day in the neighbouring markets of the city of London, whence it could be brought the short distance up the Thames in the boat kept by the pittancer and the kitchener for this kind of purpose.[24] For others less well placed, fish in any form might remain the superior food that it had been for monks in general in the early Middle Ages; and in Lent, when preserved herring were the staple food in many refectories, any other kind of fish, fresh or preserved, was no doubt acceptable as a pittance. As for drink, on a day for pittances, monks who normally drank ale might look forward to mead or wine, and those who normally drank wine could hope for a superior kind, or perhaps for spiced wine.[25]

Pittances and eating meat

Flesh meat,[26] the food which St Benedict had forbidden to all except the sick, is a special case, since for many years, every kind was to healthy monks a superior food. The role of pittances in this case was to ease the forbidden food, in some form or other, into the mainstream of monastic diet; but we shall not realize how early this happened if we consider only the dishes called pittances at the time. The consumption of flesh meat on four days in the week outside the fast seasons, and in an appointed place other than the refectory, was finally authorized for black monks by Pope Benedict XII in 1336.[27] His proviso, that even on days when this dispensation was used, not less than half the total number of monks in the monastery in question should eat as usual in the refectory, enshrined rotas,

[22] For spices, see Threlfall-Holmes (2005: 59–64); Jaritz (1985: 58–9). For wine, see Zimmermann (1971: section I/171).

[23] West (1932: no. 190) (where the pittance is referred to as a *caritas*); Slade and Lambrick (1990–1: i, nos. L 188, L 340, and ii, no. C 28); Watzl (1978: 124, no. 111). The *pain de ville* purchased for the feast of the Assumption of the BVM at Saint-Magloire in 1294 was presumably bread purchased on the day of baking: Terroine and Fossier (1966–98: ii. 546).

[24] Maunde Thompson (1902–4: ii. 76–7). For the range of fish available and changing patterns of consumption over time, see Woolgar (2000: 40–1); and Chapter 8.

[25] Chadd (1999–2002: ii. 695); Riley (1867–9: i. 76); Watzl (1978: 132, no. 124). See also Jaritz (1980: 155–6); for claret, a spiced wine, Threlfall-Holmes (2005: 65).

[26] i.e. the flesh of quadrupeds. For the flesh of birds, which was probably consumed by monks throughout the period covered by this chapter, see Harvey (1997: 620–1).

[27] Wilkins (1737: ii. 609).

assigning monks to one room or the other, in arrangements for eating. But the consumption of flesh meat by healthy monks, in a designated room and on occasions often referred to as 'recreation', was an established practice in both black monk and Cluniac houses in the north of Europe, if not more widely, well before the end of the twelfth century. These are the 'refections' of flesh meat at 'fixed times' which Gregory IX prohibited in 1235, and his reference to 'fixed times' seems to show that the occasions were now becoming meals with some of the formality of regular meals taken in the refectory.[28] Moreover, the distinction, universally accepted in the age of Benedict XII and subsequently, between *carnis*, the muscle tissue of the animal, cooked for the first time, which might not be consumed in the refectory, and *carnea*, dishes containing other parts of the animal or pre-cooked meat, which were permissible there, was in use among black monks by 1200, and pittances of the latter kind are mentioned in the thirteenth century. At Abingdon Abbey, two abbots of this period in succession decreed that the pittance on their anniversaries—an occasion when all monks would have eaten together in the refectory—should include rissoles, or meat balls.[29] Since abbots made a point of endowing the best possible anniversaries, we can assume that rissoles represent a superior form of the permissible meaty dishes at this time.

Among Cistercians, the enduring influence of St Bernard perhaps made it harder than it might otherwise have been to establish meat eating as a normal practice. In the 1180s, the gossipy Walter Map, who was no friend to the Order, hinted at nothing worse than the consumption of pig's offal and fowl.[30] Yet by the 1220s and 1230s, when, on more than one occasion, the General Chapter of the Order found it necessary to reaffirm the provisions of the Rule on the subject of flesh meat, it was evidently proving difficult to hold the line; and by the early fourteenth century, the Chapter aimed no higher than to prevent a weekly indulgence of this kind.[31] For some communities Benedict XII's Statutes of 1335, confining the consumption of flesh meat by healthy monks to the abbot's chamber and the infirmary, may have seemed a little restrictive.[32]

The content of pittances and dietary change

To obtain quality, it might be necessary to sacrifice quantity, and in consequence the messes, or portions, in which pittances were served were quite frequently smaller than those used for the common dishes. Only the common pittance, the name often given to the third dish at dinner on days of the week other than the

[28] Auvray (1896–1955: ii. 323); and for the relaxed practice in Cluniac houses *c.*1200, see Charvin (1965–82: i. 45 (26–7)).

[29] Slade and Lambrick (1990–1: i, nos. L 188, L 340, and ii, C 312). For rissoles, see also Terroine and Fossier (1966–98: ii, no. 348); and for *carnis*/*carnes* and *carnea*, Harvey (1993: 40).

[30] James, Brooke, and Mynors (1983: 76).

[31] Canivez (1933–41: ii. 14 (7), 85 (8); and iii. 330 (2)). Cf. Schreiner (1982: 107–9).

[32] Canivez (1933–41: iii. 423–5).

three fasts, seems to be spoken of as 'large'. In the common pittance, however, as the name implies, the food was the same for everyone to whom it was served, whereas choice was often permitted when smaller pittances were served. Above all, even in periods when monasteries relied heavily on their own estates for staple foodstuffs, and especially for cereals, pittances brought the variety and sophistication of market products to monastic tables, for the best were purchased. Spices, for example, which were consumed as pittances not only during Lent but also on feast days, whatever the season, fresh marine fish, and good wine—we can assume that these items were normally purchased, whatever the location of the monastery. When, moreover, the sacrist of Bury St Edmunds provided a pittance including twenty-four pike, each 22 inches long, counting head and tail, as he was bound to do on the Feast of Relics, we can be confident that he relied on a fishmonger and not the abbey fishpond for his needs.[33] Much earlier, *c.*1080, the Customs of Cluny permitted the cellarer to buy fish for the third dish of cooked food at dinner on Sundays and Thursdays if he could do so at a fair price.[34] The growth of pittances reflects the concurrent growth of both internal and long-distance marketing in western Europe and reminds us of the early date at which these developments took off.

Engagement in the market enlarged the range of available comestibles, including drink, but also exposed monasteries to fluctuations and long-term changes in supply, for better or for worse, which they might be able to do little or nothing to control. These are sometimes conspicuous in the case of drink. On his visit to Christ Church, Canterbury, Gerald of Wales observed not only a very large number of dishes, but also a wide variety of drinks, and since the day was a major feast, we can assume that many of these were pittances. There was, he says, so much wine and cider, pyment (wine, with spices, sweetened with honey), claret, must, mead, and mulberry wine (*moretum*) that no one wanted the ale, although, as he pointed out, the ale of Kent was superlatively good.[35] A century later, it is unlikely that visitors to Christ Church saw mead on the table: indeed, as a festal drink it was already declining in importance when Gerald wrote. One reason for this decline may have been a long-term rise in the cost of honey, on which mead made very heavy demands; but in England another may have been its inability to compete at this level of consumption with Gascon wine, now imported in ever increasing quantities.[36] Much later, and in response to another change in the European wine trade, sweet malmsey wines found their way into the repertoire of monastic beverages in England and elsewhere. These had a high alcohol content, and the pittance consisting of 1 gallon of malmsey to

[33] Gransden (1973: 55); and, more generally, Dyer (1994*c*). See also Chapter 8.

[34] Herrgott (1726: 147); Migne (1853: col. 761).

[35] Brewer, Dimock, and Warner (1861–91: i. 51; iv. 41). For pyment, must, and mead, see Acton and Duncan (1966: *passim*); and for claret, above, 220, n. 25.

[36] Carus-Wilson (1954: 266–9); Knowles (1963: 465). On the decline of mead, see Jaritz (1976: 208–9).

3½ gallons of Gascon wine which the monks of Westminster, then a community of about forty monks, received on Quinquagesima Sunday 1535, may tell us the proportions in which such wines were commonly served.[37]

In the case of food, the quality of dishes sometimes fluctuated, but the strong undertow of development was the tendency of foods which were superior in one age to become ordinary in another. As extra dishes, which could be small and need not contain the identical foods for everyone present, pittances were important agents in such changes. They made it possible for the abbot and the cellarer—the latter being for a long time the actual provider of food and drink in a typical monastery—to introduce new foods without commitment for the future, but also, if circumstances proved favourable, to provide them as regular features of the diet. Inevitably, when this happened, the measure of superiority changed. Although, for example, eggs and cheese were commonly served as extra dishes at dinner and supper in medieval refectories, and in quantities far exceeding those we have to envisage when these foods were merely ingredients in the common dishes, it became the practice in black monk houses to treat many of the former dishes as *extra numerum* when the number of pittances deemed permissible on the day in question was counted. The very large quantities allowed by Abbot Aethelwold (954–63) to the community at Abingdon—a wey of cheese every ten days for about forty monks—suggests that this practice was already followed in this house.[38] This view, moreover, was apparently endorsed by the Cistercian General Chapter not later than the 1220s, conservative though this body was in dietary matters.[39] Flesh meat, as we have seen, entered the diet of black monks in the decades around 1200 as a pittance or the equivalent under another name. Yet by the closing decades of the thirteenth century, and despite the extreme reticence with which monks mentioned the practice, we begin to glimpse a structure of three dishes at dinner, all of meat, for those on the day's rota for this food, only one of which is described as a pittance.[40] Major shifts of this kind in the normal provision of food and drink, and in expectation on the part of monks as consumers, do much to explain a widespread tendency to define an acceptable pittance with increasing particularity—not, for example, as fish, but as freshwater fish or the best available; not as bread or even white bread, but as wastel or simnel; not as wine, but as good wine.[41]

The structure of the meal and pittances

How, then, did the proliferation of extra dishes affect the apparently simple procedure at meals prescribed by St Benedict for his monks? Monastic customaries,

[37] Westminster Abbey Muniments 33344. I am greatly indebted to the Dean and Chapter of Westminster for permission to use their muniments. See also Threlfall-Holmes (2005: 65–6).

[38] i.e. 10 or 12 oz daily per monk, depending on the weight of the wey. See Stevenson (1858: ii. 279); and on the date of this source, see Hudson (2002–: ii, pp. xxii–xxiii). On cheese and eggs, see also Dinter (1980: 48); Maunde Thompson (1902–4: ii. 102); and Zimmermann (1971: introduction, 48).

[39] Canivez (1933–41: ii. 3 (12) (*s.a.* 1221)).

[40] Riley (1867–9: i. 450); and see above, 220–1.

[41] Above, 220; Jaritz (1985: 58).

our principal source at this point, are concerned with norms of behaviour which, more often than not, may have been modified in practice, and some have a reforming purpose tending to compromise them as guides to the status quo. Very tentatively, however, we can identify the significant features of dinner, the main meal of the day, in a medieval refectory from the twelfth century onwards.[42]

Formality is most easily achieved where numbers are large, and inevitably, in the present context, it suffered from the decline in commensality from the thirteenth century onwards. Yet all monastic meals and even the various forms of light refreshment which now existed were to some extent formal occasions if properly conducted. Many of the formalities mentioned at dinner, including a grace or benediction not only at the beginning and ending of the meal but also, in some periods and monasteries, when each main dish was brought in, had no parallel in secular households. But every ritual was well known to virtually everyone present and lent itself to a crisp performance if this was desired. Moreover, daylight was precious to those intending to read in the cloister after dinner, and seasonally, if not, in north Europe, throughout the year, the siesta was to be fitted in first. Monks, in fact, rarely needed, and could not have spared, the two hours and more that aristocratic households may have spent over the main meal of the day in the later Middle Ages;[43] nor was supper, with the evening collation (a devotional reading, often in the chapter-house) and compline still to come, a meal that could be extended without regard to time. It is in this context of meals that were neither leisurely nor hurried that we should envisage the treatment of the common dishes and the pittances.

From an early date, the rota for serving meals was dominated by junior monks, but they were now assisted by servants. Dishes were served in sequence, each mess or portion being normally for one or two monks. As in a secular household at comparable levels of society, the full sequence was not always served to everyone present—monks who were not priests, for example, might be denied some of the pittances—but the pace of the meal was that of the high table, at which everyone was probably entitled to every dish or invited to partake of his own share by the president. Each dish was normally served first at this table, but other tables were not necessarily served in order of seniority.

Despite the emphasis laid on individual dishes, and encouraged by the pittance system itself, a tripartite structure at dinner, prefiguring a later and clearer structure of three courses, is apparent from an early date in black monk and Cluniac houses, though never as clearly among the Cistercians. Some early sources place the daily pittance between the two common dishes which descended from the *pulmentaria* of the Rule—a position no doubt explained by its superior quality. This arrangement is hard to trace after the mid-thirteenth century, but may well

[42] The following account draws principally on Spätling and Dinter (1985: 54–60); Gransden (1963: 100–3); Maunde Thompson (1902–4: ii. 98–132); Sullivan (1975: 131–2, 149–50); Leccisotti and Walker Bynum (1975: 216, 221–2); Angerer (1985: 93–8); Becker (1968: 197–204); Angerer (1987: 61–70).

[43] Woolgar (1999: 88).

have persisted.[44] Other pittances, if any, were served at points in the meal varying from monastery to monastery, but within the capacity of the servers we should probably allow for a certain clustering around one or other of the common dishes or the daily pittance, the dishes which gave the meal its tripartite structure. Some of the smaller pittances, no doubt, provided the *interfercula* (*entremets*) mentioned at Abingdon Abbey and at St Swithun's Priory, Winchester.[45] In relatively wealthy monasteries, where, on most days in the year, it may have been the rule rather than the exception to have four or five extra dishes of one kind or another, it was probably easy to make do without consuming one's share of either of the inferior common dishes.[46]

Pittances, secular society, and change

The pittance system represents the monastic response to a long-term rise in standards of living outside the cloister, and as secular standards of living seeped into the cloister, so did the monastic response find imitators outside. Here, in due course, secular institutions came into existence with a regime of common dishes as the basic provision at meals which is comparable to the monastic regime, and a desire for something better on appropriate days which was sometimes expressed in the monastic way: the superior food was a pittance. In the late fifteenth century, for example, the fellows of Merton College, Oxford, enjoyed so-called pittances of superior food on major and minor feast days in their calendar.[47] In monasteries, the focus of this chapter, many dishes of this kind retained their identity as superior food to the end of the Middle Ages. After the mid-fourteenth century, however, and even in Orders which had admitted them reluctantly, they were no longer remarkable and rarely contentious. The number of new endowments for this purpose fell—in some cases concurrently with a flight of intercessory masses from monasteries to other institutions which was already under way in the early fourteenth century, and in some cases, it appears, independently of such a trend.[48] Moreover, pittances, which had long exemplified the superiority of market products over home-grown foodstuffs, were now to some extent monetized in a new sense, as benefactors, clearly prompted by the monks themselves, directed that the money available for this purpose should be distributed in whole or in part in cash. When this happened, pittance money became part of the *peculium*, or quasi-private property of the individual monk.[49] In their traditional

[44] Davril and Donnat (1984: 59–60); Migne (1853: col. 728); Maunde Thompson (1902–4: ii. 126), where the 'large pittance' (the third dish of cooked food at dinner) is described as *medium ferculum*. Cf. Canivez (1933–41: i. 17 (19) (*s.a.* 1134)).

[45] Slade and Lambrick (1990–1: i, no. L 550); Kitchin (1892: 307–8, 330–3).

[46] Cf. above, 217.

[47] Fletcher and Upton (1996: 357, 361, 365, 371, etc.).

[48] For the background, see Colvin (2000: 163–73); and for changing patterns in lay gifts to Austrian monasteries, Jaritz (1990: 29–33).

[49] e.g. B. F. Harvey (1965: 238n.); Wilson and Gordon (1908: 59, 68); Lunardon and Spinelli (1977: 272, no. xxviii). Cf. Charvin (1965–82: i. 111 (44)).

form, however, these dishes, and in particular those containing flesh meat and fish, did indeed narrow the distance placed by the Rule between monastic diets and secular diets at the higher levels of society and they gave the former a variety which they would not otherwise have possessed.

Yet the system also enhanced the differences in standards of living which inevitably existed within the monastic Order itself from an early date. These arose from different views on the permissible limits of change within the constraints imposed by the Rule, but also, and perhaps principally, from the existence of different levels of endowment, and different local and regional economies. We must remember at this point that nunneries in general attracted fewer endowments than other monasteries. Many pittances were individually endowed, and in the period extending from *c.*1100 to *c.*1350 this seems to have occurred on a far larger scale than in earlier or later periods. In this period, a new momentum in belief in Purgatory brought about a great increase in the number of men and women outside the cloister who sought the prayers of monks to ensure the safe passage of their souls through Purgatory and were persuaded to endow pittances as part of the bargain; and many of those concerned were of relatively humble status.[50] The actual number of new pittances was no doubt much larger than the number known to us today. Yet even in this period it seems likely that many dishes of this kind were a charge on the ordinary income of the monastery in question; and we can be confident that this was always true of the daily pittance, the third dish of cooked food at dinner. In these circumstances, some monasteries were able to embrace the new dietary opportunities introduced by the pittance system and others less able. Consumption of flesh meat, fish, and spices may briefly illustrate this point.

In the later Middle Ages, except in monasteries influenced by the observant movements of this period, it was considered permissible for Benedictines and Cistercians to eat meat on four days in a normal week outside the fast seasons of Advent and Lent, and the rules for Cluniacs in this respect, though less clear-cut, were no less tolerant.[51] Fish might be eaten not only on the remaining three but also daily in the fast seasons. Whether this relatively expensive food could be afforded so frequently, and, if so, what varieties could be afforded, depended on the resources of the individual monastery, and similarly in the case of meat. A rental recording arrangements at the Cluniac priory of St Denys at Nogent-le-Rotrou in 1317–18 seems to show that the monks here, who were in financial difficulties, had meat on three days in each week outside the fast seasons, bought large quantities of herrings during Lent, but at other times of the year bought fish only for principal feasts and Saturdays: if at these other times they ate fish more

[50] For the rise of Purgatory, see Southern (1982), a review of J. Le Goff, *La Naissance du Purgatoire* (Paris: Éditions Gallimard, 1981; English trans., *The Birth of Purgatory*, London: Scolar Press, 1984); and for urban donors of pittances in England in this period, Postles (2003: 183–5).

[51] Above, 220–1; Charvin (1965–82: i. 45 (27), 89 (108), and 153 (48)); and for a proposed reform at Cluny itself, Vaquier (1923: 168–70, 173).

frequently than this implies, it came presumably from their own pond or ponds.[52] Such precise evidence, however, is scarce. Some monasteries—and these included Durham Cathedral Priory—could afford to buy a wide range of exotic spices. Others, in common with the nuns of Marrick Priory, might have to confine their purchases to one or two, and at Marrick in 1414–15, these were pepper and saffron.[53] Pittances even lent a competitive edge to monastic standards of living, and did so in unlikely places. In 1423 and again in 1432, the General Chapter of the Carthusian Order forbade priors of the Order to use large pittances and good wine in order to attract recruits from other charterhouses to their own.[54] In their pursuit of higher standards of living with the aid of pittances, monastic households at first anticipated and later to some extent matched the differentiation in food and drink which became conspicuous in the households of the secular aristocracy in the later Middle Ages and one of the supports of the social hierarchy of that period.[55]

[52] Charvin (1965–82: ii. 448–50).

[53] Threlfall-Holmes (2005: 93–102); Tillotson (1989: 16, 31). Cf. on saffron, Dyer (1998*a*: 63).

[54] Hogg (1980: 135).

[55] For secular practice, see Dyer (1998*a*: 55–70); Woolgar (1999: 111–65; 1995: 19–21); Vale (2001: 20–1, 203–5).

16

Diet in Medieval England: The Evidence from Stable Isotopes

G. MÜLDNER AND M. P. RICHARDS

This chapter introduces a relatively novel method of exploring medieval food-ways, the direct measure of past human diets through stable isotope analysis of bone collagen. This analysis reflects long-term dietary averages and is used to gain information on broad categories of foods consumed by humans in the past. In medieval historical and archaeological studies there is a tendency to focus on the specifics of diet. Approaches such as isotope analysis, which provide evidence of general dietary patterns, have therefore only rarely been applied. By contrast, isotopic methods are widely used in prehistoric archaeology, as this field does not have the benefit of documentary sources.

The method is briefly outlined here and then the chapter presents results of the isotope analyses for a number of English medieval sites. It aims to demonstrate that in combination with the rich textual evidence from this period, isotope analysis can tell us new and unique information about diet and its role in medieval society.

The principles of stable isotope analysis for reconstructing diet

Stable isotope analysis as used for dietary reconstruction refers to the measurement of the ratios of the two stable isotopes of carbon (^{13}C and ^{12}C) and nitrogen (^{15}N and ^{14}N) in proteins extracted from archaeological bone samples. The ratio of these isotopes differs between broad categories of foods, such as those from the sea and those from the land. When these foods are consumed by humans, certain dietary components, for instance amino acids from the proteins, are used in the building or repair of various body tissues, including bone. Fortunately, when they are incorporated into the body, the original isotopic signature of the food is preserved, albeit somewhat altered, in the newly formed tissue, for example in the major bone protein (collagen). If bone collagen is extracted from a human bone found at an archaeological site and its isotopic signature measured, we can

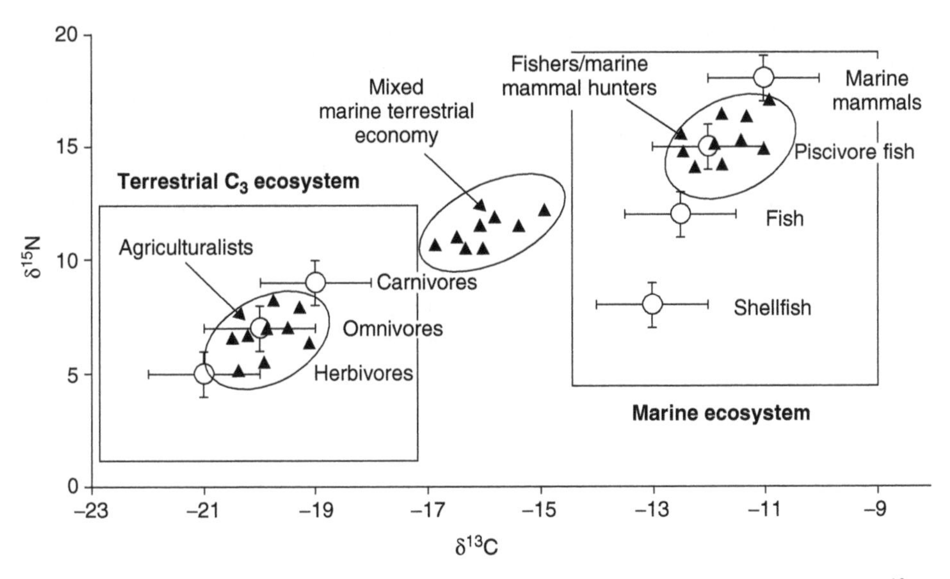

Fig. 16.1 Typical stable isotope scatterplot, showing the carbon stable isotope ratios ($\delta^{13}C$) on the x-axis, and the nitrogen stable isotope ratios ($\delta^{15}N$) on the y-axis. The open round symbols designate typical collagen stable isotope values for fauna in Holocene western Europe (with error bars $\pm$ 1 standard deviation). Note the differences in the carbon isotope ratio between the terrestrial and marine ecosystems as well as the differences in nitrogen between different trophic levels. The solid triangular symbols stand for typical human isotope values. Their mode of subsistence and the proportion of plant versus animal protein they consumed are inferred from where the human values plot in relation to the faunal background data

infer what types of foods were used to form it, and therefore what foods the human had eaten during this time.

The formation and repair of bone is a very slow process, and the signature measured in archaeological samples is the product of many instances of protein making its way into the bone from the meals consumed. The isotopic signature observed in bone collagen is therefore an average of the protein eaten by that human (or animal) over between ten and thirty years before the individual's death.

Bone collagen isotope ratios only tell us about the sources of dietary protein and can characterize diet in no more than very general terms. Currently, they can only be used to distinguish between the consumption of aquatic (marine and freshwater) and terrestrial foods, to detect the inclusion of arid C_4 plants (e.g. maize and millet) in the diet, and to estimate the proportion of plant and animal proteins consumed (Fig. 16.1). They cannot, for example, detect the occasional consumption of fish in a diet that is predominantly land based. It is also not possible to distinguish between different cuts of meat or even between the consumption of meat or milk from the same animal. Also, as bone is renewed only very slowly, a sudden dietary change is expressed as a mix between the old and the new diet, until the collagen has turned over completely and the 'signature' of the new diet is fully incorporated in the bone.[1]

Despite these limitations, and as we will demonstrate below, stable isotope analysis is a very powerful tool for reconstructing past diets. Like many methods, it is most effective when used in conjunction with other types of evidence, such as plant and animal remains or, if available, historical documents. Over these more traditional sources of information, isotopic data has the considerable advantage that it allows us to target specific sites, groups, and even individuals, and provides *direct* evidence about their long-term diet. Isotopic results are directly comparable in diachronic investigations and are mostly unaffected by biases arising from changes in deposition rites, differential preservation, or the availability and character of historical texts.

First applications to medieval archaeology

Despite its potential to address questions concerning diet in the Middle Ages, stable isotope analysis has only rarely been used by medieval archaeologists. With the exception of a small-scale study on medieval Besançon,[2] Mays was the first to apply the method to problems surrounding medieval diet.[3] In order to assess the importance of marine foods in medieval subsistence, he analysed carbon stable isotope ratios for sixty-seven individuals from five lay and monastic sites in north-east England, including the city of York and the deserted medieval village of Wharram Percy. His results suggest that terrestrial foods formed the bulk of the diet for all groups he investigated, but that there were at least smaller contributions of marine protein at some of the monastic and coastal sites.

The rural settlement of Wharram Percy also provided the samples for the second application of stable isotope analysis to medieval England, a study designed to compare historical information on infant feeding practices with scientific data on the age of weaning.[4] By analysing bone collagen of children of different ages of death and inferring from the isotopic data whether they had been weaned or not, it is possible to reconstruct weaning practices and estimate the cultural age of weaning in a population.[5] For Wharram Percy, the isotopic study shows that the children were fully weaned by two years of age. This is in keeping with recommendations by medieval medical writers and demonstrates that their advice—or a parallel practice—was followed even in the remote English countryside.[6]

Isotope data for the adult population of Wharram Percy, which was obtained in the same study, indicates a rather uniform diet of C_3 plants (probably cereals) and animal (meat or dairy) protein. Although the excavation of the settlement

[1] Schwarcz and Schoeninger (1991); Ambrose (1993); Katzenberg (2000).
[2] Bocherens, Fizet, Mariotti, Olive, Bellon, and Billiou (1991). [3] Mays (1997).
[4] Richards, Mays, and Fuller (2002). [5] Herring, Saunders, and Katzenberg (1998: 430–7).
[6] Richards, Mays, and Fuller (2002: 209); Mays, Richards, and Fuller (2002: 655–6).

had produced the remains of some marine fish and molluscs, there was no measurable contribution by marine foods to the isotopic signal, suggesting that they formed no regular or only a very minor part of peasant diet.[7] Isotopically, overall subsistence at Wharram Percy appears very similar to that of earlier, prehistoric populations in Britain.

Complete dietary change in medieval England?

Following these case studies, there was a need for a more comprehensive approach to isotope analysis and medieval diet. In order to explore whether the dietary variation indicated in historical sources is reflected in the isotopic record, fifty humans from three later medieval sites in northern England were chosen to represent three different social groups. This pilot study comprised individuals from the low-status rural hospital of St Giles by Brompton Bridge near Richmond, friars and wealthy townspeople from inside the church of the Augustinian friary in Warrington, and a group of men who lost their lives in the Battle of Towton in 1461.[8]

Contrary to expectations, the isotopic differences between the three sites were almost negligible. They indicate that at least in terms of the relative consumption of plant and animal protein from different sources, the diet of the three groups was very similar.[9] Nevertheless, the observed values are significantly different not only from those obtained from Wharram Percy, but also from the isotopic data available for other historical as well as prehistoric populations in Britain (Fig. 16.2).[10] The data from the three sites were subsequently found to be quite typical for later medieval populations in England and also have parallels on the Continent.[11] These results are important as they suggest that the diet of large parts of the medieval population was completely different from that of preceding periods.

Explaining medieval diet

The medieval isotope data are very unusual and explaining the signal in terms of the underlying diet is not straightforward. The observed nitrogen isotope ratios are uncommonly high, and, as such, would normally be associated with the consumption of marine foods. The associated carbon values, however, suggest a diet based almost entirely on terrestrial resources with only minor contributions

[7] Ryder (1974: 44).

[8] Cardwell, Chundun, Costley, Hall, Hanson, Huntley, Jamfrey, Maxwell, Speed, Simpson, and Stallibrass (1995); Lupton (2000); Fiorato, Boylston, and Knüsel (2000).

[9] Müldner and Richards (2005).

[10] Richards, Mays, and Fuller (2002); Privat, O'Connell, and Richards (2002); Richards, Hedges, Molleson, and Vogel (1998); Richards (2000).

[11] Müldner and Richards, unpublished data; Bayliss, Shepherd Popescu, Beavan-Athfield, Bronk Ramsey, Cook, and Locker (2004); Polet and Katzenberg (2003).

Fig. 16.2 Human stable isotope data for three later medieval sites, St Giles by Brompton Bridge, Warrington, and Towton, in comparison with Wharram Percy and other human isotope data from Britain. All values are means with the error bars indicating ± 1 standard deviation. Note that the $\delta^{15}N$ ratios for the three later medieval sites are significantly different from those of the other populations

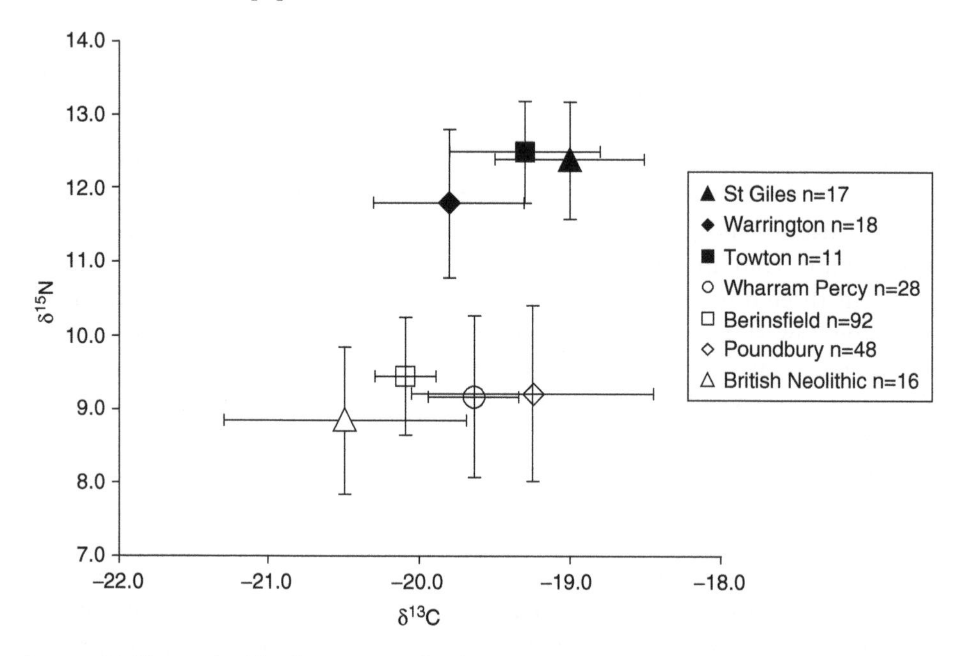

Sources: Müldner and Richards (2005); Richards, Mays, and Fuller (2002); Privat, O'Connell, and Richards (2002); Richards, Hedges, Molleson, and Vogel (1998); Richards (2000).

from marine protein. Therefore, two main possibilities need to be considered: the consumption of omnivore protein, such as that from domestic pigs, or else the inclusion of freshwater fish in diet.

The first scenario is that the unusual isotopic values are indicative of a terrestrial diet with a high proportion of animal, and in particular omnivore, protein. When omnivorous animals, such as pigs or birds, are fed mostly on animal products, their tissue $\delta^{15}N$ ratios are elevated close to those of carnivores. Due to the trophic level shift between diet and consumer, humans eating omnivore flesh gain even more enriched nitrogen values, conceivably in the region of those observed in the medieval samples. References to pigs roaming the streets of medieval towns and feeding on scraps and waste are well known from historical and iconographic sources.[12] Isotopic studies can corroborate this evidence by showing that a significant proportion of pigs from medieval York were omnivorous.[13] Although changes in pig-feeding practices in the historical period could account to some extent for the raised human nitrogen values, it seems

[12] Rixson (2000: 115); see Chapter 6. [13] Müldner, unpublished data.

nevertheless unlikely that pork consumption alone is responsible for the substantial enrichment observed in the later medieval samples. The zooarchaeological evidence suggests that pigs, compared with cattle, played only a secondary role in the urban meat supply.[14] The unusual isotope data have also been observed not only in town populations but also in rural contexts, such as at the hospital of St Giles, where most pigs were demonstrably kept on a plant-based diet.[15]

The alternative explanation for the medieval isotope data is a contribution from aquatic foods, in particular freshwater fish, to the diet. The isotopic signatures of both lacustrine and riverine fish are extremely variable, but in general the carbon ratios tend to be within the terrestrial range, while the nitrogen values are rather high, similar to marine fish.[16] When stable isotope results for freshwater and migratory fish from the Dominican Priory in Beverley are plotted with the human data and other contemporaneous fauna from northern England, it is apparent that a mixed diet of terrestrial, freshwater, and marine foods offers a convincing explanation for the medieval human isotope values (Fig. 16.3). Eels in particular, which are known to have been widely available even to the lower classes, match the data well.[17]

Fig. 16.3 Later medieval humans from St Giles by Brompton Bridge, Warrington, and Towton in comparison with contemporaneous fauna from northern England. Note how the human $\delta^{15}N$ ratios are far more enriched than those of terrestrial carnivores, but how both carbon and nitrogen values plot between the ranges of terrestrial animals, freshwater, and marine fish

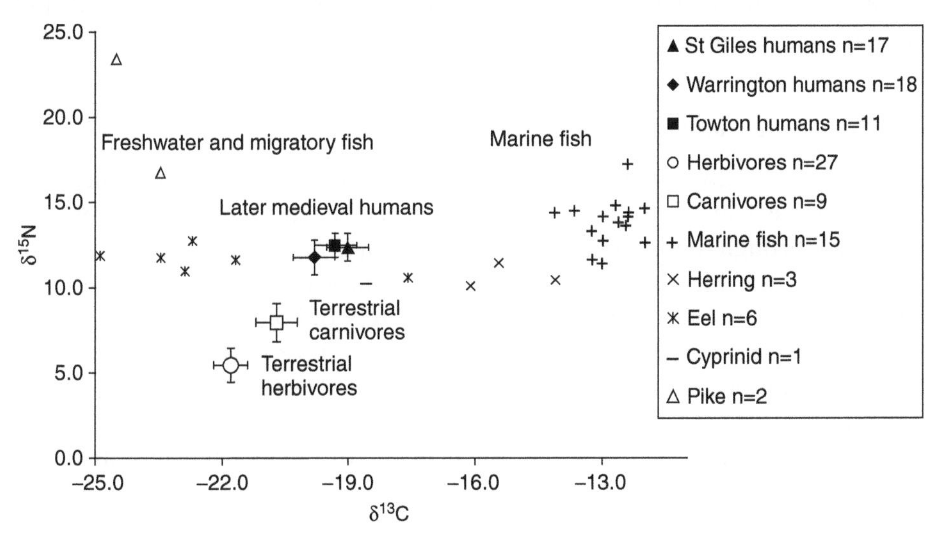

Source: Müldner and Richards (2005).

[14] See Chapter 6. [15] Müldner and Richards (2005).
[16] Dufour, Bocherens, and Mariotti (1999: 622–5). [17] Woolgar (2000: 39); see Chapter 8.

Diachronic trends?

If one accepts a combination of freshwater fish and marine fish as an explanation for the unusual isotope values, then it would appear from the data that aquatic protein made up a significant proportion of the overall diet. This must almost certainly be seen in the context of the dietary rules enjoined by the Church, which proscribed the consumption of meat on nearly half the days of the year. The fact that the medieval isotope signal is so very different from the data for earlier populations may therefore illustrate the impact of increased fish consumption.

There are still too few data to assess whether these findings indicate a genuine diachronic trend or if they are merely an artefact of how little isotope work has been carried out on historical populations in Britain. So far, the very enriched nitrogen values which make the medieval isotope data so remarkable have only been observed from the eleventh century onwards, a date which coincides with the point at which herring and cod began to be traded and eaten widely, as outlined in Chapter 8. While there was some consumption of aquatic foods in the post-Roman and early medieval periods, stable isotope and fish bone data suggest it was much less.[18] A similar situation could be suggested by the limited isotopic evidence for continental Europe.[19] Documentary sources indicate a decline in fish consumption in the fifteenth century and especially during the Reformation, when the rules about fasting lost most of their significance, although some major elements of fish consumption were sustained.[20] The isotopic results do not reveal this trend so far, although the chronology for most of the sites does not allow for fine distinctions within the later medieval period, and there are hardly any data for post-medieval England.[21]

Social differences and group diets

Stable isotope analysis only allows a very general characterization of diet, and can recognize few of the many different ways in which social difference is expressed through food.[22] When significant isotopic differences between groups in a society are identified, the underlying dietary differences must be very substantial. In the case of the late Roman cemetery of Poundbury in Dorset, for example, isotope analysis revealed that the diet of most individuals was based on terrestrial foods, while only those buried in stone mausolea consumed a significant proportion of marine fish. In this way, the upper classes set themselves apart in life, by their diet, just as they did in death, through their elaborate burial architecture.[23]

[18] O'Connor (2000: 56); Müldner, unpublished data.

[19] Ervynck, Strydonck, and Boudin (1999/2000); Herrscher, Bocherens, Valentin, and Colardelle (2001); Polet and Katzenberg (2003).

[20] Woolgar (2000: 44). See also Chapter 8.

[21] Richards (2002).

[22] Gumerman (1997: 117–24).

[23] Richards, Hedges, Molleson, and Vogel (1998: 1249–50).

High and low status The differences between the rural population of Wharram Percy and the three other sites, St Giles, Warrington, and Towton, are the key examples that significant isotopic variation can be found within late medieval British society. The villagers of Wharram Percy were mostly ordinary peasants and their mixed diet of terrestrial plant and animal protein, in contrast to the aquatic foods consumed at the other sites, fits in well with historical evidence that the lower classes often had dairy products rather than fish during the fasting periods.[24] While these findings are promising with regard to the future potential of using isotopes to define upper- and lower-class diet in the Middle Ages, there is no convincing example so far where isotope analysis has revealed dietary differences within one site, for example by contrasting males and females or individuals in different burial locations (for instance inside or outside a church).[25] Dietary differentiation in medieval England therefore may have often been more subtle than usually thought, or, alternatively, of a kind that stable isotope analysis cannot easily detect, such as the distinction between meat and dairy consumption or between different cuts of meat from the same animal. The results nevertheless justify additional work, preferably on a sample for which it is reliably known that there were substantial social differences between individuals.

Town and country While the considerable isotopic differences between the rural Wharram Percy and the urban Warrington group could be interpreted in terms of a better and more varied diet of the town-dwellers, the obvious similarities between the individuals from Warrington and the inmates of the rural hospital of St Giles illustrate that the situation is not that simple. In addition, the Warrington sample represents a mixed, high-status and monastic group and cannot be regarded as a cross-section of society in medieval Warrington. A much more representative sample for an urban population is available in the form of the individuals buried in two plague pits which were recently identified in Hereford Cathedral Close.[26] Since the Black Death would have eliminated most of the social boundaries that usually govern medieval burial rites, it is probable the individuals in these mass graves came from all tiers of society in Hereford in the mid-fourteenth century.

Isotopic measurements, which are currently available for a small sample only, indicate that the nitrogen values of the Hereford group plot at intermediate points between those for Wharram Percy and Warrington (Fig. 16.4). The differences in subsistence between the inhabitants of medieval Hereford and Wharram Percy were therefore not as substantial as those between Wharram Percy and the three sites in the pilot study. The $\delta^{15}N$ ratios for Hereford are nevertheless still rather high. They suggest that the townsfolk consumed relatively

[24] Dyer (1998*a*: 157–8); see Chapter 7.

[25] Herrscher, Bocherens, Valentin, and Colardelle (2001: 485–6); Müldner (2001: 84–93); Müldner, unpublished data.

[26] Stone and Appleton-Fox (1996).

Fig. 16.4 Box-and-whisker plot of $\delta^{15}N$ ratios for Warrington, Hereford, and Wharram Percy. The whiskers indicate the overall range of values in the sample, the box represents the interquartile range, and the bar the median value. Note how the values for the Hereford group are intermediate between Warrington and Wharram Percy

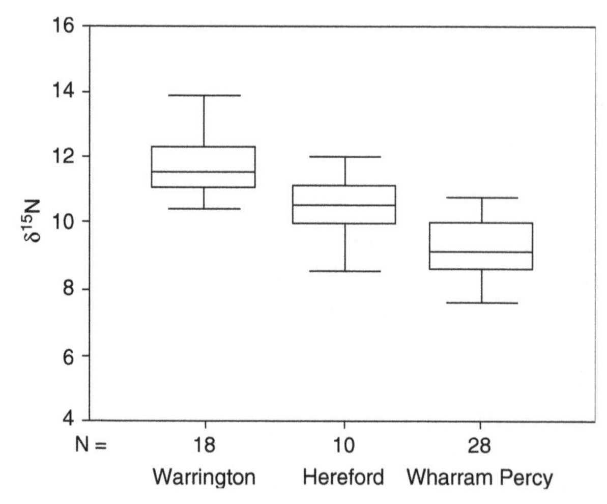

Sources: Müldner and Richards (2005); Müldner (2003); Richards, Mays, and Fuller (2002).

more animal protein than the Wharram Percy peasants and probably also some aquatic foods, although not in the same quantities as the inhabitants of Warrington and those of St Giles.[27] These data are consistent with historical information about diet in medieval towns, where the markets provided a constant supply of fresh foods in great variety and ensured that standards of living were generally higher than in the countryside.[28]

Monastic diet

The obvious similarities between the isotopic data from the rural, low-status St Giles samples and the urban, high-status Warrington group are at first surprising. They appear less unusual, however, when it is considered that both places were effectively religious institutions. The inmates of the hospital lived together under a common dietary regimen, just like the friars of Warrington, and fasting regulations were probably followed much more closely than in the population at large.

With this in mind, it may be significant that nearly all medieval sites where very high $\delta^{15}N$ ratios, indicative of fish consumption in greater quantities, have been observed, are monastic in character.[29] In the mixed monastic and lay cemetery at Warrington, where only the burials of women and children can be reliably identified as those of lay people, the peak nitrogen values are exhibited by males,

[27] Müldner (2003). [28] Dyer (1998*a*: 196–7).

[29] Polet and Katzenberg (2003); Müldner and Richards (2005); Müldner and Richards, unpublished data.

one of them a priest, distinguished by the inclusion of chalice and paten in his grave. Two priests from the St Giles cemetery display similarly enriched nitrogen ratios. One of them had a severe bone pathology, almost certainly acquired in childhood, which makes it likely that he spent most of his life at the hospital.[30] Consequently, and even though in most cases it cannot be established whether an individual had lived under a dietary regimen associated with a religious order or clerical status long enough for their bone to reflect their monastic diet, there are at least grounds to suspect that much of the unusual medieval isotope data is biased by members of religious communities and their very specific diet, their stricter observance of fasting and greater reliance on fish during times of abstinence.[31]

Although very high nitrogen values seem therefore likely to be displayed by members of religious communities, these are unfortunately not isotopic signatures that could serve to identify such individuals in the burial record. The men from the Battle of Towton, who were also part of the pilot study, are thought to have come from a variety of social backgrounds, yet were almost certainly lay people.[32] Their enriched $\delta^{15}N$ ratios nevertheless are much more similar to the isotopic profile observed for St Giles and Warrington than that for Hereford or Wharram Percy.[33] While these results do not prove that the Towton individuals were a homogeneous group of long-serving, professional soldiers enjoying a privileged diet, they illustrate that humans whose diets are isotopically almost identical may have fulfilled very different roles in society.

Location Finally, we need to consider briefly the link between location and everyday diet. In the predominantly market-based economy of the later Middle Ages, where food might be preserved and transported over long distances, the local availability of resources was less crucial than in earlier periods, but is still likely to have had some impact. When Mays investigated the relative importance of marine resources in north-east England, the data suggested a greater importance of marine foods at some, but not all coastal locations.[34] Only carbon isotopes were analysed for his study, however, and it is possible that additional information from $\delta^{15}N$ values, which proved crucial for the interpretation of other medieval isotope data, would allow a reassessment.

It is perhaps unsurprising that the sites where isotopic evidence suggests the consumption of freshwater fish are all located directly on rivers: Warrington on the Mersey, St Giles on the Swale, and Hereford on the Wye. Wharram Percy, on the other hand, is situated in the remote Yorkshire Wolds, and although there was a small stream and later a fishpond to the south of the church, these would not have been a significant source of food for the common villager.[35] Location is

[30] Knüsel, Chundun, and Cardwell (1992); Müldner (2001: 95–7).
[31] Bond (2001); Harvey (1993: 34–71).
[32] Knüsel and Boylston (2000: 170–2).
[33] Müldner and Richards (2005).
[34] Mays (1997: 565).
[35] Beresford and Hurst (1990: 65–7).

therefore almost certainly another factor contributing to the pronounced isotopic differences between Wharram Percy and the other sites.

Research prospects

This review illustrates that, while far from being a universal key to questions surrounding diet and nutrition in medieval England, stable isotope analysis of human bone nevertheless deserves its place among the more established methods of dietary reconstruction. Isotopic evidence is currently scarce, and the differences and trends suggested here, whether diachronic, regional, or social, offer an enticing, but unfortunately not a comprehensive tale. The need for much more work to be done is most apparent. Suitable assemblages need to be identified to target specific periods of time, sites, or groups of interest, to test interpretations, and to answer particular questions.

Isotope analysis is one of the fastest developing fields in archaeological science. Multi-isotopic approaches, that is, with the isotopic analysis of additional elements, such as sulphur or oxygen, to complement the data from carbon and nitrogen, will be of particular interest to studies of medieval diet. They offer the prospect of teasing out of the isotopic signal more detailed information about different foods.[36] In the future, they may help to resolve the many interpretative issues surrounding the unusual medieval isotope values and to estimate proportions of omnivore protein, freshwater, and marine fish in the diet. Even in advance of this, stable isotope analysis is likely to have a significant impact on the study of medieval diet, because it can make available a large body of previously unused evidence—the chemical information contained in the bones of the vast numbers of human skeletons surviving from the medieval period.

[36] See for example Clementz and Koch (2001); Richards, Fuller, Sponheimer, Robinson, and Ayliffe (2003).

17

Medieval Diet and Demography

P. R. SCHOFIELD

There is no simple and straightforward association to be made for the Middle Ages between diet and demographic behaviour. We can and do risk certain assumptions regarding this relationship, however. These assumptions include correlations between the behaviour of population and the overall availability of food, and the consequences for demography and health of limited food availability, in terms of quantity and also of range and quality. We might also argue, with some justification, that large-scale population movements were not occasioned by diet. Background mortality and crisis mortality, the consequences of endemic and epidemic disease, may have had significantly greater roles to play in the demographic history of the period.

This chapter reviews some of the ways in which diet could have effected changes in the demographic regime in medieval England. This involves an assessment both of the extent to which the availability of food in general shifted over the period *c.*1200 to *c.*1500, with some reference to earlier centuries, and the degree to which the consumption of food altered, in quality and type of foodstuffs, topics covered in varying ways in many of the preceding chapters. In that respect, we should comment on entitlement to food and its demographic consequences, including the ways in which that entitlement might be influenced by wealth, age, and gender. Before we do so, however, we need to rehearse some of the limits which apply in approaching these sorts of relationships.

Historians and demographers have for some time dwelt upon the complex and rather opaque relationship between diet and demographic behaviour. Thomas Malthus elaborated a model of population change which was consequent upon availability of resources. Contending 'that population cannot increase without the means of subsistence, is a proposition so evident, that it needs no illustration', Malthus argued that, given the logarithmic increase in population and the arithmetic increase in resource, a positive check on population (in other words, a check that was driven by mortality) could only be occasioned through misery and deprivation.[1] Malthus also suggested a fertility-driven explanation, that population could be checked preventively, through reduced nuptiality.

Malthus's views have informed some of the most important empirical and theoretical work on European population history. The majority of commentators on pre-industrial Europe have posited a close association and inverse relationship between food resource and mortality. In particular, as Fogel has described, historical demographers established a broad consensus that mortality crises had a general effect of lowering population and that they were the product of crises in subsistence, either directly, through starvation, or indirectly, through dearth- and famine-related epidemics.[2] It was, as he also notes, only the publication by Wrigley and Schofield of the mammoth *Population History of England and Wales* which, through a wealth of empirical data on early modern England, established an alternative hypothesis. This hypothesis had two interpretative consequences. The direct consequence was to question any immediate relationship between, on the one hand, mortality crises arising from harvest failure and a collapse in food supply and, on the other, long-term movements in population. Wrigley and Schofield, along with Ronald Lee, employed data derived from parish records to show not only that short-term variations in mortality were only weakly correlated with grain (wheat) price movements, but that long-term trends in mortality showed not even the weakest correlations.[3] The indirect consequence was to throw emphasis upon background mortality and, in particular, the demographic consequences of chronic malnutrition. To quote Fogel, reflecting on early modern Europe and its emergence into a modern era of significantly reduced mortality rates, 'the elimination of chronic malnutrition played a large role in the secular improvement in health and life expectation'.[4] In drawing such a conclusion, Fogel has consciously attempted to find common ground for both the argument that crisis mortality arising from subsistence crises had little or no long-term effect on population movement and the alternative thesis that subsistence crises and their associated epidemics induced long-term population shifts. While accepting entirely the arguments propounded by Wrigley and Schofield for the limited impact of crisis mortality on total mortality and the extremely limited consequence of famine for the same, Fogel considers that the decline of mortality rates was largely dependent upon a significant improvement in average nutritional status. This improvement might be measured through changes in height and body mass indices. The demographic transition was, for Fogel, a consequence of improved agriculture and associated technologies which permitted an advance in per capita consumption.[5] McKeown has argued, along broadly similar lines, that it was an improvement in nutrition prompted by agricultural advances rather than other endogenous developments, and notably medical

[1] Malthus (1966: 37). [2] Fogel (1992: 243–4); see also Smith (1993).
[3] Wrigley and Schofield (1981: chapter 9); also Fogel (1992: 244–5). [4] Fogel (1992: 280).
[5] Fogel (1992: 280).

developments, that first reduced mortality in the modern era. Essential to McKeown's argument is an association between nutrition and infection. He argues for a close association between nutritional status and diseases so that 'malnutrition contributes largely to the high level of infectious deaths'.[6] Chronic malnutrition creates the sort of conditions that ensure a constant and vital influence of infection upon populations.[7]

While the arguments of the classical economists and those, such as Fogel and McKeown, who have applied or extended them have a forceful logic, their acceptance is not universal. In particular, some historians and demographers have contested any close and clear association between a failing or lowered allocation of resources per capita and population behaviour. In a succinct polemic on European demographic history, Massimo Livi-Bacci, while acknowledging 'that nutritional variations were of great importance in the history of European populations', has argued that 'the mechanisms which link demography and nutrition are, to say the least, ambiguous'.[8] In a controversial thesis, which reviews evidence of the body's ability to adapt to privation, Livi-Bacci argues that a decline in resource is to be measured less in terms of deaths than in changes in human physiology: 'the human phenotype may assume body size compatible with limits imposed by its environment.'[9]

Livi-Bacci recognizes that the adaptability of human populations and of individuals to periods of want is finite and that extreme periods of constraint on resources lead to death, not to adaptation; but he emphasizes the potential of societies to absorb changes in diet and in the availability of food without consequence for their survival.[10] In developing this thinking, Livi-Bacci suggests that evidence of adaptability challenges an immediate and negative correlation between the availability of food and the movement of population, of a kind advanced by Fogel and McKeown. In the first place, he argued that the association between food shortage and disease only occurs 'at high levels of malnutrition' and is 'triggered more by social dislocation than by malnutrition'. Secondly, since (*a*) elite groups do not appear to have enjoyed better chances of survival than did their poorer contemporaries and (*b*) the nutritional experience of the wealthy and the relatively poor as infants was similar or the same, that is, before they were weaned, mortality levels were as or more likely to be determined by environment, culture, or society than they were by nutrition. Thirdly, in historical populations, there is an observable increase in mortality at points of apparent and general improvement in standard of living. Here Livi-Bacci makes explicit reference to the century and a half after the first outbreak of plague in the mid-fourteenth century. Finally, when mortality does begin to decline, there is no clear evidence of an improvement in standard of living; in fact, evidence,

[6] McKeown (1976: 135). [7] McKeown (1976: 136); also McKeown (1983).
[8] Livi-Bacci (1991: 22). [9] Livi-Bacci (1991: 113). [10] Livi-Bacci (1991: 114).

especially that relating to height, indicates the contrary.[11] Many of Livi-Bacci's observations cannot be easily accommodated with the views of some of the commentators discussed earlier.[12] His work encourages caution in extrapolating causative explanations from apparently compelling sets of data.

As will already be evident, much of the debate about the relationship between nutrition and mortality has been conducted between demographers and historians working on population histories of the early modern and modern periods.[13] Relatively little work has been attempted for the Middle Ages, largely for practical reasons. The sorts of cumulative data available to historical demographers working on later centuries—registration of baptisms, marriages, and deaths, information on morbidity, health statistics, and the like—do not exist, or only in the rarest of caches, for the Middle Ages. That said, there are sources for medieval England which might allow us to approach some of the questions that have engaged the attention of modernists. Medievalists do have information on local populations, the types of resources available to them, the allocation of food, and, in price and wage data, indices of change in supply and demand. Archaeology offers some limited information on health and morbidity, while narrative sources discuss the more significant moments of crisis mortality, including famines in the thirteenth and early fourteenth centuries. It behoves us at least to assess some of the ways in which we might approach the relationship between nutrition and mortality for this period.

Diet and the availability of food per capita

The principal food for the population of England in the Middle Ages was cereals. Campbell, using extensive research on grain and livestock production, has estimated the total food allocation for the population of England *c.*1300 and *c.*1375. Around 1300, the arable area in England was *c.*10.5 million acres; this had reduced by about 25 per cent by 1375 and was roughly equivalent to the cultivated acreage around the time of Domesday Book.[14] Accounting for a range of factors including the costs and methods of production, the extent of fallow, and the types of grain sown, Campbell calculated the total net grain output in 1300 as 7.39 million quarters, declining to 4.79 million quarters *c.*1375. Extrapolating from this, he estimated that the net output in 1086 was between 3.4 and 3.9 million quarters.[15] By estimating the extent of the employment of the output, in terms of, for instance, proportion of grain brewed as ale, fed to livestock, milled, and so on, Campbell proposed an 'extraction rate'—the

[11] Livi-Bacci (1991: 115–17).

[12] Though there is much struggling with the same issues in their work and a significant degree of uncertainty; see, for instance, the discussion of aristocratic mortality rates in relation to those of a more general population, McKeown (1976: 139–42).

[13] For a recent assessment of significant elements of this work, see Landers (2003: 30–5).

[14] Campbell (2000: 388–90). [15] Campbell (2000: 392–6).

percentage extracted for consumption in relation to the total net output—of 50–6 per cent in 1086, 58 per cent in 1300, and 48 per cent in 1375. Further calculation from this produces a daily kilocalorie output for England at these same points: 2,518–3,383 million kcal in 1086; 6,569 million kcal in 1300; and 3,503 million kcal in 1375. Campbell proposed that, if we assume a daily minimum requirement per person of 1,500 kcal derived from grain, then the population which could be sustained in England was in 1300, 4.38 million, and, in 1375, 2.34 million.[16]

Campbell's estimate of 1,500 kcal per individual can be tested against those few sources which offer particular insight into the general diet of the population. While sources for certain groups within medieval society, such as monks and members of high-status, secular households, suggest that they frequently consumed far in excess of this base figure,[17] it is also evident, where we can extrapolate from sources for low-status populations, that a majority of the population consumed levels of food that came closer to 1,500–2,000 kcal per day. Dyer's examination of peasant maintenance agreements, which detail grain allocations provided, typically by family members, for older relatives in retirement, indicates an amount close to Campbell's base figure. A sample of 141 agreements, over two centuries, reveals that the majority of payments were of 9 to 16 bushels of grain per annum, with an average of 12 bushels (1.5 quarters). Dyer estimates 12 bushels of wheat and barley would produce a daily allowance of about 1.5 to 1.75 lb of bread, or 1,700–2,000 kcal.[18] Analysis of payments to harvest workers, recorded in manorial accounts, while tending to offer information on much larger lump sums which must have been subsequently distributed amongst the labour force and its dependants, also indicates daily levels of a similar proportion.[19]

Long-term patterns of food availability in medieval England

The general availability of food altered significantly across the high and late Middle Ages. Dyer's work has done most to establish a broad chronology of food availability, describing a pattern generally consistent with long-term shifts in population and entitlements. In particular Dyer's examination of payments to harvest workers and *famuli* indicates ways in which a peasant diet dominated by cereals gave way, in the second half of the fourteenth century and in the fifteenth century, to a diet which was significantly higher in protein.[20] These general patterns were not representative of all patterns of consumption. The availability of food types was dependent upon status and the extent to which food availability altered across the period was also dependent upon wealth and status. In

[16] Campbell (2000: 396–9). See Chapter 2 and Table 2.3 for a revision of the calculation of the size of population.
[17] Harvey (1993: 66–7). See also Waldron (2001: 92–4).
[18] Dyer (1998*a*: 152–6). See also Dyer (1983: 197–206; 1998*b*: 56).
[19] Dyer (1994*b*: 25–9).
[20] Dyer (1998*a*: 157–60; 1983: 210; 1994*b*).

particular, the diet of the wealthy, especially members of social elites, did not alter dramatically relative to changes in the diet of their social inferiors.[21] In fact, it was the more pronounced variation in the diet of the poorer members of society which attracted most attention and, as the diets of those of lower status drew closer to patterns of high-status consumption, censure.[22]

Diet did not create social difference in this period but it was evidence of social difference and, in demographic terms, it helped perpetuate that difference. The majority of the population throughout the period consumed a diet that was principally cereal based; elites, however, consumed diets that were relatively, and often extensively, rich in protein. Thus the monks of Westminster Abbey ate, outside Lent, a diet highly rich in protein.[23] Those further down the social scale also sought to replicate, albeit to a limited degree, the diets of their social superiors. While the poorest members of society could not hope to enhance their diet unless through charity, wealthy urban and country dwellers, including peasant elites, could vary their diet. As we have seen in earlier chapters, wealthier villagers purchased meats and fish, as well as rearing their own livestock and fish for sale and consumption.[24]

The opportunities brought by wealth had dietary implications which we may detect in other demographic indicators. For instance, we know that the families of the wealthier peasants were often larger than those of their poorer neighbours.[25] In many respects, this is an indication of the potential for the reallocation of resources within the family and the opportunities for early marriage thereby presented in a system of neo-local household formation. But this also suggests that members of these wealthier families may have been better nourished and particularly able to provide for themselves, their offspring, and their elderly relatives.[26] In discussing the changing pattern of food availability in this period, therefore, we need to acknowledge that the general patterns of availability were not universal patterns and that individual entitlements to food varied significantly, with differing proportions of the population more or less able to gain an appropriate quantity and quality of food.

Diet and demography in high and late medieval England

We might reasonably suppose that this pattern of food availability and its changing nature occasioned its own demographic consequences. While the central elements of a demographic model of population movement have been and remain

[21] Dyer (1983: 196–7; 1998*a*: 55–70). Note, however, Woolgar (1999: 132–4).

[22] For an example of that censure, see John Gower, *Mirour de l'omme*, cited in Hatcher (1994: 16): 'The labourers of olden times were not accustomed to eat wheat bread; their bread was made of beans and of other corn, and their drink was water. Then cheese and milk were as a feast to them; rarely had they any other feast than this . . . Then was the world of such folk well-ordered in its estate.'

[23] Harvey (1993: 66); Dyer (1983: 193–4); Woolgar (1999: 111–35).

[24] Dyer (1983: 209).

[25] Razi (1980: 83–8, 140–6).

[26] Though, on the possible adverse effects of wealth and infant health, see also below, 251.

predicated upon the relationship between population and resources, it is also clear that, at least for the later Middle Ages, it is the relationship between mortality, and especially crisis mortality, and fertility which has exercised the attention of historical demographers. In neither the instance of mortality nor fertility is diet wholly absent, but population movement in this period, essentially post-1350, has been described in terms of the positive check of disease in relation to the preventive check of fertility, the latter described in terms that tend to eschew diet.

In order to review the relationship between food availability and population movement, we can employ a series of subcategories which also provide us with a roughly chronological engagement with the population history of high and late medieval England. The broad parameters of development are essentially these: in the late eleventh century, around the time of Domesday Book, population in England stood at approximately one third to one half of the population two centuries later.[27] Rising to a possible peak of 6 million by 1300, after a sustained period of dramatic growth in the late twelfth and thirteenth centuries, population had again begun to decline by the middle years of the fourteenth century.[28] The arrival of plague in 1348 and its recurrence in the last decades of the fourteenth century may then have reduced the population by over 50 per cent; and it failed to recover throughout the remainder of the period. It was only in the late fifteenth and early sixteenth centuries that the population began to rise once again.[29] How important was food availability to this demographic history?

Demographic growth, c.*1100*–c.*1300* From the eleventh century through to the end of the thirteenth century, we are in a period where the models of classical demography appear to be easily applied; as the demand for labour increased, so population, 'a passive variable', also increased. In the longer term, it is changes in such variables as institutional structures, climate, economic opportunities, technology, and social organization which have the greatest effect upon population movements.[30] Historians have identified, for the high Middle Ages, a period of economic expansion, with a growing urban sector, a blossoming of markets, and, by the early thirteenth century, a direct engagement in grain production by landlords which also encouraged some limited technological enhancement of production. Allied to these developments, the emergence of sophisticated legal mechanisms presented opportunities for exchange and capital accumulation based upon appropriate security.[31] This economic expansion, which also included, where it could be attempted, a widescale colonization of new

[27] Estimates of population vary considerably: Campbell (2000: 403, table 8.06) provides a very useful summary of population estimates and their references. See also Snooks (1995: 32–4).

[28] But note Campbell's lower estimate for population *c.*1300, as discussed above, 243.

[29] For a recent account of this population history, see Dyer (2003: 235).

[30] See, for instance, Lee (1986: 75–6).

[31] For general discussions of the expansion of the economy in the twelfth and thirteenth centuries, see Dyer (2003: part 2); Britnell (1993: part 2; 1995).

agricultural land, furnished the population with resources and thereby encouraged an expansion of that population.

The greatest proponent of this model for medieval England was M. M. Postan who, through a series of papers and two larger studies, presented a thesis of population movement which is essentially Malthusian in its implications and which continues to dominate, though not without modification and objection, our general view of the behaviour of population in this period.[32] What distinguishes the work of Postan from those who have commented within the last decade or two on economic growth in the twelfth and thirteenth centuries is a willingness on the part of Postan's successors to identify a relationship between economic developments, notably an increase in commerce and urbanization, and the potential for population to sustain itself at such relatively high levels as 5 to 6 million by *c.*1300.[33] For Postan's thesis, the essential feature is not that per capita food intake increased greatly in this period, but that food availability increased alongside population. The bulk of historical information on the diet of the general population survives from no earlier than the mid-thirteenth century and, from the earliest sources, including monastic corrodies, we gain a sense of a diet, even for the relatively wealthy peasant, which was dominated by grains.[34] In that sense, the food availability of the general population for this period has been perceived by historians as essentially a passive agent of the demographic model. It was acted upon by institutional factors, including the demands of lordship and the state, as well as by largely exogenous forces, notably climate and the fertility of the soil.[35] A consequence of population expansion encouraged by economic expansion was vulnerability, with those increasingly dependent upon trade and commerce exposed to changing economic conditions.[36]

Demographic crisis, c.*1300–*c.*1380* Even before the end of the period of demographic growth, the population of the high Middle Ages had suffered a

[32] Postan (1973*b*; 1973*c*; 1973*e*; 1973*f*); Postan and Titow (1973). See also Postan (1966: 600 ff.; 1972). For a recent assessment of the Postan thesis and its legacy, see Hatcher and Bailey (2001).

[33] See, for instance, the comment in Newman, Boegehold, Herlihy, Kates, and Raaflaub (1990: 116): 'Many Europeans in this large population doubtless went hungry, but the community successfully maintained its size until plague overwhelmed it. This Malthusian deadlock might have held on indefinitely within Europe. The plague broke its grip . . .' See also the references in n. 31 above.

[34] See, for instance, Hallam (1988*d*: 826–7).

[35] The significant counter-thesis to that expounded by Postan was a Marxist one. Marxists such as Brenner have much to say on the subject of resources, but this is principally in relation to the ways in which lordship restricts control over resources and inhibits productivity and population growth. Thus, 'the determination of the impact of the pressure of population on the land—who was to gain and who to lose from a growing demand for land and rising land prices and rent—was subject to the prior determination of the qualitative character of landlord/peasant class relations': Brenner (1985: 22, also 31); see also Hatcher and Bailey (2001: 110–11) and Smith (1991: 34–5). For examples of work in relation to exogenous and endogenous factors, see the following: lordship and the state: Brenner (1985); Hilton (1976: 17–18); Maddicott (1975); climate: Titow (1960); yields: Titow (1972); on the economic potential for technological advances in agriculture, in this period see, for instance, Campbell (1995: 192).

[36] Britnell (1995: 23).

series of assaults. From at least the early thirteenth century, we can chart a series of very poor harvests, interspersed with bountiful years.[37] Harvest failure culminated in the Great Famine of the second decade of the fourteenth century, in which it has been suggested 10 per cent of the population died, but the spectres of harvest failure, dearth, and famine haunted the land for decades after that. Nor was the threat of famine wholly removed by the halving of population by the plague in the mid-fourteenth century.[38] This period of agrarian boom and bust left a significant proportion of the population at risk of severe shortage and even starvation.[39] A noted dependency upon grains is illustrated by marked fluctuations in price, with bread grains sometimes doubling in price over consecutive years. These changes illustrate the complexity of the relationship between price and those factors that acted upon it, but the quality of the harvest and the interplay of supply and demand are undoubtedly of central importance.[40] It is striking that other foodstuffs failed to display the same pattern of fluctuation. Livestock prices did not follow grain prices in the worst harvest years. Instead, the price of, for instance, pigs and of cheese remained low, a feature which may illustrate the relative cost and, hence, unavailability of these foodstuffs for the majority of the population.[41]

If, in the thirteenth and the early fourteenth centuries, a predominantly carbohydrate-dependent population did succumb to starvation in years of crisis, as local studies of individual manors and evidence from tithing-penny payments suggest it did, the demographic consequences of such crises are likely to have been reasonably muted.[42] We know, from the work of historical demographers of early modern England, both that the statistical relationship between grain price movements and long-term shifts in population is extremely weak, and that population showed no marked decline in the decades after the most severe harvest failures of the period.[43] This is not to suggest, of course, that there were no long-term consequences, either demographically or socio-economically, but they are less than easy to detect. While there is fairly unequivocal evidence for infectious disease following in the wake of harvest failure in this period, the impact of shortage or associated infection is largely hidden from us.[44]

In fact, the social and demographic consequences of the famines of the thirteenth and early fourteenth centuries stand in marked contrast to the plagues of the mid- and late fourteenth century. While the direct consequences of plague

[37] Titow (1972: appendix C); Farmer (1988: 737–8); Campbell (2000: 309 ff).

[38] Kershaw (1973*b*); Razi (1980: 39–40); on dearth and famines in the mid-fourteenth century, see Mate (1991: 90–103); Frank (1995); on famine in the fifteenth century, see Pollard (1990: 46–8; 1989); Goldberg (1988: 43); Dyer (1998*a*: 267–8); Walter and Schofield (1989*b*: 28 ff).

[39] See, for instance, Riley (1866: 93–8).

[40] Farmer (1988: 735, and more generally, 733–9).

[41] Farmer (1988: 753–4); see also Chapter 14.

[42] Razi (1980: 39–40); Poos (1985).

[43] See above, 240; note, however, the evidence of early fourteenth-century tithing-penny data: Poos (1985).

[44] See, for a description of an epidemic at the time of the Great Famine, Riley (1866: 94); for the behaviour of population in the decades after the Great Famine, see Poos (1985).

have been an object of debate for generations of historians,[45] it is for the greater part accepted that plague's initial impact had little association with food availability. There is a growing body of literature in which historians and epidemiologists have set out to identify alternatives to *Yersinia/Pasteurella pestis* as the Black Death disease, but in all cases the incidence of such a devastatingly virulent soft tissue disease, capable of the eradication of perhaps 50 per cent of a population (in England, possibly in the region 2.5 million people), appears not to have been dependent upon the nutritional status of the victim.[46]

That said, there has been a suggestion that the incidence of plague was at its strongest where the population or sections of the population had undergone severe periods of dearth or famine.[47] If this were at all likely, then we might expect that the generation which had survived the Great Famine and which may have carried the physiological burden of that experience with them for three decades was especially vulnerable in 1348–50.[48] In other words, where we can find or extrapolate evidence for age-specific mortality in the first plague years, we might expect to find that those who had experienced the Great Famine as children, and who were therefore aged between 40 and 50 when plague struck, were more or even most likely to succumb to the disease.[49] Razi's examination of age-specific mortality at Halesowen suggests that the most vulnerable age cohort in the first outbreak of plague was indeed those aged between 40 and 60, and Lock, too, who argues for high mortality rates amongst the young, also acknowledges, though is sceptical of, 'the disproportionate number of those over forty among the Black Death victims'.[50]

Background mortality and endemic disease: the long fifteenth century While crisis mortality remained a feature of the late medieval demographic regime, its heyday had passed by the end of the fourteenth century. Undoubtedly epidemic disease and, to a much lesser extent, famine occasioned significant surges in mortality in the fifteenth century, but the impact of these crisis events was never of the severity of the major national outbreaks of the previous

[45] Hybel (1989) analyses a series of debates which, beginning in the second half of the nineteenth century, explored the relationship between the Black Death and change in late medieval society.

[46] Recent attempts to discuss the identity of the Black Death include Twigg (1984; 2003: 48–50); Scott and Duncan (2001); Cohn (2002*a*; 2002*b*).

[47] Jordan (1996: 186).

[48] Research currently being conducted by Daniel Antoine on the remains, and especially the teeth, of victims of the Black Death excavated from the plague cemetery at Smithfield includes those who were juveniles at the time of the Great Famine. These show clear signs of the traumatic impact of the famine and, possibly, of subsequent crises in the 1320s. It seems, at this stage, unlikely that the sample will permit discussion of final achieved height and the likely focus will be on the immediate effects of famine in the early fourteenth century rather than the longer-term physiological effects of famine-induced trauma. I am most grateful to Professor Derek Keene for this information.

[49] But note the apparent negative association between iron deficiency and the replication of the plague bacillus, below, Chapter 18. On the long-term developmental consequences of food shortage in infancy and childhood, see Rivers (1988: 61–2), cited in Jordan (1996: 186).

[50] Razi (1980: 108–9); Lock (1992: 318–20). See also Chapter 18.

century.[51] Instead, those historians who argue for a demographic regime driven by mortality present evidence for high levels of background mortality in the fifteenth century. For certain communities in late medieval England, evidence hints at very high levels of mortality occasioned by a potent combination of endemic and epidemic disease. Work by Harvey, who employed wage accounts and infirmarers' accounts to reconstruct the demographic experience of the monks of Westminster Abbey from the fifteenth century, and by Hatcher, using profession and obituary lists of the monks of Christ Church Canterbury for the same period, has uncovered a distinctive range of diseases which afflicted the brethren of these houses and which also indicate very low life expectancy levels.[52] While Harvey's research also suggests that monks suffered significantly from fluctuations in diet, a product of the demands of the round of the liturgy, there is no clear indication that endemic disease and low life expectancy were in any respect the result of diet and/or general food availability. In fact, for these relatively privileged members of late medieval society, an abundance rather than a shortage of food was as likely to occasion health problems.[53] What we cannot also tell from this material—and this evidence undoubtedly is among the very best that survives[54]—is the extent to which a low life expectancy was the product of a dietary burden inflicted upon individuals in infancy. Nor can we, in the absence of census data or other meaningful demographic data on infancy and childhood, assess the extent to which diet, or poverty of diet, affected individual responses to endemic disease. Instead, we can make some reasoned inferences regarding the care of infants and, rather more concretely, identify general improvements or adjustments in diet in the later Middle Ages which might allow us to judge the role of diet in its effect upon the demographic regime of the long fifteenth century. Was diet a factor in the purported low life expectancy of this period or did background mortality, as we would assume was the case for epidemic disease, play upon the population irrespective of and perhaps despite the quality of the late medieval diet?

The protein content in individual diets increased in this period. While this may have brought benefits in terms of health, even a minor increase in the predominant cereal-based diet, allied to a generally improved living environment, could have brought as great or greater advantages to the population. As modern studies of cereal-dependent populations have shown, there does not have to be a dramatic increase in diet or a change in diet for individuals to benefit significantly in health and nutritional terms. What is important, in order to ensure an appropriate protein intake from a cereal-based diet, is an effective restriction of

[51] See, for instance, the list of plague years in Hatcher (1977: 17), and the evidence gathered by Creighton (1965); also Gottfried (1978).

[52] Harvey (1993: 127–9); Hatcher (1986). On morbidity, see also Harvey and Oeppen (2001).

[53] Harvey (1993: 109); Hatcher (1986: 34). See also Waldron (1985).

[54] Hatcher has also worked upon the obedientiary material for Durham Priory for the same period; the results of this research are still to be published.

energy.[55] As Gopalan and others have described, inadequacy of diet is not an absolute, but is aggravated by conditioning factors, such as the presence of endemic disease, the prevalence of breast-feeding, and the physical demands of the individual's living and working environment(s).[56] In late medieval England, an improved availability of food, even basic foodstuffs, allied to an enhanced labouring environment—the fifteenth century has been described as the 'golden age of the English labourer'—may have brought significant demographic advantages.

If then, as all major indicators including wage and price data suggest,[57] the general condition of the population improved in terms of food availability, we may be inclined to explain the population stagnation of the late Middle Ages in terms of a high-pressure regime of both high mortality and high fertility.[58] An improving diet and general entitlement, revealed by lowering prices and increasing wages, could have created a healthy population, the growth of which was held in such severe check only by repeated outbreaks of disease, both endemic and epidemic. In other words, if population did not increase in this period, if there was no mortality transition, it was as a result of a failure to respond to the challenge of the exogenous factor of disease in what has been described as a 'golden age of bacteria'.[59] Not all demographic historians would subscribe to this interpretation. Those who prefer to argue for a low-pressure regime operating in late medieval England point to the high incidence of female employment, especially life-cycle employment as live-in servants, which delayed marriage and household formation and, hence, the period of fecundity.[60] In discussion of both high- and low-pressure regimes, the role of diet and food availability is of indirect consequence relative to, respectively, mortality and fertility. That said, while mortality may have acted as a variable largely independent of food quality and availability, neither regime can be entirely comprehended without some recognition of diet.

In terms of high-pressure regimes, characterized by high mortality and high fertility, we might suppose that infant mortality, as a result of infectious disease, was a key factor in the very low life expectancies suggested for this period, though there is little hard evidence for this. Rising background levels of mortality in the fifteenth century, as indicated by monastic data, suggest life expectancy levels—at the age of 20, the average expectation for monks at Christ Church Canterbury would have been a further twenty-eight years—lower than those typically to be found in early modern England.[61] Since, for early modern England, we also assume that infant mortality was a significant feature of the mortality regime, we might also suppose the same to be true for the later Middle

[55] Gopalan (1992: 25–30).
[56] Gopalan (1992: 19–20).
[57] Phelps-Brown and Hopkins (1962); Postan (1973*d*); Penn and Dyer (1994); Hatcher (1994).
[58] Harvey (1993: 145); Hatcher (1977: chapter 4); Razi (1980: 124–39); Bailey (1996).
[59] Thrupp (1965).
[60] Goldberg (1986; 1992*a*); Poos (1991).
[61] Hatcher (1986: 32–3); Harvey (1993: 128–9).

Ages. While infectious disease could have caused high infant mortality irrespective of the nutritional status of infant and mother, these levels could have been worsened by either the nutritional status of the mother or the duration of breast-feeding. We know, for instance, that opportunities for highly paid irregular work, including heavy agricultural work, were presented to women in the second half of the fourteenth century.[62] This situation did not last and women were forced from such lucrative opportunities as landlords retreated from direct management of their demesnes in the last decades of the fourteenth century.[63] Briefly, however, we can envisage a regime where women, alongside men, enjoyed a considerably improved earning capacity in work that was irregular and not dependent upon life cycle—in other words, in the third quarter of the fourteenth century, women could both have worked for high wages and, at almost or exactly the same time, borne and reared children. In a period of repeated national plague outbreaks, we do not have to search hard for reasons for infant mortality. We might, however, at least anticipate that infant mortality, and therefore the overall demographic regime, could have been accentuated by a labouring environment which, while it presented new opportunities for women and also may also have encouraged early marriage and household formation,[64] may also have weakened the nutritional status of infants. Judging by modern studies of lactation and breast-feeding, the different demands of heavy agricultural work are unlikely to have impaired the mother's ability to provide a nutritious diet for her infant, but it is at least conceivable that patterns of irregular employment disrupted customary childrearing practices.[65] In particular, a regime which presented opportunities for early household formation may also have encouraged large families, provided the nutritional resources to enhance fecundity, and, thereby, reduced the period of breast-feeding for each infant. An infant who has not been breast-fed lacks immunological protection; however, breast-fed children become increasingly vulnerable to disease from the point they are weaned.[66]

Low-pressure regimes, where fertility and mortality both operate at relatively low rates, display similar associations with diet and nutritional status. These also can be applied to the situation for all or part of the later Middle Ages in England. It is possible that a high-pressure regime, which may have included a high rate of infant mortality, gave way, in the late fourteenth and fifteenth centuries, to a regime which was lower in its intensity. In this case, we would also anticipate, of course, a decline in infant mortality. Life-cycle service, especially

[62] Penn (1987).

[63] Penn (1987: 13) describes this as 'an exceptional period in which an acute labour shortage served to enhance the importance and value of female labour'.

[64] If we apply a 'real wages' or 'proletarian model' of household formation; see, for instance, Poos (1991: 141–8).

[65] Livi-Bacci (1991: 75–6).

[66] Livi-Bacci (1991: 76–7): 'The age of weaning is thus an important variable in infant survival.' On the seasonality of weaning and its association with labour, Landers (2003: 30), citing Breschi and Livi-Bacci (1997).

the employment of women as live-in servants, delayed age at marriage and may have reduced the size of households. In these circumstances we should anticipate that an improving general standard of living, relatively small family units, and expendable capital all worked to reduce mortality, and notably infant mortality. Opportunities for breast-feeding which extended over more than just the first weeks or months of infancy are likely to have been more prevalent in such conditions. We might also expect that a greater proportion of the population could afford to employ wet-nurses.[67]

The early stages of population recovery in the late fifteenth and sixteenth centuries can and have been explained in terms of either declining mortality or increasing fertility.[68] Diet and nutritional status serve in either model as foundations upon which the other key variables can act. The assumption associated with an upswing in population is that, in a state of relative nutritional advantage, population responded to either mortality or fertility.

Conclusion

In assessing the importance of diet and food availability in the Middle Ages, we are left with a series of imponderables which apply in relation both to the population stagnation of the later Middle Ages and to the earlier period of growth. Ultimate relativities of diet are hidden from us; more than one chronicler tells us that the poor ate 'unclean foods' in the worst harvest years, but we do not know the extent to which diets could be supplemented.[69] Instead, our view is driven to foodstuffs which were the major objects of account, notaby cereals.[70]

Most importantly, we can only estimate the proportion of the population that was left vulnerable to food shortages in this period. In reflecting upon the demographic consequences of economic expansion in the high Middle Ages, Britnell has estimated that 'among the costs of commercialization was a more precarious life for perhaps a fifth or more of the population'.[71] Since the relationship between alternative demographic measures (wages and prices) and the very poor can be extremely muted,[72] we cannot be entirely confident that nutritional status and food entitlements did not have significant demographic consequences, especially in the period before the mid-fourteenth century, for a meaningful proportion of the medieval population. The same applies for other significant hidden elements, notably the nutritional status of the very young.[73]

[67] On possible archaeological evidence for breast-feeding in the medieval village, see Fuller, Richards, and Mays (2003).

[68] See, for instance, Bailey (1996: 15–17).

[69] For instance, Denholm-Young (1957: 70).

[70] The archaeology of cooking vessels and the investigation of stored traces of brassicas may help in this respect.

[71] Britnell (1995: 23). See also Campbell (2000: 388); Livi-Bacci (1991: 116), for recognition of the problem.

[72] Wrigley (1989).

[73] Smith (1991: 61).

While we cannot identify the potential consequences of general and changing food availability throughout the medieval population, we are given sufficient indication that diet was a variable capable of effecting change in population movement in this period. The minute and particular causative features are not, however, open to our scrutiny.[74]

[74] 'Of other more fundamental biological changes we know nothing, and I doubt whether anything about them worth knowing will ever be discovered': Postan (1973*b*: 12).

18

Nutrition and the Skeleton

T. WALDRON

It is manifestly *not* the case that we are what we eat, but normal growth and development of the skeleton certainly do depend upon a diet that will ensure that sufficient calories and the correct balance of carbohydrate, protein, and fat are provided. In addition, other essential elements are necessary; in the case of the skeleton, the most important are calcium and phosphorus, which are both components of bone mineral. Finally adequate supplies of vitamins are required, vitamins C and D being the most significant in the case of the skeleton. If any of the components of the diet is deficient, skeletal metabolism may be adversely affected, but it should also be noted that untoward effects may result from the presence of other substances in excess.

The growth of the skeleton

The long bones grow in length from birth until the end of puberty, when the growing ends—the epiphyses—fuse with the shaft and the individual's final height is achieved. Skeletal growth is not a steady phenomenon but has two periods of accelerated growth—or spurts—imposed upon it. These occur from birth to about the age of 2, and from the onset of puberty until the epiphyses fuse. The pubertal growth spurt is the one with which people are most familiar but, in fact, the maximum rate of growth occurs at the earlier age; from birth to age 1, growth takes place at the rate of 18–25 cm per year, from one to two years at 10–13 cm per year, while the pubertal growth spurt takes place at the relatively stately rate of 6–13 cm per year. Maximum achieved height is dependent largely on genetic factors as is readily demonstrated by the fact that tall parents tend to have tall children, and short parents, short children.[1] Each individual has a genetic potential for height,[2] but the most important factor that determines

[1] It is usually noted that children of tall parents are, on average, shorter than their parents while those of short parents are, on average, taller. This tendency is referred to as a regression to the mean.

[2] See, for example, Preece (1996).

whether or not that potential is achieved is the state of nutrition during the periods of active growth.[3]

There has been a very considerable interest by both historians and physical anthropologists in the height of past populations, and the popular conception has grown up that people in the past were much shorter than nowadays. Studies on the skeleton, however, tend to suggest that a substantial increase in the mean height of the population is a very recent phenomenon.[4] For example, the heights of a large skeletal assemblage excavated from St Peter's Church in Barton-upon-Humber show no change in mean height over the nine hundred years that the cemetery was in use (Table 18.1).[5] The mean heights of two other medieval assemblages, from Hertford (Jewson's Yard) and Kellington, are virtually identical with those from Barton (Table 18.2) and suggest that the nutritional status of adolescents growing up in at least these three parts of England was—on average—comparable and, in the case of Barton, not dissimilar to those of the population that lived after them.

One of the disadvantages of using the mean height of an assemblage as a surrogate for nutritional status is that short-term fluctuations are almost invariably obscured because height is the result of very many years' experience. Thus, it is

Table 18.1. Mean height (m) of skeletal assemblage from St Peter's Church, Barton-upon-Humber

Date	Male		Female	
	Mean	Range	Mean	Range
1700–1850	1.71	1.58–1.85	1.59	1.47–1.70
1500–1699	1.72	1.70–1.74	1.56	1.52–1.60
1300–1499	1.71	1.64–1.81	1.59	1.51–1.70
1150–1299	1.70	1.57–1.76	1.58	1.53–1.62
950–1149	1.69	1.54–1.83	1.61	1.54–1.68

Note: For comparison, the mean heights of contemporary males and females in Great Britain are 1.76 and 1.62 m, respectively: Ruston, Hoare, Henderson, Gregory, Bates, Prentice, Birch, Swas, and Farron (2004).

[3] Diseases that may cause poor absorption from the gut—such as chronic diarrhoea—will also have the same effect. For example, see Checkley, Epstein, Gilman, Cabrera, and Black (2003). In this study, deficits in the height of children—compared with a WHO standard—were proportional to the prevalence of diarrhoea; one may assume that a similar situation obtained in the medieval period.

[4] I once took part in a round table conversation in the company of historians, economists, and archaeologists and when one of the historians remarked that populations were shorter in the past, I commented that there did not seem to be much fluctuation judged by the skeletal evidence. 'In which case,' he retorted, without a trace of irony, 'the skeletons are wrong.'

[5] Maximum achieved height is estimated from the skeleton by substituting the maximum length of individual long bones of the arm or leg into regression equations that have been derived from the study of modern subjects. Those most commonly used were published by Trotter (1970). The heights shown in the tables in this chapter were derived using the maximum length of the femur which gives a more reliable estimate: Waldron (1998).

Table 18.2. Mean height (m) of skeletal assemblages from the medieval cemeteries at Jewson's Yard, Hertford, and St Edmund's Church, Kellington

Site	Male		Female	
	Mean	Range	Mean	Range
Jewson's Yard	1.70	1.68–1.81	1.58	1.49–1.77
St Edmund's	1.71	1.59–1.91	1.61	1.50–1.72

Source: Waldron (1998).

reasonable to assume that the children who were actively growing and approaching puberty during the period of the Great Famine in the early fourteenth century would have been shorter than their contemporaries who either finished their growth earlier, or whose growth spurt did not start till later. An opportunity to test this hypothesis presented itself during the examination of skeletons excavated from the Black Death plague pit at the site of the Old Mint in London. An analysis was made of mean height by presumed year of birth, calculated from the estimated age at death. Four birth cohorts were defined, those born before 1303, and those born 1303–12, 1313–22, and 1323–33. In the event, there was not the smallest difference between the mean heights in either sex, irrespective of their year of birth. There are a number of explanations for this unexpected result, the most obvious being that those who were born early in the century were not affected by food shortages to the extent that their final achieved height was impaired. It is possible that their growth may have been curtailed during the famine years but if they survived, and food became plentiful again, their growth spurt started again and they caught up; we know from other evidence that children probably did not stop growing until a later age than is the case nowadays and most likely did not achieve their final height until their early twenties.[6] The numbers within each cohort were small, however, and there is likely to have been some error when assigning age at death (and hence year of birth); these two factors may easily have combined to obscure any real difference that there may have been.

Some effects of status on final achieved height are apparent at the Old Mint site, however (Fig. 18.1). Here a cumulate frequency distribution is shown for male skeletons excavated from within the church of St Mary Graces and from the lay cemetery associated with the abbey.[7] The mean height of the men from the church was 4 cm greater than those from the lay cemetery: 80 per cent of those buried within the church were over 1.65 m, but only about 55 per cent of

[6] Tanner (1981).

[7] St Mary Graces was the last Cistercian foundation in Britain, established by Edward III in 1350 on the site of a plague pit. It was the only foundation of the Order to be established on an urban site and was demolished five years after the Dissolution in 1539.

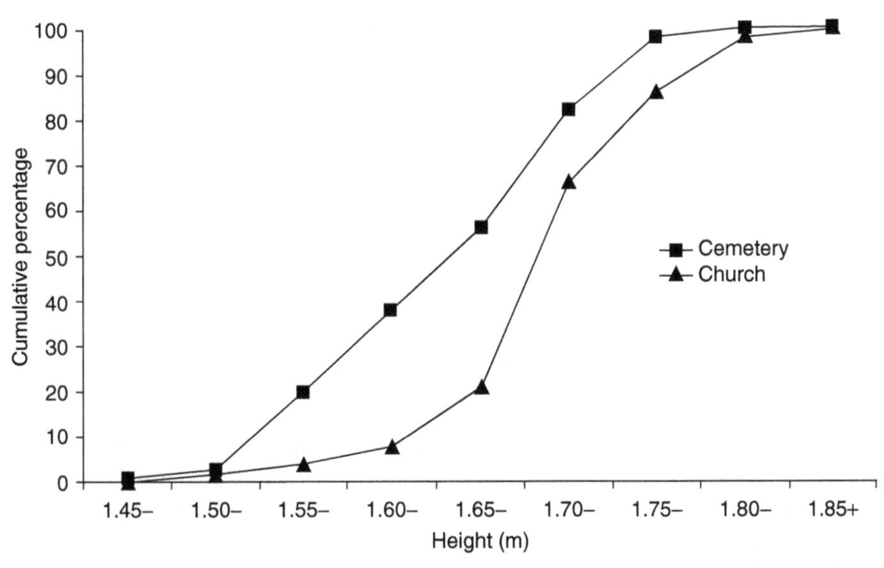

Fig. 18.1 Cumulative percentage distribution of heights (m) of male skeletons from within the church and in the lay cemetery at the Old Mint site, London. The skeletons from the church—presumably of monks or high-status individuals—show a marked shift to the right, indicating that a higher proportion of this group achieved greater heights than those buried in the lay cemetery. The difference in mean heights is 4 cm

those in the lay cemetery. We assume that those who were buried within the church were either the monks or abbots, or benefactors of the abbey and other high-status lay individuals, and further, that their nutritional status was better during their adolescence than those who came to be buried in the cemetery outside.

The growth of children Although there are no generally agreed formulae for determining the heights of children from their skeletons, use can be made of long bone measurements to study their heights for age by comparison with a modern sample.[8] When the length of the femur of children who died at Barton between the ages of 1 and 15 was compared with those of modern children of the same age, it was found that the great majority (*c.*88 per cent) were 'short for dates', that is, had shorter femurs than their modern counterparts. These data indicate that the rate of growth at all ages was slower in the past, presumably reflecting a relatively poorer state of nutrition during this period than enjoyed by modern American children, perhaps a not very surprising conclusion.

[8] The age at death of children can be established with much greater certainty than that of adults and is done by considering the state of formation and eruption of the dentition. If radiographs of the jaws are used, then juvenile skeletons can be aged to within a year or two at most. For further details, see Goode, Waldron, and Rogers (1993). The modern comparative data were published by Maresh (1955).

Table 18.3. Calculated weight (kg) of population from St Peter's Church, Barton-upon-Humber

	Male	Female
Mean	72.9	62.5
Range	59.1–88.7	53.6–72.1

Note: For comparison, the mean weights of contemporary males and females are 84 and 69 kg, respectively: Ruston, Hoare, Henderson, Gregory, Bates, Prentice, Birch, Swas, and Farron (2004).

Body weight

A number of formulae have been proposed from which to estimate body weight from the skeleton. Some utilize the thickness of the cortex of the femoral shaft just below the lesser trochanter, others use the maximum diameter of the femoral head. Using measurements of the femoral head,[9] estimates of body weight could be determined for both males and females who died during the medieval period at Barton (Table 18.3). The mean weights are by no means implausible and are several kilograms lighter than the modern-day mean. In order to sustain a weight of 73 kg, a male undertaking a moderate amount of physical activity would require approximately 3,400 kcal a day, while a female of 62 kg also undertaking moderate physical activity would require about 2,500 kcal a day. These daily requirements are greater than one might suppose were actually achieved by the bulk of the medieval labouring population and suggest that the estimate of weight is probably too high. This is also borne out from an estimation of the Body Mass Index (BMI)[10] of the Barton assemblage, utilizing the appropriate mean height; for both males and females the BMI approximates to 25, which would tend to suggest that they were all rather portly and which is not at all likely.

Dietary deficiencies

Dietary deficiencies produce a wide range of harmful effects depending upon whether there is an inadequate supply of the so-called macro-elements of the diet—carbohydrate, protein, and fat—or of micro-elements, such as minerals or vitamins. A completely inadequate amount of calories in the diet will result in starvation which, in the most extreme case, will lead to premature death. Less dire effects include weight loss, susceptibility to infection, lassitude, and, in the case of females, amenorrhoea and loss of fertility. None of these consequences can be deduced from the skeleton, but there is one effect which could, in theory,

[9] In order to use the formulae which relate to cortical thickness, radiographs of the femur have to be taken; this was not done routinely at Barton, hence reliance on the femoral head diameter, which can be measured directly. The formulae used are to be found in Auerbach and Ruff (2004).

[10] BMI = weight in kg/(height in metres)^2; anything less than 25 is now deemed to be acceptable, although values in the range 20 to 22 are considered to be most desirable.

be detected and that is osteoporosis. It is conceivable that osteoporosis in the skeleton of a child or a young adult could have been the result of starvation, but there have been no such reports from medieval contexts. This may have been because there are no agreed criteria by which to make the diagnosis, or because bone specialists have not been alert to the possibility, expecting to find the condition only in the elderly.

Among the conditions that result from the deficiency of specific nutritional factors, scurvy and rickets certainly produce skeletal abnormalities and the most common of all, iron deficiency, is widely thought to do so.

Scurvy and rickets Scurvy and rickets both produce alterations in the growing skeleton and are both caused by vitamin deficiency. Scurvy is caused by a lack of vitamin C which is plentiful in fresh fruit and vegetables, whereas rickets is due to the lack of vitamin D which is mainly derived from the action of sunlight on the skin.[11] Both scurvy and rickets would have been present in children living in towns during the medieval period in the long northern winters in Europe; those living in the country were probably better provided with fruit and vegetables from gardens or farms, as we have seen in earlier chapters. Both conditions are reversible when fresh fruit and vegetables come back to the table, and when the sun starts to warm the air sufficiently to enable little children to play outside again. Both affect the growing ends of the bones but, in the case of scurvy, the changes may not be very great to the naked eye; there are very characteristic changes on X-ray, and radiography is needed in order to make a definitive diagnosis.[12] By contrast, the changes in the bones in rickets are obvious and, in addition, if the child is walking, the bones of the leg tend to bend under his or her weight and this distortion of the femur and the tibia does not right itself even when the other changes are reversed; it will persist into adulthood, when it is—again—easily recognized.

A major difference between the two conditions is that scurvy is a killing disease whereas rickets is not. It is somewhat of a paradox, then, that scurvy is almost never reported in skeletal assemblages whereas rickets is relatively common, and cases are found in many medieval assemblages. The explanation for the apparent deficit in cases of scurvy is probably because children's skeletons are not routinely X-rayed and so cases are missed. And the reason why rickets *is* found may be that it occurs in the context of gastro-intestinal disease with malabsorption which is likely to have been the cause of death of many young children in the medieval period.

Iron deficiency Iron deficiency is the most common deficiency by far in contemporary populations and there is no reason to suppose that this was not the case during the medieval period. Women are particularly prone because of

[11] Ultra-violet light from the sun forms a precursor of vitamin D from cholesterol in the skin. This compound is then activated by chemical reactions in the liver and in the kidney. The only significant natural dietary sources of vitamin D are found in oily fish and fish oils.

[12] Fain (2005).

their inevitable blood loss during menstruation and childbirth. When the body stores of iron become very depleted, anaemia will supervene, which gives rise to a considerable morbidity but little mortality. In recent years it has come to be believed that iron deficiency can be recognized in the skeleton by the presence of cribra orbitalia—small pits in the superior roof of the orbit—which may or may not be accompanied by a thickening and porosity of the parietal bones, so-called parietal hyperostosis.

Cribra is relatively common in most medieval skeletal assemblages, especially in children, and this is usually interpreted as indicating that they suffered from iron deficiency anaemia. In fact, this interpretation is almost certainly incorrect and has arisen because of the confusion between the different kinds of anaemia. There are, roughly speaking, two kinds of anaemia, iron deficiency and haemolytic. The haemolytic anaemias, of which the best-known examples are sickle cell anaemia and thalassaemia, are all characterized by a very short red cell lifespan—normally about 120 days. Because of this, the red bone marrow, where the red cells are produced, enlarges greatly in volume in the shafts of the long bones, the vertebral bodies, and the skull which may become considerably thickened with—*inter alia*—porotic hyperostosis. The haemolytic anaemias do not occur in northern Europe except in immigrants, so when porotic hyperostosis is found it is frequently attributed to iron deficiency anaemia. Finding cribra in association with porotic hyperostosis gives rise to the conclusion that the two have the same cause, so when cribra are found as an isolated phenomenon it is still assumed to be a reaction to iron deficiency anaemia. Unfortunately for this hypothesis, the red cell lifespan is not materially shortened in iron deficiency anaemia, the bone marrow is little expanded, and on clinical grounds there is no reason to suppose that cribra and iron deficiency are related to each other. Until clinical validity for the connection can be established, the notion should be discounted.

Malnutrition and tuberculosis The association between malnutrition and tuberculosis is well known and it persists into the present day. Infection may be acquired via airborne droplets or from infected milk and dairy products. In the first case the infectious agent is *Mycobacterium tuberculosis* and in the second, *M. bovis*. The former bacterium gives rise to the human form of the disease, which is primarily a lung disease, the bacteria inhaled from the air settling in the lung and giving rise to the possibility of extension to other organs. The disease spreads particularly quickly among those who are poorly nourished and who live in close proximity to each other. Mothers may pass the infection to their children, husbands to wives, and neighbour to neighbour. The disease that results from the ingestion of infected cow's milk—the bovine form—affects the lymph glands in the gut in the first instance but it too may spread to distant organs. Unlike the human form, however, there is no direct person-to-person infection.

Nowadays, tuberculosis is the most common cause of death from a single infectious agent and it has been estimated that approximately a third of the world's population is infected.[13] We have no accurate account of the prevalence of tuberculosis in the medieval period but, during the early modern period, deaths from 'consumption' and 'cough' accounted for at least a fifth of all deaths in non-plague years, to judge from the London *Bills of Mortality*. Deaths during this period were reported by searchers with no special medical knowledge and according to John Graunt, what they reported were bodies 'very lean and worn away'.[14] Although other diseases that cause severe weight loss, such as malignant disease, diabetes, and thyrotoxicosis, would all have been present at the time and would have been inevitably fatal, it is probable that the majority of deaths included under this rubric were, in fact, due to tuberculosis. There is little reason to suppose that, at least in the towns and cities of medieval England, tuberculosis would not also have been rife and a substantial cause of death.

One of the features of tuberculosis, whether the human or bovine form, is a propensity to affect the skeleton, which it may do in up to 30 per cent of cases.[15] In about half the cases of skeletal tuberculosis the spine is involved, giving rise to a characteristic deformity of the spine that is sometimes known as Pott's disease. Apart from the spine, almost any skeletal element can be involved, although it is unusual for more than a single extra-spinal lesion to occur. Tuberculosis may also involve one of the larger joints—the hip, knee, or wrist most commonly—in which case they almost always fuse[16] and this may cause considerable disability.

The appearances of tuberculosis in the skeleton—especially when the spine is involved—are sufficiently characteristic that they are not easily confused with any other condition. Although it is of considerable antiquity, the disease is not commonly diagnosed in medieval populations, which is at odds with its presumed high prevalence.[17] One reason for the discrepancy may be that the individuals with tuberculosis died before the skeleton was involved, or perhaps the expression of the disease was different from the way it is at present, so that there were few cases in which the skeleton was involved.

Two cases of tuberculosis were found in the lay cemetery from the Old Mint site, one a young male of between 15 and 24 at the time of death in whom the disease had affected the spine. In addition there was an elderly male (aged at least 45 when he died) whose left wrist was entirely fused. In both cases the disease was confirmed by the extraction of mycobacterial DNA from the affected bones.[18] The prevalence of tuberculosis in the assemblage from this medieval cemetery was 0.85 per cent, which seems low in the light of our prior expectation: it would be of very great interest to know what the true prevalence was in

[13] Tuberculosis accounts for *c*.3.1 million deaths per year, about the same number as die from HIV/AIDS; malaria accounts for between 1.5 and 1.7 million deaths: data from www.ph.ucla.edu/epi/Layne/epidemiology.

[14] Graunt (1662: 14).

[15] Liyanage, Gupta, and Cobb (2003).

[16] Sequira, Co, and Block (2000).

[17] Pálfi, Dutour, Deák, and Hutás (1999).

[18] Taylor, Crossey, Saldanha, and Waldron (1996).

the medieval period, particularly to see whether it was higher or lower than in the succeeding early modern period. Apart from looking for the morphological changes, it is possible to extract both bacterial DNA[19] and mycolic acids[20] from the skeleton. Both will be present in the body during an infection and if it were possible to analyse a large number of skeletons from medieval contexts this should provide a better estimate of the likely prevalence of the disease than reliance on macroscopic lesions.[21] The results would almost certainly conform with studies of assemblages from ancient Egypt, which showed a high prevalence of infection in skeletons that were apparently normal.[22]

Plague Plague provides an interesting example of how malnutrition may sometimes provide protection against an infectious disease. In modern epidemics of plague it is usually found that young men are more commonly affected than any other sector of the population.[23] One explanation that has been given for this is that the very young, the very old, and women are more likely to be iron deficient than young men and that since the plague bacterium requires a plentiful supply of iron to replicate, while it may be able to colonize the tissues of those who are iron deficient, it may not be able to multiply sufficiently rapidly to produce an overwhelming infection.[24] There was an excess of males among the skeletons excavated from the Black Death plague pit at the Old Mint site which conforms to expectation from modern studies. There may, therefore, have been at least *some* advantage in being a female during the medieval period.

The effects of plenty

So far as the diet is concerned, it is by no means always the case that more is better. Indeed, in contemporary developed societies, the most important nutritional problems seem to be related either to total excess or imbalance, and there has been a dramatic increase in the number of individuals labelled as obese.[25] Obesity was not entirely unknown in the medieval period: the fat monk is a stereotype that has persisted to the present day,[26] and Henry VIII was also

[19] See, for example, Taylor, Goyal, Legge, Shaw, and Young (1999); Konomi, Lebwohl, Mowbray, Tattersall, and Zhang (2002); Gernaey, Minniken, Copley, Dixon, Middleton, and Roberts (2001). With modern techniques, it is possible to determine the species of bacterium from the DNA recovered; to date, in all cases that have been studied from the UK, the infecting organism has been found to be *M. tuberculosis*.

[20] Mycolic acids are present in the fatty coat which surrounds the tubercle bacillus and each species has a unique combination of acids which permits them to be differentiated one from the other.

[21] Of the two techniques, mycolic acid would be preferable since it is not so readily subject to contamination as DNA analysis: Tanaka, Annop, Leite, Cooksey, and Leite (2003).

[22] Zink, Grabner, Reischl, Wolf, and Nehrlich (2003).

[23] Boisier, Rahalison, Rasolomaharo, Ratsitorahina, Mahafaly, Razfimahefa, Duplantier, Ratsifasoamanana, and Chanteau (2002).

[24] Ell (1984).

[25] That is, those whose BMI is greater than 30; those whose BMI is greater than 40 are considered to be morbidly obese.

[26] Patrick (2004).

notoriously fat in middle and old age, but it is not likely that obesity would have been prevalent among the urban or rural poor.

One disease that is easily recognized in the skeleton and which is in some cases related to obesity is osteoarthritis. There is good epidemiological evidence that obesity is strongly related to the development of osteoarthritis of the hands and the knee, but only weakly to osteoarthritis of the hip.[27] It is not difficult to understand how osteoarthritis of the knee might be related to body weight; it is not so simple to understand why arthritis of the hands is similarly weight related, or why osteoarthritis of the hip is not. The results of the epidemiological studies suggest that a metabolic factor is responsible[28] but give no clue to the reason why some joints and not others are affected. The results of a study of the weight of individuals with and without osteoarthritis of the hand or knee buried at some medieval monastic sites in London suggest that those with osteoarthritis were indeed heavier than those without, indicating that the association between obesity and some forms of joint disease is far from being a recent phenomenon.[29]

DISH DISH—diffuse idiopathic skeletal hyperostosis—is an interesting condition that is characterized by an exuberant formation of new bone into the anterior longitudinal ligament of the spine and by the ossification of extra-spinal entheses and ligaments (Plate 18.1). Although changes may appear in the spine at any level, they are often most prominent in the thoracic region where they are confined to the right-hand side of the vertebral bodies.[30] In time the ossification leads to fusion of variable numbers of vertebrae, but the intervertebral disc spaces and the facet joints are normal in the absence of any other pathology. DISH is rarely found in individuals under the age of 40 but becomes increasingly common in older age groups and is more common in males than in females.

In a paper in the *British Medical Journal* in 1985 I suggested, rather tongue in cheek, that DISH might be an occupational disease for those following the monastic way of life on the basis of an apparently elevated prevalence in burials from Merton Priory.[31] The suggestion was taken sufficiently seriously for others also to note an increased frequency of the condition at other monastic or high-status sites.[32] Data from two sites in particular give strong support to the notion,

[27] See Coggon, Reading, Croft, McLaren, Barrett, and Cooper (2001); Sowers (2001); Livense, Bierma-Zeinstra, Verhager, Baar, Verhaar, and Koes (2002); Dawson, Juszczak, Thorogood, Marks, Dodd, and Fitzpatrick (2003); Sayer, Poole, Cox, Kuh, Hardy, Wadsworth, and Cooper (2003).

[28] Once candidate is leptin, a hormone that is derived from adipose tissue. It is found in joints affected by osteoarthritis in concentrations that are correlated with the BMI: Dumond, Presle, Terlain, Mainard, Loeuille, Netter, and Pottie (2003).

[29] Patrick (2004).

[30] It is thought that the reason that the changes are seen only on the right side of the thoracic vertebrae is because the pulsation of the descending aorta (which lies on the left) prevents ossification of the anterior longitudinal ligament on that side. If the aorta lies on the right side of the vertebrae, as it does in the rare condition known as transposition of the great vessels, then the ossification occurs on the *left* side in the thoracic region.

[31] Waldron (1985).

[32] See, for example, Janssen and Maat (1999); Maat, Mastwijk, and Sarfatij (1998); Mays (1991).

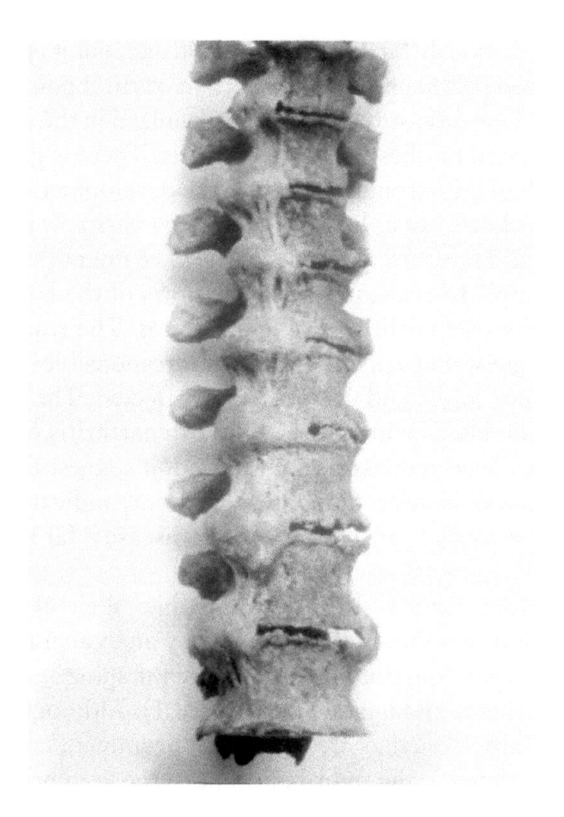

Plate 18.1 Diffuse idiopathic skeletal hyperostosis (DISH) in the thoracic region of the spine of a medieval female from Barton-upon-Humber. Note the flowing new bone on the right-hand side of the vertebrae. Photograph: T. Waldron.

Wells Cathedral and the Old Mint site in London.[33] At Wells Cathedral a large number of skeletons was excavated, including some from the thirteenth-century Lady Chapel and the sixteenth-century Stillington's Chapel. In both cases the burials found there were those of priests or lay benefactors. The prevalence of DISH in males from these two sites is shown in Table 18.4. DISH occurred much more frequently in the presumed high-status burials; the differences at the Old Mint site are highly significant ($p = 0.001$), although those at Wells are not ($p = 0.12$). When the data from both sites are combined, however, the difference is again highly significant ($p = 0.001$), which suggests that it is very unlikely to have arisen by chance.

The reason for the condition occurring so frequently in monastic or high-status burials is probably to do with differences in nutrition. In the present day DISH has been found to be associated with obesity and late onset diabetes, and may also be related to other metabolic disorders, including alterations in vitamin A metabolism.[34] We know that those living in the monasteries were likely to enjoy a diet with an extremely high calorific value,[35] and even if not all the food

[33] Rogers and Waldron (2001).

[34] Julkunen, Karava, and Viljanen (1966); Denko, Boja, and Moskawitz (1994); Abiteboul and Arlet (1985).

[35] Harvey (1993); above, Chapters 7 and 8, for other examples of the calorific content of high-status diet.

Table 18.4 Prevalence of DISH in male burials at Wells Cathedral and the Old Mint site

Site	Number of males	Number with DISH	Prevalence (%)	95%CI
Wells lay cemetery	93	6	6.5	3.0–13.4
Lady Chapel	15	2	13.3	3.7–37.9
Stillington's chapel	13	3	23.1	8.2–50.3
Mint lay cemetery	99	0	0	0–3.7
Church and chapels	52	6	11.5	5.4–23.0
Total lay	192	6	3.1	1.4–6.6
Total high status	80	11	13.8	7.9–23.0

Note: 95%CI = 95% confidence intervals.
Source: Rogers and Waldron (2001).

provided in the daily allowance was eaten by the brothers, it seems highly likely that their consumption was sufficient to render at least some of them obese. The frequent consumption of oily fish and offal may also have contributed to the high prevalence of DISH—both, as we have seen in earlier chapters, prominent in the diet of those for whom the eating of flesh meat was restricted.

On this evidence it seems that DISH provides a good marker on a population level if not for obesity, then for a rich diet. Since the condition may occur in those who are not obese, or whose diet is not excessive, it is not possible to argue, as some have tried to do, that a single individual with DISH must necessarily have been of high status;[36] the argument holds true only for groups or populations, not for individuals.

Dietary intoxicants

Toxic substances may enter into food and drink accidentally or deliberately, but few will directly affect the skeleton. Two which are of some interest in this respect are lead and fluoride.

Lead poisoning Exposure to lead has been common during many periods in the past and during the medieval period it would have been considerable. The main route of exposure for the general population would have been through contaminated food and drink; the main sources of lead were poorly glazed earthenware, poor-quality pewter vessels, and contaminated wine. Lead poisoning would have produced a considerable morbidity and it may have been a cause of death in children. There are no morphological signs of lead poisoning in the skeleton, but lead interferes with the normal growth of the skeleton and produces a dense white line at the growing ends of the long bones which can be seen on X-ray, the so-called lead line. Although not absolutely pathognomonic

[36] Bruintjes (1987).

of lead poisoning, other causes are either extremely rare or readily differentiated, and it is not difficult to make the diagnosis from an X-ray.[37] A radiological survey of the long bones of children from a medieval assemblage would be instructive in determining the extent to which they had been exposed to lead; any found to have lead lines would most probably have died from lead poisoning, but up to this point no study has been carried out.

Fluoride Fluoride is taken up avidly by the skeleton and the teeth and when in excess may cause brown mottling of the teeth and widespread production of new bone, especially at the site of entheses—the points at which muscle tendons enter the bone—and in ligaments,[38] and fluoride poisoning may sometimes be confused with DISH. Fluoride is present in sufficient concentration in some ground waters in England—for example in Essex and Yorkshire—to cause tooth mottling. Judging from the state of the teeth in medieval skeletons, however, few were exposed to fluoride in the water in sufficient quantities to protect them from dental caries.

Conclusion

The composition of the diet is to some extent reflected in our skeletons and an examination of the remains of our ancestors provides the most direct evidence by which to assess at least some components of it. Other chapters in this volume show that there was a substantial discrepancy between the diets of the rich and the poor, and the ill effects of poor nutrition must have been evident to those who chose to observe them; they would certainly have been evident to those who experienced them. There seems no reason to suppose that sections of the population would not have achieved their full height potential and would have been subject to cyclical episodes of rickets and scurvy.

There remains much to be done, however, before we can know the true extent of some of the deficiency diseases; provide data by which valid comparisons can be made of the final achieved heights of past populations; and study growth rates in children. By working in concert, bone specialists and historians will be able to maximize their efforts in this field and, to some extent, compensate each other for the deficiencies in their own disciplines.

[37] Woolf, Riach, Derweesh, and Vyas (1990).
[38] Kleerekoper (1996).

19

Conclusion

C. M. WOOLGAR, D. SERJEANTSON, AND T. WALDRON

This examination of food and drink has ranged widely, from the day-to-day diet of individuals to the macrocosm of demography, from analyses of the role of individual commodities to studies exploring the complexity and variety of consumption. Prominent among the themes of this book are long-term changes in diet, defined in terms of the overall levels of availability of food, the range of foodstuffs, and the effects of diet in terms of nutrition. Some changes were widespread; others affected some groups of society or regions to a greater degree. On a general level, medieval diet contrasted with the prehistoric period and with some aspects of post-medieval consumption.

There were other long-term patterns of importance, related to seasonality, religion, and status. In this, causality and dietary change can be seen to have had a complex interaction. Advances in the technology of agriculture or fishing, the relationship between the animal economy and humans, the rise in urbanization with its demands for provisioning of urban centres, and cultural links with new trading patterns, all show close connections between the economy and diet. Cultural factors—and personal choice—had significant potential for influence on diet. Adherence to the teachings of the Church brought varying dietary implications, and so might links to ethnic or other groupings, the imitation of patterns of consumption, or sumptuary restrictions.

The cross-disciplinary nature of the subject brings to the fore questions of methodology and evaluation of the evidence. There are recurring discussions of the relationship between the historical, archaeological, and scientific record, especially about what is evidence for food production and what illuminates consumption. Assumptions drawn from historical sources need to be tested carefully against the archaeology. Bones may be better evidence for consumption than historical records where the latter are dominated by information about production. In contradistinction, the nuances of consumption apparent from documentary sources indicate something of the range of interpretation that we must consider for purely archaeological and scientific material.

The availability of food

Between the departure of the Romans and the mid-sixteenth century, long-term changes in food availability pivot around four points: the Middle Saxon period, with its new economic, social, and political structures, the development of towns, and the expansion of agriculture; the mid-eleventh to early twelfth centuries, with some aspects perhaps linked to the Norman Conquest, others to more general cultural change; the late thirteenth and early fourteenth centuries, encompassing the Great Famine of 1315–17 and the peak levels of medieval population; and the period following the Black Death, in the wake of its catastrophic mortality.

Of the factors that had a major impact on the availability of food, the level of population was outstandingly significant. It grew through the Saxon period and the economy moved to a greater element of cereal production, pushing out the margins of cultivation. We have little direct evidence of cereal production at this period, but the changes found their reflection in different patterns of animal husbandry. As the population increased, quantities of livestock per capita diminished between the eleventh century and the end of the thirteenth. This decreased the productivity of agricultural land and animal husbandry, restricting for many both the variety of diet and amounts of food—and at some points, leading to starvation. The population, already falling at the start of the fourteenth century, was further checked by famine and the Black Death—reversing many of the constraints on the general availability of food. After the late fourteenth century famine was a far less significant element in England and the pattern of diet was considerably improved, with a greater availability of meat and dairy products.[1]

The growth of population required systems of production and distribution to sustain the availability of food. Urbanization, from the development of *wics* to the much larger towns of the early fourteenth century, commercialization, and development of systems of production for the market, all had an impact on the range, quantity, and quality of foodstuffs in town and country. The increasing percentage of the population which lived in towns was dependent on reliable supplies of cereals, the market for which is well documented after 1200. Changes in butchery patterns demonstrate how meat from cattle and sheep developed into a market-based commodity. The supply of some foodstuffs, such as garden produce or domestic fowl, although present in the urban market, was never fully commercialized. It nonetheless formed an important element in sustaining both the variety of diet and the economy of town and countryside and may have been a significant mitigation of the diet of the peasantry, particularly at the extreme levels of population. On the other hand, the market-based economy was the key to the expansion of the range of commodities within late medieval diet, through the import of marine fish, spices, dried and exotic fruits, and wine. Equally

[1] Dyer (1998*b*: 70–1).

significant in terms of the food supply and range of diet were the efficiency of agriculture and livestock husbandry, both peasant and seigneurial, the development of marine fisheries, and access to wild resources.

The diversity of foodstuffs

Food in England became increasingly diverse at two points in the medieval period: in the eleventh century, and in the fifteenth. Choice in terms of foodstuffs is, on one level, a question of resource: those without the means cannot obtain variety. But the potential for variety—or a notion that there should be other possibilities—has to occur first, whether it be through trade, exchange, the development of specialized industries, or facets that become culturally desirable. The change in the eleventh century is characterized by the appearance of significant quantities of marine fish in both fresh and preserved form, a new pattern of consumption for game, and the advent of a range of exotic spices. It had an especial impact at high-status sites, although it was not exclusively felt there: the alteration in urban diet was also considerable. The Norman Conquest and the arrival of an Anglo-Norman aristocracy influenced these changes. New hunting practices were linked directly to this group; but other aspects of dietary change, such as the growth in fish consumption, particularly in upper-class households and the monasteries, may have reflected a wider pattern across Europe and coincided with the widening impact on dietary patterns made by the Church—although the expansion in the consumption of fish was a development that had started in pagan communities in northern Europe and it was also a particularly convenient resource for the population of the burgeoning towns. Trading connections brought in new commodities, and diet in England, at least at an elite level, came to conform broadly to a wider pattern of European cuisine. For the first time since the end of the Roman Empire, one can point to a European-wide style of cookery, in which exotic spices, strong tastes, and thin, acidic sauces were prominent. A common milieu for this development can be found in the contacts with the southern and eastern Mediterranean. While this may have been an elite cuisine, it was one that had its reflection at other layers within society. There was at the same time a good deal of variation between countries and within countries, but it provided a common background to the investment in food as part of social display and was to form part of a common European culture throughout the later Middle Ages.[2]

The impact of the second change, which occurred in the fifteenth century, especially in towns, was less far-reaching. The range of game and birds (but not fish, which were already present in considerable variety) that is regularly recovered on archaeological sites for this period is noticeably wider than before.

[2] Mennell (1985: 40–61); Woolgar (in press).

One might point, for example, to the presence of rabbits and wild birds, a broadening of urban diet.

Other patterns

Status and social stratification had important consequences for diet throughout the Middle Ages. From the feasting patterns recorded in the bone assemblages from Saxon sites to the restrictions on access to game imposed after the Conquest, sumptuary regulation of the consumption of roast meat or wild birds in the later Middle Ages, the use of expensive spices, or simply the quantities of food available, diet and status were inextricably linked. Social aspiration found its expression in food: what was considered a luxury in one generation became routine in the next, even in a monastic setting. The rise of meat eating was not solely a phenomenon of the period after the Black Death; its subtle growth within monastic diet, for example, in emulation of upper-class diet, can be seen throughout the high and late Middle Ages in both the historical and archaeological record. At the same time, high-status diet might be rejected by others, themselves of high rank.

Seasonality had a major impact on all areas, from matching the patterns of agricultural production and animal husbandry to those of consumption, to the arrangements for preserving food. We can see this from the historical evidence for the late medieval period and, prima facie, it must have been significant at an earlier period as well. Part of the seasonal arrangements were connected to Christian observance, but we should note that the pattern of diet that was implicit in religious observance was by no means uniform, that there was considerable variation in detail across social groupings and temporally. Models of monastic diet, derived from the Benedictine Rule, were mitigated and varied by both ecclesiastical influence and secular habit; the diet of the laity might be influenced in turn by religious exemplar. This concern for diet and spiritual well-being was matched by medieval notions of physical health, which created further patterns in consumption. The historical evidence shows diet to be both a highly personal matter and also one that might have a corporate dimension.

Changes in nutrition

The diet of the peasantry The range of evidence for the diet of the peasantry is greater than might have been expected, given the lack of archaeological material from rural sites and the focus of documentation on other parts of the economy. It is important to recognize, however, that there were significant divisions within rural society in terms of land and livestock, resources such as gardens, and access to wild foods. Of the long-term shifts in the balance of nutrition, however, one of the most significant was the growth of a dependence on cereals. At the upper end of the scale, the wealthier peasantry had strips in the common fields for growing cereals, in the period before the Black Death, some wheat, but mostly barley, oats or mixed corn, and beans. They would also have been able to grow garden produce and to keep a few animals, but the level of livestock in the countryside was very low by the thirteenth century. It may have been much higher earlier, particularly in the Saxon period: the advanced age of many animals from rural sites in the twelfth century points to this decline—these were animals at the end of their working lives, not raised primarily for meat. By the thirteenth century the peasantry can have consumed little by way of meat and dairy products: this must have led to a decline in fat and protein in diet for the majority of the population. While some sources suggest the poor rarely ate beef, zooarchaeology indicates that it was consumed, along with mutton, although the quantities cannot be established. Pig keeping may have been important among the peasantry, but swine may have been kept as much for sale as for consumption, since the peasants appear to have eaten few of them. The peasantry was partly within the money economy and partly without. Some foodstuffs would have been purchased, in particular wheat and other cereals if the household could not grow enough—and the sale of other goods, such as pigs and cheese, would have facilitated this. At the same time, peasants had hens, geese, and sometimes ducks—and some specialized in keeping domestic fowl for their own consumption, as renders to the landlord, and to sell. Eggs were another important source of protein and very substantial quantities were produced. Peasants were less likely to keep geese, although many did.

Fishing on a small scale, from streams and rivers, may have augmented diet, but there are few fish bones on rural sites to confirm this. For those who lived inland, marine fish, both preserved and fresh, could only be obtained by purchase. Of these, herrings must have been the most affordable, but the archaeological evidence for their consumption is scant, even after the eleventh century when fish constituted a greater element in diet in some places. Remains also suggest that little stockfish was eaten in the villages. Rural sites also produce very little game or small birds, although we know from historical evidence that these were caught—perhaps indicating that there was more value in catching them for sale than for domestic consumption.

Those in the countryside—or in the towns—with little or no land must have had the most restricted diet of all. Some may well have kept chickens, although they would have had no cattle, sheep, or pigs. Some wild food, especially plant foods, was available in the countryside, but after the Conquest the poor were impeded by legal restrictions from killing most wild animals, and there is little evidence from archaeological remains that they did so. Some of the rural population ate horseflesh, at least if we believe the evidence from some rural sites where there was a surprisingly high percentage of horse bones butchered in a fashion similar to cattle bones. Presumably it was the poorest who ate horseflesh, though paradoxically it is likely that they were too poor to have owned a horse.[3]

[3] Albarella and Davis (1994); Albarella (1997).

The poor also had access to charity, the alms or leftovers from monastic houses and the houses of the nobility, although this would have had no more than an occasional impact.

Most of the population would have had ample food during and immediately succeeding the grain harvest. Just as horses and draught cattle had to be fed more at the times of year when they had to provide the largest amount of work, the human labour force was treated in the same way, with most food provided during the period of harvest. The leanest time of the year fell in the months preceding the grain harvest, particularly in May and June. The diet of the wealthy peasantry, at least, was reasonably good, particularly when supported by the products of horticulture.

Urban diet Urban diet differed from that in the countryside in a number of respects. First, the amount of food produced in the town was usually limited to that from a comparatively restricted number of gardens, orchards, and livestock. There was probably at the same time a greater reliance on foods that were preserved, or could be maintained in a state fit for sale. Meat (and possibly also cereals) for the first Saxon town-dwellers may not have been obtained through the market, but through the organization and patronage of the lords who established the towns. The earliest towns or trading ports had a meat diet which was unvaried compared with the later period, although it was not necessarily more or less varied than that in the countryside. There is debate about how much livestock there was in towns in the high Middle Ages and whether what was raised there was in fact destined for consumption within the town. The presence of butchers indicates that towns did get supplies of fresh meat. By the later Middle Ages this was both varied and specialized, with cattle and sheep fattened and driven to urban slaughterhouses. Some foods that preserved well—such as hams, sausages, and fish—may have been peculiarly well adapted to the urban market. Large, marine fish, for example, seem to have been consumed more in towns than in high-status residences or the countryside.

A crucial element in medieval urban life was the maintenance of a supply of cereals—and hence bread—a subject of considerable interest to both municipal and national authorities. In the late medieval period, the use of mechanisms to ensure constant pricing, but with variation in the size of loaf dependent on market fluctuations, enshrined in theory the position that some bread might always be within the financial reach of the majority. Urban markets had a wide range of produce, including game and wildfowl by the end of the Middle Ages. Garden produce brought to the town for sale supplemented the supply of fresh vegetables and fruit from urban horticulture; dairy produce, domestic fowls, and their eggs were similarly brought in from surrounding villages. At the same time, there must have been very clear contrasts between what was available to the rich and the poor in the towns.

Upper-class diet Upper-class diet differed from that of most of the population, but it was not unchanging. It had always contained substantially more meat than that of all other sections of society. In the Saxon period, there was beef; and pork and beef were eaten in the eleventh to thirteenth centuries. In late medieval England there were changes, with a decline in the consumption of pork, a concentration on lighter meats and birds for the highest echelons, but with very substantial quantities of beef and mutton for the bulk of the household. Cattle, sheep, and sometimes pigs predominate in the remains from wealthy households and these meats must be what was fed to most of the household members and servants. The impact of game—and especially venison—for display and other elements of conspicuous consumption are well known. Elaborately decorated and spiced food, especially after the Conquest, with wine rather than ale, marked out the quality of the establishment.

Fish also constituted an important element in upper-class diet from at least the twelfth century: it was here that both the greatest range of species was eaten and in the largest amounts. Household bread was of the highest quality and ale was available in substantial quantities, but we can be less sure about the proportion of fruit and vegetables eaten. They were highly regarded in some circles and, along with dairy produce, they may have been drawn from home farms, gardens, and orchards, escaping documentation.

Monastic diet The original Benedictine diet of the sixth century was designed for a Mediterranean environment, where vegetables were varied, plentiful, and had a long growing season, and olive oil was available. The modification of this regime in England is evident particularly after the Norman Conquest, when monastic diet was demonstrably a variant of upper-class diet. Monasteries began to expand the meat-based components of diet, as well as embracing fish in substantial quantities. This dietary regime extended well beyond the abbot or prior at the head of the house, to the monks.[4] The level of wealth of different monasteries affected what appeared on the table. The poorest houses in late medieval England might serve water rather than ale, whereas the richest had the finest wines. Monastic consumption before the Conquest is more opaque. Roe deer are particularly in evidence on Late Saxon monastic sites; but there is a gap in our knowledge of monastic fish eating in the Saxon period, which may be filled by sites such as the Anglian abbey at Flixborough.[5]

Drink In terms of nutrition, consumption of ale (and subsequently beer) was almost as important as eating bread or making pottage with cereals, although it was not as efficient in transferring energy from the cereal to the consumer. Other beverages—mead, cider, and perry—were probably as significant only in

[4] Harvey (1993: 34–71). [5] Loveluck (2001).

restricted areas. Wine was a hallmark of upper-class consumption, much as whey as a beverage marked out the peasant producer. Milk was important in the diet of the young and, along with cream, an element in the diet of the female upper classes. Some may also have been available to the peasantry during the summer. Verjuice, like vinegar, was used most for tart sauces or for preserving foods.

Regional differences There are some indications of variation in diet on a regional basis, but it is less marked in these studies than might have been expected. The most important difference lay in cereal production, although it is not clear that this pattern was mirrored in terms of consumption. Wheat requires good soils and a long growing season. This restricted its availability in the north and west, particularly in Scotland and Wales, but there must have been some effects in England. This would have made a difference to the quality of the diet, but its impact on the peasantry, who grew more barley and rye and much less wheat in the period before the Black Death, would not have been as great. At the same time, the aristocratic households of medieval England could transport their best corn and flour as they moved. There were, however, observable differences in terms of price: distance from London and the competing pressures of the great urban markets made some commodities, such as ale, more affordable in the countryside. Some areas exhibited specialities in terms of production, for example, in concentrations of orchards.

Market forces worked to some extent to even out regional differences. In terms of cattle production, for example, livestock, raised in pastoral areas, in Wales or in the north, was moved on the hoof to centres of consumption. Pastoral areas likewise produced a greater volume of dairy products—although they will not necessarily have been consumed there. Some areas may have specialized, for instance, in producing goats. The consumption of fish had a stronger regional impact: those closer to the coast ate considerably more fish and its presence in diet in south-west England, with the development of the fisheries there, is more marked than elsewhere at the end of the Middle Ages. The range of fish also differed on a regional basis. The fenlands, not only around the Wash, but also in more restricted areas in Somerset and in East Sussex, were important sources of freshwater fish, as well as wildfowl. Although these might seem obvious resources for local consumption, both freshwater fish and wildfowl were probably of more value to the peasantry as commodities for sale than for consumption.

Imported food The market economy, from the Middle Saxon period onwards, brought food into England. At the great fairs and urban markets of medieval England, preserved foodstuffs from overseas, such as stockfish and herring, were traded. Customs accounts document traffic in a wide range of foodstuffs, including vegetables, conger eels, dairy products, and cereals. Some of these goods were staples of consumption, particularly in urban markets; but many of them were redolent of high status. The most distinctive feature of the trade was the goods of high value, particularly spices, dried fruits, condiments and confections from the Mediterranean, and wine. Preserved venison and other commodities might be moved by a household from its estates in Ireland to England. Imported goods, however, are barely visible in the archaeological record beyond a few botanical specimens, the evidence for preserved fish, and pottery containers that may have been associated with particular goods, such as vessels for wine.

Was England distinctive in its diet?

If the evidence for diet in England is contrasted with that of the Continent, it can be seen that there are dietary patterns in common, particularly with north-west Europe, even before the changes of the eleventh and twelfth centuries. The environment here would have offered opportunities similar to England for agriculture and animal husbandry, while the Mediterranean lands had a different tradition of agriculture and food. Some of the most comprehensive zooarchaeological studies of early medieval towns have been carried out on material excavated from the emerging urban centres of Saxon and Viking north-west Europe. Research on this material, such as the fauna from Haithabu in northern Germany[6] and sites in the Low Countries, is now making an important contribution. Fewer assemblages from France have been published relative to the importance of that country, although the first syntheses of archaeological finds of animals in the Middle Ages were drawn from French material.[7]

The most significant contrast between Britain and the rest of Europe was in the part played by pigs in the economy and diet. In northern and central France and Germany more pigs were kept and eaten than further north and east, which were mainly areas where cattle predominated.[8] In northern France, as in England, numbers of pigs declined relative to sheep in the later Middle Ages,[9] but in France and Germany they continued to be more frequent than elsewhere. Britain is alone in northern Europe in having a high percentage of sheep, a species which was otherwise numerically most important in the countries around the Mediterranean. English reliance on wool exports dictated consumption to a greater extent than in neighbouring countries. The average age at which sheep were culled rises over time elsewhere in Europe, as it did in England, reflecting the importance of wool production.[10]

A wider range of meats and other foodstuffs was consumed at sites inhabited by wealthy households than at rural settlements. In Flanders in the late Middle Ages social distinctions in terms of consumption may have been at least as clearly defined as in England.[11] In France wild boar as well as deer featured in the diet of

[6] e.g. Reichstein and Pieper (1986).
[7] Audoin-Rouzeau (1993; 1997).
[8] Audoin-Rouzeau (1997: 73).
[9] Audoin-Rouzeau (1992: 85, fig. 1); Clavel (2001: 188).
[10] Chapter 5; Sykes (2001: fig. 15).
[11] Ervynck (2004).

the wealthy: the scarcity of wild boar in medieval England presages their complete extinction in Britain in the seventeenth century. The hunting and eating of wild birds in France mirrored the English pattern: the main species were almost the same, and the quantity and range of species increased in the later Middle Ages.[12] Finds from rural sites confirm that in the later Middle Ages the peasantry was raising and selling chickens, as they did in England.[13]

There were similarities in fish consumption between the Low Countries and England, as well as some differences. Sites have quantities of flatfish, together with herring and eels—which might be expected—and more cyprinids were eaten. The earliest good evidence for carp rearing has been found at the castle of Ename in Flanders. Away from the coast, stockfish penetrated everywhere. Otherwise, at sites distant from the coast, unsurprisingly there is a stronger emphasis on freshwater fish, especially cyprinids and eels. Between the eleventh and thirteenth centuries fish consumption was quite high in French towns, with amounts of marine fish increasing through to the sixteenth century.[14]

The range of fruit and vegetables which could be grown in north-west Europe was much restricted compared to the countries around the Mediterranean. As in England, fruits such as figs and raisins had to be imported, while the range of local produce is similar to English finds.[15]

Essentially in the animal economy of north-west Europe there were many correspondences with England, governed both by the constraints of the environment and also by the broadly similar patterns of landed organization—although in the thirteenth and fourteenth centuries the direct exploitation of estates by English landlords, rather than leasing out their property, has produced vastly more documentation for the rural economy and evidence for local practice. The comparisons between habits in England and France following the Conquest reveal a not unexpected community of culture between the two countries, with some features appearing first in England. Fish, other than the ubiquitous herring, eels, and stockfish, may have varied more between different regions than other foods. There are also sites in France—as in England—where remains of fish are unexpectedly rare compared with places that are otherwise apparently similar.[16] Local taste and preference no doubt had some influence, but the broader picture shows similar trends everywhere.

Historical evidence points to other unifying factors. The evidence for patterns of feasting, the exchange of food as gifts, in both pagan and Christian contexts,

[12] e.g. at La Charité-sur-Loire and Vatteville: Audoin-Rouzeau (1987); Sykes (2001: table 8).

[13] Audoin-Rouzeau (1992: 88).

[14] Ervynck and Neer (1994); Neer and Ervynck (1996: 161–2); Heinrich (1994); Sykes (2001: 53–5); Clavel (2001); Querrien (2003).

[15] e.g. at Sint-Salvator's Abbey at Ename in Flanders, where quantities of fig and raisin pips were found in a latrine: Ervynck, Cooremans, and Neer (1994: 315).

[16] Sykes (2001: 55) noted that there were unexpectedly few fish from the castles at Vatteville and Mayenne in Normandy, interestingly similar to their virtual absence at Carisbrooke Castle.

demonstrated for Merovingian Gaul, has many similarities to practices in Early and Middle Saxon England.[17] In the high and late Middle Ages, the dietary regime established by the Church was a major influence on patterns of consumption across Europe, in terms of both conformity and delineating heresy. Equally there was a strong association between patterns of consumption and status, although the detail of consumption varied with local products.[18] The contrast between the Mediterranean lands and northern Europe is, however, very apparent. Although the evidence is scattered, a study of food and drink in southern Italy between approximately 900 and 1200 demonstrates a similar pattern to the archaeological record, noting the presence of pigs, sheep, and goats, but fewer cattle, and the consumption of fish, especially freshwater and lacustrine fish.[19] For the later medieval period, when there is substantially more documentation, from cookbooks to domestic accounts, it can be readily seen that cookery around the Mediterranean had distinctive elements within the overall pattern of cuisine: a close association with Arab styles of cuisine, a use of oils rather than dairy products as a source of fat, and a wider range of spices.[20] But even with a similar pattern of food production, there might be nuances in consumption, in local varieties and cooking practices, readily observable to the consumer. At the level of the upper-class meal in the fourteenth and fifteenth centuries, for example, there were differences in the structure of menus: eating in France would have been a different gastronomic experience from a formal meal in England even if many of the basic elements of diet and some of the dishes were similar.[21] After the Black Death, one important contrast with the Continent is the robustness of the food supply in England. Cereal prices might fluctuate, but overall productivity was normally sufficient to avoid famine; the supply of meat was equally sustained.

Interpretation, methodology, and themes for research

The very considerable range of work that has come to fruition over the last twenty years has done much to expand our knowledge of medieval diet. The outline that was established by Christopher Dyer can now be taken back through the Saxon period.[22] Although the different evidence will not allow us to estimate calorific content of diet for the earlier period, it does show us the range of consumption. It is also possible to fill in detail for later medieval diet and to contrast it with consumption elsewhere in Europe. There are, however, many areas where further work might be done to clarify particular questions and methodologies.

[17] Effros (2002).
[18] e.g. Dembińska and Weaver (1999: 71–135); Crossley-Holland (1996: 61–112); Riera i Melis (2000).
[19] Skinner (1997: 3–21).
[20] Grewe (1992); Santich (1992); Weiss Adamson (2002); Flandrin (1983).
[21] Flandrin (1992).
[22] Dyer (1983).

The discussion of food and related subjects in medieval England requires a close understanding of how historical evidence, archaeological remains, and the results of archaeological science can be brought together. Typically, any analysis of food must take into account the relationship between production and consumption: here the sources divide. Historical records often concentrate on production or trade, much less commonly or directly on consumption; the archaeological remains of bones provide direct evidence of consumption, although they may also provide some information about production.[23] The dilemma at its simplest can be seen in interpretations of the increased consumption of veal in the later Middle Ages. Was this driven by a demand for the meat, either as conspicuous consumption or as a matter of taste? Was it driven by the production of an excess of male calves, since these are a concomitant of intensive dairying to meet demands for milk, cream, butter, and cheese? Similar questions arise with fish, especially herring and other species that were preserved. Was their consumption driven by demand, to meet the dietary pattern established by the Church? Did the availability of a cheap source of protein to feed the urban population allow more rigorous adherence to this dietary pattern than might otherwise have been the case?

The different categories of evidence provide a range of answers to these questions. Historical records can reveal the quantities of meat and fish consumed, allowing us even to estimate levels of calories in individual diet, and to establish some, at least, of the motives for consumption. It is not possible to use excavated remains in this way, although they may provide evidence for consumption by a whole range of individuals who may, or may not, be covered by the historical record.

Archaeological remains have other limitations: bones provide more direct evidence for consumption than the remains of seeds; the remains of larger animals are a better witness to consumption than those of smaller creatures, including most birds and especially fish. Historical sources offer a potentially bewildering range of terminology for food in its fresh and preserved states; archaeology cannot distinguish the remains of salt meat from fresh meat, although it can sometimes make a distinction between fresh and preserved fish. Even where assemblages have survived well, and bone fragments have been counted in hundreds or even thousands, it is still difficult to estimate directly the quantity of meat provided.

Comparisons between larger mammals and smaller mammals, and between birds, fish, and mammals, have been based on the number of fragments identified. For individual sites, fragment numbers are often converted to numbers of animals, although this cannot be done without reservation for sites where it was usually joints or parts of animals that were eaten, and not—except in cases like chickens—the whole animal.[24] Sometimes the comparison of the weight of

[23] Albarella (1999). [24] O'Connor (1989: 194).

mammal, bird, and fish bone may provide a more realistic indication of the quantity of meat which would have been provided by each, using the rule of thumb that the ratio of meat to bone is comparable for each class of animals; but weighing bone is a time-consuming method of analysis which few workers use.[25] At Stert Street, Abingdon, birds made up 3 per cent of all mammal, birds, and fish by weight in the twelfth to thirteenth centuries. The percentage by weight of bird bones in the unsieved deposits there was the same—3 per cent—in the fifteenth century as in the earlier Middle Ages, although in the sieved deposits from one well of fifteenth- to sixteenth-century date, it reached 18 per cent.[26] Differences of this degree help to point to work that might be done to clarify our understanding of medieval diet.

To assist in comparisons, the studies in this book have used standard meat weights for animals.[27] These weights, however, may be deceptive. The archaeological evidence suggests that animals became progressively smaller in at least the two centuries before the Black Death. They then increased in size, almost certainly because more food and grazing were available; and they increased in size again after the Dissolution.[28] This increase in size may also have led to better-quality meat. Here again is an area where we need to understand more about the consequences of generalization.

The contribution to the study of diet that has been made by archaeological science is in some aspects the least developed: its methodologies are both comparatively new and promise to add information about areas of consumption that are less well understood, covering both pre- and post-Conquest material in a similar way and offering us insight into the dietary regimes of rural society or the urban poor. Isotopic and osteological analyses of human bone have important contributions to make to this debate, providing direct evidence of the consequences of nutrition and offering them over a much wider area than can be covered by historical documentation. They may also add significantly to the study of gender and diet. Historical documentation indicates that dietary patterns based on the gender of the consumer may have had some impact in the later Middle Ages, particularly in relation to the upper classes and in religious settings. Differential impacts on health based on gender are to be expected, but the link to diet is at this stage unclear and may be supplied by direct, osteological evidence.

Our knowledge of cereals, the largest constituent in diet for the majority from the Middle Saxon period onwards, is drawn principally from historical evidence and is therefore at its best for the period after 1200. This is another area where the development of models based on one strand of evidence, the historical, might be used to take forward the interpretation of another, the archaeobotanical,

[25] Wilson (1979); Bramwell and Wilson (1979).

[26] Wilson (1979).

[27] Based on Stouff (1970) and Harvey (1993).

[28] e.g. Albarella and Davis (1996: 43–6).

perhaps focusing on in-depth studies of limited regional areas using both sources. A greater integration between the archaeological and historical evidence might open the door to the application of statistical methods in the interpretation of medieval cereal assemblages. At the same time, the late medieval evidence for the diet of the peasantry, particularly its involvement in gardening and poultry farming, requires us to look more closely at the archaeology of earlier periods.

There is a good deal more research that might be done on historical records of consumption—some are very detailed and substantial—and also on the records of production and sale. Together these have the potential to add further perspectives to the comparatively small number of regional studies that we have at present. This work is likely to bring out local nuances of consumption, which may be less apparent in archaeological analyses. Some documentation also offers the possibility of looking closely at the diet of a particular individual, or small group, over a period of some years. But historical records of consumption are not so numerous that they are going to provide widespread, in-depth coverage for the late medieval period. Literary texts have much to say on diet, from the place of gluttony in the Seven Deadly Sins and the vices of carnality to the well-known but perhaps not fully understood passages in Langland and Chaucer; these offer means of developing our understanding of the motivation for patterns of diet.

While we can see that diet had a range of consequences for the individual in medieval England, its impact at the level of population remains a matter for debate—although in the critical case of famine, historians can point to episodes of deprivation that had significant, but probably shorter-term, consequences than the effects of epidemic disease. Into this pattern we need to fit the archaeological evidence, especially the direct evidence from human bones, both in terms of what it tells us of consumption and in terms of disease—and in turn, this may allow us to modify the methodological debates about food and the dynamics of population in the later medieval period and to extend them back into earlier periods.

To broaden the range of sources employed for the study of diet in this and other ways offers the best opportunities for developing our knowledge and for creating models for interpretation. The last are a source of both questions and inspiration for all the disciplines involved, applicable well beyond the case of diet—benefits that have come especially from the study of the medieval period.

Bibliography

MANUSCRIPTS: UK

Cambridge: University Library
EDC 7/4
EDR D8/1/1-D8/3/11, 16–30; D8/4/1–3; D10/2/5

Chelmsford: Essex Record Office
TB 124

Dorchester: Dorset Record Office
D204/A5-A10

Gloucester: Gloucestershire Record Office
GDR, 40/T2

Ipswich: Suffolk Record Office
HA6:51/4/4.7

Kew: National Archives
C 47
DL 29
DL 30
E 32
E 36
E 101
SC 6

Lincoln: Lincolnshire Archives Office
2 Anc. 2/2/116

London: British Library
Add. MS 32050
Add. MS 34213
Sloane MS 686

London: Westminster Abbey Muniment Room
WAM 12186
WAM 26873
WAM 33344

Longleat House
MS 6369
MS Misc. IX

Manchester: John Rylands University Library
Latin MS 235

Northampton: Northamptonshire Record Office
Finch Hatton MS 519
Westmorland (Apethorpe) 4. xx. 4

Norwich: Norfolk Record Office
HARE 2199, 194 x 4
MCR/B/26
NH 8
NRS 10193

Oxford: Bodleian Library
MS DD Weld c. 19/4/2–4

Oxford: Merton College
Muniments 4546

Stafford: Staffordshire Record Office
D641/1/3/8
D1734/2/1/427

Stratford-upon-Avon: Shakespeare Birthplace Trust Record Office
BRT 1/3/40

Taunton: Somerset Record Office
DD/L P37/7, 10A-C, 11, 12

Truro: Cornwall Record Office
AR12/25

Winchester: Hampshire Record Office
5M50/69
11M59/B1/102–50
21M65/A/1/19

Winchester: Winchester College Muniments
WCM 1
WCM 11504
WCM 22213

Windsor: St George's Chapel Muniments
XV. 53. 35. v (1)

Worcester: Worcester Cathedral Library
Muniments of the Dean and Chapter, E13

Worcester: Worcestershire Record Office
899:95 BA 989/2/33

York: Borthwick Institute of Historical Research
Dean and Chapter wills

MANUSCRIPTS: OVERSEAS

France: Nantes: Archives départementales, Loire-Atlantique
E206/1 and 3

United States of America: Chicago University Library
Bacon 405–510

The Vatican: Biblioteca Apostolica Vaticana
MS Lat. Vat. 4015

PRINTED WORKS, REPORTS, AND THESES

Abiteboul, M., and Arlet, J. (1985). 'Retinonol-Related Hyperostosis', *American Journal of Roentgenology*, 144: 435–6.

Acton, B., and Duncan, P. (1966). *Making Mead: A Complete Guide to the Making of Sweet and Dry Mead, Melonell, Metheglin, Hippocras, Pyment and Cyser*. 2nd impression. Andover: Amateur Winemaker.

Albala, K. (2002). *Eating Right in the Renaissance*. Berkeley and Los Angeles: University of California Press.

Albarella, U. (1997). 'Size, Power, Wool and Veal: Zooarchaeological Evidence for Late Medieval Innovations', in De Boe and Verhaeghe (1997: 19–30).

——(1999). ' "The Mystery of Husbandry": Medieval Animals and the Problem of Integrating Historical and Archaeological Evidence', *Antiquity*, 73: 867–75.

——(2004). 'Mammal and Bird Bones', in H. Wallis (ed.), *Excavations at Mill Lane, Thetford, 1995*. Norwich: East Anglian Archaeology, 108: 88–99.

——(2005). 'Meat Consumption and Production in Town and Country', in K. Giles and C. C. Dyer (eds.), *Town and Country 1100–1500*. London: Society for Medieval Archaeology, Monograph Series, 22: 131–48.

——Beech, M., and Mulville, J. (1997). *The Saxon, Medieval and Post-Medieval Mammal and Bird Bones Excavated 1989–1991 from Castle Mall, Norwich (Norfolk)*. London: English Heritage, AML Report 72/97.

——and Davis, S. J. M. (1994). *The Saxon and Medieval Animal Bones Excavated 1985–1989 from West Cotton*. London: English Heritage, AML Report 17/94.

———— (1996). 'Mammals and Birds from Launceston Castle, Cornwall: Decline in Status and the Rise of Agriculture', *Circaea*, 12: 1–156.

——and Thomas, R. (2002). 'They Dined on Crane: Bird Consumption, Wildfowling and Status in Medieval England', *Acta zoologica Cracoviensia*, 45: 23–38.

Alcock, N. W. (1970). 'An East Devon Manor in the Later Middle Ages. I: 1374–1420, The Manor Farm', *Reports and Transactions of the Devonshire Association*, 102: 141–87.

Allen, T. G. (ed.) (1994). 'A Medieval Grange of Abingdon Abbey at Dean Court Farm, Cumnor, Oxon.', *Oxoniensia*, 59: 219–447.

Almond, R. (2003). *Medieval Hunting*. Stroud: Sutton.

Ambros, C. (1980). 'The Mammal Bones', in P. Wade-Martins (ed.), *Fieldwork and Excavation on Village Sites in Launditch Hundred, Norfolk*. Norwich: East Anglian Archaeology, 10: 158–9.

Ambrose, S. H. (1993). 'Isotopic Analysis of Paleodiets: Methodological and Interpretive Considerations', in M. K. Sandford (ed.), *Investigations of Ancient Human Tissue: Chemical Analyses in Anthropology*. Langthorne: Gordon and Breach, 59–130.

Amherst, A. M. T. (1894). 'A Fifteenth-Century Treatise on Gardening by Mayster Ion Gardener', *Archaeologia*, 54: 157–72.

Angerer, J. F. (ed.) (1985). *Caeremoniae regularis observantiae sanctissimi patris nostri Benedicti ex ipsius regula sumptae secundum quod in sacris locis, scilicet Specu et monasterio Sublacensi, practicantur*. Siegburg: F. Schmitt, Corpus Consuetudinum Monasticarum, 11/1.

——(ed.) (1987). *Breviarium caeremoniarum monasterii Mellicensis*. Siegburg: F. Schmitt, Corpus Consuetudinum Monasticarum, 11/2.

Anon. (1790). *A Collection of Ordinances and Regulations for the Government of the Royal Household, Made in Divers Reigns from King Edward III to King William and Queen Mary*. London: Society of Antiquaries.

——(1885). *The Great Roll of the Pipe for the Seventh Year of the Reign of King Henry the Second AD 1160–1161*. London: Pipe Roll Society, 4.

——(1981). *Red Deer Management*. Inverness: HMSO for Red Deer Commission.

Armitage-Smith, S., Lodge, E. C., and Somerville, R. (eds.) (1911–37). *John of Gaunt's Register*. 4 vols. London: Royal Historical Society, Camden, 3rd series, 20, 21, 56, 57.

Arthur, J. R. B. (1960). Untitled note referring to Chillington Manor House, *Archaeologia Cantiana*, 74: 195–6.

——(1961). Untitled note referring to Bicknor Court, *Archaeologia Cantiana*, 76: 192–3.

Ashby, S. (2002). 'The Role of Zooarchaeology in the Interpretation of Socioeconomic Status: A Discussion with Reference to Medieval Europe', *Archaeological Review from Cambridge*, 18: 18–59.

Ashdown, R. (1993). 'Avian Bones', in Williams (1993: 154–8).

Ashley, W. (1928). *The Bread of our Forefathers*. Oxford: Clarendon Press.

Astill, G., and Grant, A. (eds.) (1988). *The Countryside of Medieval England*. Oxford: Blackwell.

Aston, M., and Bond, C. J. (eds.) (1988*a*). *Medieval Fish, Fisheries and Fishponds*. 2 vols. Oxford: BAR, British Series, 182.

————(1988*b*). 'Warwickshire Fishponds', in Aston and Bond (1988*a*: ii. 417–34).

————(1988*c*). 'Worcestershire Fishponds', in Aston and Bond (1988*a*: ii. 435–55).

Attenborough, F. L. (1922). *Laws of the Earliest English Kings*. London: Cambridge University Press.

Audoin-Rouzeau, F. (1987). 'Medieval and Early Modern Butchery: Evidence from the Monastery of La Charité-sur-Loire (Nièvre)', *Food and Foodways*, 2: 31–48.

——(1992). 'Approche archéozoologique du commerce des viandes au Moyen Âge', *Anthropozoologica*, 16: 83–92.

——(1993). *Hommes et animaux en Europe de l'époque antique aux temps modernes*. Paris: CNRS.

——(1997). 'Les Ossements du cheptel médiéval', *Ethnozootechnie*, 59: 69–78.

Auerbach, B. M., and Ruff, C. B. (2004). 'Human Body Mass Estimation: A Comparison of "Morphometric" and "Mechanical" Methods', *American Journal of Physical Anthropology*, 125: 331–42.

Ault, W. O. (1972). *Open-Field Farming in Medieval England*. London: George Allen and Unwin.

Austin, D. (ed.) (1982). *Boldon Book: Northumberland and Durham*. Chichester: Phillimore, Domesday Book Series, 35.

Austin, T. (ed.) (1888). *Two Fifteenth-Century Cookery Books*. London: N. Trübner, EETS, os 91.

Auvray, J. (ed.) (1896–1955). *Les Régistres de Grégoire IX*. 4 vols. Paris: Bibliothèque des Écoles Françaises d'Athènes et de Rome.

Ayres, K., Locker, A., and Serjeantson, D. (2003). 'Mammal, Bird and Fish Remains and Oysters. Phases 2f–4a: The Medieval Abbey: Food Consumption and Production', in Hardy, Dodd, and Keevill (2003: 360–406).

Backhouse, J. (2001). *Medieval Birds in the Sherborne Missal*. London: The British Library.

Bailey, M. (1989). *A Marginal Economy? East Anglian Breckland in the Later Middle Ages*. Cambridge: Cambridge University Press.

——(1990). 'Coastal Fishing off South East Suffolk in the Century after the Black Death', *Proceedings of the Suffolk Institute of Archaeology and History*, 37: 102–14.

——(1996). 'Demographic Decline in Late Medieval England: Some Thoughts on Recent Research', *EconHR*, 2nd series, 49: 1–19.

——(ed.) (2002). *The English Manor, c.1200–c.1500*. Manchester: Manchester University Press.

Baillie-Grohman, W. A., and Baillie-Grohman, F. (eds.) (1909). *The Master of Game by Edward, 2nd Duke of York: The Oldest English Book on Hunting*. London: Chatto and Windus.

Barr, H. (1993). *The Piers Plowman Tradition*. London: J. M. Dent.

Barrett, J. H. (1997). 'Fish Trade in Norse Orkney and Caithness: A Zooarchaeological Approach', *Antiquity*, 71: 616–38.

——Locker, A. M., and Roberts, C. M. (2004*a*). ' "Dark Age Economics" Revisited: The English Fishbone Evidence AD 600–1600', *Antiquity*, 78: 618–36 and appendices at http://antiquity. ac. uk/ProjGall/barrett/.

————— (2004*b*). 'The Origins of Intensive Marine Fishing in Medieval Europe', *Proceedings of the Royal Society: Biological Sciences*, 271: 2417–21 and appendix at www.journals.royalsoc.ac.uk/.

——NICHOLSON, R. A., and CERÓN-CARRASCO, R. (1999). 'Archaeo-Ichthyological Evidence for Long-Term Socioeconomic Trends in Northern Scotland: 3500 BC to AD 1500', *Journal of Archaeological Science*, 26: 353–88.

BAYLISS, A., SHEPHERD POPESCU, E., BEAVAN-ATHFIELD, N., BRONK RAMSEY, C., COOK, G. T., and LOCKER, A. (2004). 'The Potential Significance of Dietary Offsets for the Interpretation of Radiocarbon Dates: An Archaeologically Significant Example from Medieval Norwich', *Journal of Archaeological Science*, 31: 563–75.

BAZELEY, M. L. (1921). 'The Extent of the English Forest in the Thirteenth Century', *Transactions of the Royal Historical Society*, 4th series, 4: 140–72.

BECKER, P. (ed.) (1968). *Consuetudines et observantiae monasteriorum sancti Mathiae et sancti Maximini Treverensium, ab Johanne Rode abbate conscriptae*. Siegburg: F. Schmitt, Corpus Consuetudinum Monasticarum, 5.

BENNETT, J. M. (1996). *Ale, Beer, and Brewsters in England: Women's Work in a Changing World, 1300 to 1600*. Oxford: Oxford University Press.

BENOIST, J. O. (1984). 'Le Gibier dans l'alimentation seigneuriale (xie–xve siècles)', in D. Menjot (ed.), *Manger et boire au Moyen Âge: actes du colloque de Nice (15–17 octobre 1982)*. 2 vols. Nice: Publications de la Faculté des Lettres et Sciences Humaines de Nice, i. 75–87.

BENSON, L. D. (1988) (ed.). *The Riverside Chaucer*. 3rd edition. Oxford: Oxford University Press.

BENT, D. C. (1977–8). 'The Animal Remains', in P. Liddle (ed.), 'A Late Medieval Enclosure in Donnington Park', *Leicestershire Archaeological and Historical Society*, 53: 8–29, at 14–15.

BERESFORD, M., and HURST, J. (1990). *Wharram Percy: Deserted Medieval Village*. London: Batsford/English Heritage.

BIDDICK, K. (1984). 'Field Edge, Forest Edge: Early Medieval Social Change and Resource Allocation', in K. Biddick (ed.), *Archaeological Approaches to Medieval Europe*. Kalamazoo: Western Michigan University, Medieval Institute Publications, 105–18.

——(1989). *The Other Economy: Pastoral Husbandry on a Medieval Estate*. Berkeley and Los Angeles: University of California Press.

BIRRELL, J. R. (1982). 'Who Poached the King's Deer? A Study in Thirteenth-Century Crime', *Midland History*, 7: 9–25.

——(1990–1). 'The Forest and the Chase in Medieval Staffordshire', *Staffordshire Studies*, 3: 23–50.

——(1992). 'Deer and Deer Farming in Medieval England', *Agricultural History Review*, 40: 112–26.

——(1994). 'A Great Thirteenth-Century Hunter: John Giffard of Brimpsfield', *Medieval Prosopography*, 15(2): 37–66.

——(1996). 'Peasant Deer Poachers in the Medieval Forest', in R. H. Britnell and J. Hatcher (eds.), *Progress and Problems in Medieval England: Essays in Honour of Edward Miller*. Cambridge: Cambridge University Press, 66–88.

——(2001). 'Aristocratic Poachers in the Forest of Dean: Their Methods, their Quarry and their Companions', *Transactions of the Bristol and Gloucestershire Archaeological Society*, 119: 147–54.

Blyke, R., Strachey, J., Pridden, J., and Upham, E. (eds.) (1783). *Rotuli parliamentorum; ut et petitiones et placita in parliamento*. 6 vols. London: Record Commission.

Boardman, S., and Jones, G. (1990). 'Experiments on the Effects of Charring on Cereal Plant Components', *Journal of Archaeological Science*, 17: 1–11.

Bocherens, H., Fizet, M., Mariotti, A., Olive, C., Bellon, G., and Billiou, D. (1991). 'Application de la biogéochimie isotopique (^{13}C, ^{15}N) à la détermination du régime alimentaire des populations humaines et animales durant les périodes antique et médiévale', *Archives des sciences, Genève*, 44: 329–40.

Boisier, P., Rahalison, L., Rasolomaharo, M., Ratsitorahina, M., Mahafaly, M., Razfimahefa, M., Duplantier, J.-M., Ratsifasoamanana, L., and Chanteau, S. (2002). 'Epidemiologic Features of Four Successive Annual Outbreaks of Bubonic Plague in Mahajanga, Madagascar', *Emerging Infectious Diseases*, 8: 311–16.

Bond, C. J. (1988). 'Monastic Fisheries', in Aston and Bond (1988*a*: i. 69–112).

——and Chambers, R. A. (1988). 'Oxfordshire Fishponds', in Aston and Bond (1988*a*: ii. 353–70).

Bond, J. (2001). 'Production and Consumption of Food and Drink in the Medieval Monastery', in G. Keevill, M. Aston, and T. Hall (eds.), *Monastic Archaeology: Papers on the Study of Medieval Monasteries*. Oxford: Oxbow Books, 54–87.

Bossard-Beck, C. (1984). 'Le Mobilier ostéologique et botanique', in J.-M. Pesez (ed.), *Brucato: histoire et archéologie d'un habitat médiéval en Sicile*. Rome: Collection de l'École Française de Rome, 615–71.

Bourdillon, J. (1988). 'Countryside and Town: The Animal Resources of Saxon Southampton', in D. Hooke (ed.), *Anglo-Saxon Settlements*. Oxford: Blackwell, 177–95.

——(1993). 'Animal Bones', in S. M. Davies (ed.), *Excavations in the Town Centre of Trowbridge*. Salisbury: Wessex Archaeological Reports, 134–43.

——(1994). 'The Animal Provisioning of Saxon Southampton', in J. Rackham (ed.), *Environment and Economy in Anglo-Saxon England*. York: CBA, Research Report 89: 120–5.

——(1998). 'The Faunal Remains', in Poulton (1998: 139–74).

——and Coy, J. (1980). 'The Animal Remains', in P. Holdsworth (ed.), *Excavations at Melbourne Street, Southampton 1971–76*. London: CBA, Research Report 33: 79–120.

Bourne, W. R. P. (1981). 'The Birds and Animals Consumed when Henry VIII Entertained the King of France and the Count of Flanders at Calais in 1532', *Archives of Natural History*, 10: 331–3.

Bowie, G. (1979). 'Corn Drying Kilns, Meal Milling and Flour in Ireland', *Folk Life*, 17: 5–13.

Boyd, B. (1986). 'Minor Finds of Cereals at Two Medieval Rural Archaeological Sites in North-East Scotland', *Circaea*, 4: 39–42.

Bramwell, D. (1977). 'Bird Bone', in Clarke and Carter (1977: 399–409).

——and Wilson, R. (1979). 'The Bird Bones', in M. Parrington (ed.), 'Excavations at Stert Street, Abingdon, Oxon.', *Oxoniensia*, 44: 20–1.

Brenner, R. (1985). 'Agrarian Class Structure and Economic Development in Pre-Industrial Europe', in T. H. Aston and C. H. E. Philpin (eds.), *The Brenner Debate: Agrarian Class Structure and Economic Development in Pre-Industrial Europe*. Cambridge: Cambridge University Press, 10–63.

BRESCHI, M., and LIVI-BACCI, M. (1997). 'Month of Birth as a Factor in Children's Survival', in A. Bideau, B. Desjardins, and H. P. Birgnolo (eds.), *Infant and Child Mortality in the Past*. Oxford: Oxford University Press, 157–73.

BREWER, E. (1992). *Sir Gawain and the Green Knight: Sources and Analogues*. Cambridge: Cambridge University Press.

BREWER, J. S., DIMOCK, J. F., and WARNER, G. F. (eds.) (1861–91). *Giraldi Cambrensis opera*. 8 vols. London: Rolls Series.

BRIDBURY, A. R. (1955). *England and the Salt Trade in the Later Middle Ages*. Oxford: Oxford University Press.

BRITNELL, R. H. (1966). 'Production for the Market on a Small Fourteenth-Century Estate', *EconHR*, 2nd series, 19: 380–7.

——(1986). *Growth and Decline in Colchester, 1300–1525*. Cambridge: Cambridge University Press.

——(1993). *The Commercialisation of English Society 1000–1500*. Cambridge: Cambridge University Press.

——(1995). 'Commercialisation and Economic Development in England, 1000–1300', in Britnell and Campbell (1995: 7–26).

——and CAMPBELL, B. M. S. (eds.) (1995). *A Commercialising Economy: England 1086 to c. 1300*. Manchester: Manchester University Press.

BRITTON, C. E. (1937). *A Meteorological Chronology to AD 1450*. London: HMSO, Meteorological Office Geophysical Memoirs, 70.

BRITTON, E. (1977). *The Community of the Vill*. Toronto: Macmillan of Canada.

BROWN, A. E. (ed.) (1999). *Garden Archaeology*. London: CBA, Research Report 78.

BROWN, D. (2002). *Pottery in Medieval Southampton c.1066–1510*. York: CBA, Research Report 133.

BROWN, T. A. (2001). 'Ancient DNA', in D. R. Brothwell and A. M. Pollard (eds.), *Handbook of Archaeological Sciences*. Chichester: John Wiley, 301–11.

BRUINTJES, T. J. D. (1987). 'Diffuse Idiopathic Skeletal Hyperostosis (DISH): A 10th Century AD Case from the St Servaas Church at Maastricht', *Bone*, 1: 23–8.

BULLOCK, A. E. (1998). 'The Fish Remains', in Poulton (1998: 164–71).

BURNETT, D. P. (1992). 'Animal Bone from Great Linford Village', in D. C. Mynard and R. J. Zeepvat (eds.), *Great Linford*. Buckingham: Buckinghamshire Archaeological Society, Monograph Series, 3: 231–9.

BYERLY, B. F., and BYERLY, C. R. (eds.) (1977). *Records of the Wardrobe and Household 1285–1286*. London: HMSO.

——(eds.) (1986). *Records of the Wardrobe and Household 1286–1289*. London: HMSO.

CAMPBELL, B. M. S. (ed.) (1991). *Before the Black Death: Studies in the 'Crisis' of the Early Fourteenth Century*. Manchester: Manchester University Press.

——(1995). 'Measuring the Commercialisation of Seigneurial Agriculture, *c.*1300', in Britnell and Campbell (1995: 132–98).

——(2000). *English Seigniorial Agriculture, 1250–1450*. Cambridge: Cambridge University Press.

——GALLOWAY, J. A., KEENE, D., and MURPHY, M. (1993). *A Medieval Capital and its Grain Supply: Agrarian Production and Distribution in the London Region c.1300*. London: Institute of British Geographers, Historical Geography Research Series, 30.

Campbell, J. (2002). 'Domesday Herrings', in Harper–Bill, Rawcliffe, and Wilson (2002: 5–17).

Canivez, J. M. (ed.) (1933–41). *Statuta capitulorum generalium Ordinis Cisterciensis ab anno 1116 ad annum 1786*. 8 vols. Louvain: Bureaux de la Revue.

Cantor, L. M. (1982). 'Forests, Chases, Parks and Warrens', in L. M. Cantor (ed.), *The Medieval English Landscape*. London: Croom Helm, 56–85.

——and Moore, J. S. (1963). 'The Medieval Parks of the Earls of Stafford at Madeley', *North Staffordshire Journal of Field Studies*, 3: 37–58.

Cardwell, P., Chundun, Z., Costley, S., Hall, J., Hanson, J., Huntley, J., Jamfrey, C., Maxwell, R., Speed, G., Simpson, R., and Stallibrass, S. (1995). 'The Excavation of the Hospital of St Giles by Brompton Bridge', *Archaeological Journal*, 152: 109–245.

Carlin, M. (1998). 'Fast Food and Urban Living Standards in Medieval England', in Carlin and Rosenthal (1998: 27–51).

——and Rosenthal, J. T. (eds.) (1998). *Food and Eating in Medieval Europe*. London: Hambledon Press.

Carruthers, W. (1991). 'Plant Remains Recovered from Daub from a 16th-Century Manor House, Althrey Hall, near Wrexham, Clwyd, U. K.', *Circaea*, 8: 55–9.

——(2003). 'The Charred and Waterlogged Plant Remains', in M. Rylatt and P. Mason (eds.), *The Archaeology of the Medieval Cathedral and Priory of St Mary, Coventry*. Coventry: Coventry City Council, 120–5.

Cartledge, J. (1983). 'Mammal Bones', in B. Ayers and P. Murphy (eds.), *A Waterfront Excavation at Whitefriars Street Car Park, Norwich 1979*. Norwich: East Anglian Archaeology, 17: 30–2.

——(1985). 'The Animal Bones', in M. Atkin (ed.), 'Excavations at Alms Lane', in *Excavations in Norwich 1971–8 part II*. Norwich: East Anglian Archaeology, 26: 144–265.

——(1988). 'Mammal Bones', in B. Ayers (ed.), *Excavations at St Martin-at–Palace Plain, Norwich, 1981*. Norwich: East Anglian Archaeology, 37: 111–13.

——(1989). 'The Animal Bone', in L. Butler and P. Wade-Martins (eds.), *The Deserted Medieval Village of Thuxton, Norfolk*. Norwich: East Anglian Archaeology, 46: 52–4.

Carus-Wilson, E. M. (1954). 'The Effects of the Acquisition and of the Loss of Gascony on the English Wine Trade', in E. M. Carus-Wilson, *Medieval Merchant Venturers*. London: Methuen, 265–78.

Cazier, P. (ed.) (1998). *Isidorus Hispalensis sententiae*. Turnhout: Brepols, Corpus Christianorum, Series Latina, 111.

Chadd, D. (ed.) (1999–2002). *The Ordinal of the Abbey of the Holy Trinity, Fécamp*. 2 vols. Woodbridge: Boydell Press, Henry Bradshaw Society, 111–12.

Chapman, N. G., and Chapman, D. I. (1982). *The Fallow Deer*. London: HMSO, Forest Record, 124.

Charvin, G. (ed.) (1965–82). *Statuts, chapitres généraux et visites de l'Ordre de Cluny*. 9 vols. Paris: E. de Boccard.

Checkley, W., Epstein, L. D., Gilman, R. H., Cabrera, L., and Black, R. E. (2003). 'Effects of Acute Diarrhea on Linear Growth in Peruvian Children', *American Journal of Epidemiology*, 157: 166–75.

Cherryson, A. (2002). 'The Identification of Archaeological Evidence for Hawking in Medieval England', *Acta zoologica Cracoviensia*, 45: 307–14.

Chew, H. M., and Weinbaum, M. (eds.) (1970). *The London Eyre of 1244*. London: London Record Society, 6.

Chibnall, M. (ed.) (1951). *Select Documents of the English Lands of the Abbey of Bec*. London: Royal Historical Society, Camden, 3rd series, 73.

—— (ed.) (1982). *Charters and Custumals of the Abbey of Holy Trinity Caen*. London: British Academy, Records of Social and Economic History, new series, 5.

Childs, W. R. (ed.) (1986). *The Customs Accounts of Hull 1453–1490*. Leeds: Yorkshire Archaeological Society, Record Series, 144.

—— and Kowaleski, M. (2000). 'Fishing and Fisheries in the Middle Ages', in Starkey, Reid, and Ashcroft (2000: 19–28, 243–5).

Cigman, G. (ed.) (1989). *Lollard Sermons*. Oxford: Oxford University Press, EETS, os 294.

Clark, P. (1983). *The English Alehouse: A Social History, 1200–1830*. London: Longman.

Clarke, H., and Carter, A. (eds.) (1977). *Excavations in King's Lynn, 1963–1970*. London: Society for Medieval Archaeology, Monograph Series, 7.

Clavel, B. (2001). *L'Animal dans l'alimentation médiévale et moderne en France du nord (xiie–xviie siècles)*. *Revue archéologique de Picardie*, special number 19.

Clementz, M. T., and Koch, P. L. (2001). 'Differentiating Aquatic Mammal Habitat and Foraging Ecology with Stable Isotopes in Tooth Enamel', *Oecologia*, 129: 461–72.

Clutton-Brock, J. (1984). 'The Master of the Game: The Animals and Rituals of Medieval Venery', *Biologist*, 31: 167–71.

Clutton-Brock, T., Guinness, F. E., and Albon, S. D. (1982). *Red Deer: Behaviour and Ecology of Two Sexes*. Chicago: University of Chicago Press.

Coggon, D., Reading, I., Croft, I., McLaren, M., Barrett, D., and Cooper, C. (2001). 'Knee Osteoarthritis and Obesity', *International Journal of Obesity and Related Metabolic Disorders*, 25: 622–7.

Cohen, A., and Serjeantson, D. (1996). *A Manual for the Identification of Bird Bones from Archaeological Sites*. London: Archetype Press.

Cohn, S. K. (2002*a*). 'The Black Death: End of a Paradigm', *American Historical Review*, 107: 703–38.

—— (2002*b*). *The Black Death Transformed: Disease and Culture in Early Renaissance Europe*. London: Arnold.

Colley, S. M. (1984*a*). 'Some Methodological Problems in the Interpretation of Fish Remains from Archaeological Sites in Orkney', in N. Desse-Berset (ed.), *2nd Fish Osteoarchaeology Meeting*. Valbonne: CNRS, 117–32.

Colley, S. M. (1984*b*). *Fish Remains from Hamwic (Saxon Southampton) Six Dials Variability Study*. London: Department of the Environment, AML Report 4576.

Colvin, H. M. (2000). 'The Origin of Chantries', *Journal of Medieval History*, 26: 163–73.

Constable, G. (ed.) (1967). *The Letters of Peter the Venerable*. 2 vols. Cambridge, Mass.: Harvard University Press, Harvard Historical Studies, 78.

Corran, H. S. (1975). *A History of Brewing*. Newton Abbot: David and Charles.

COX, J. C. (1905). *The Royal Forests of England*. London: Methuen.
——(1907). 'Forestry', in *VCH Yorkshire I*. London: Archibald Constable and Co., 501–23.
COY, J. (1980). 'The Animal Bones', in J. Haslam (ed.), 'A Middle Saxon Iron Smelting Site at Ramsbury, Wiltshire', *Medieval Archaeology*, 24: 1–68, at 41–51.
——(1981). 'The Animal Bones', in R. J. Silvester (ed.), 'An Excavation on the Post-Roman Site at Bantham Ham, South Devon', *Proceedings of the Devon Archaeological Society*, 39: 106–10.
——(1983). 'Animal Bone', in S. M. Davies (ed.), 'Excavations at Christchurch, Dorset, 1981 to 1983', *Dorset Natural History and Archaeological Society Proceedings*, 105: 43–5.
——(1985). 'Fish Bones', in Cunliffe and Munby (1985: 256–61).
——(1987). *Animal Bones from Abbots Worthy (Itchen Abbas Road), Hampshire*. London: English Heritage, AML Report 156/87.
——(1989). 'The Provision of Fowls and Fish for Towns', in Serjeantson and Waldron (1989: 25–40).
——(1996). 'Medieval Records versus Excavation Results: Examples from Southern England', *Archaeofauna*, 5: 55–63.
——(in press). 'Late Saxon and Medieval Animal Bone from the Western Suburbs', in Serjeantson and Rees (in press).
CRABTREE, P. J. (1989). *West Stow, Suffolk: Early Anglo-Saxon Animal Husbandry*. Norwich: East Anglian Archaeology, 47.
——(1991). 'Zooarchaeology and Complex Societies: Some Uses of Faunal Remains for the Study of Trade, Social Status and Ethnicity', *Archaeological Method and Theory*, 2: 155–205.
——(1994). 'The Animal Bone Remains from Ipswich, Suffolk, Recovered from Sixteen Sites Excavated between 1974 and 1988', Unpublished report.
——(1995). 'The Symbolic Role of Animals in Anglo-Saxon England: Evidence from Burials and Cremations', in K. Ryan and P. J. Crabtree (eds.), *The Symbolic Role of Animals in Archaeology*. Philadelphia: University of Pennsylvania, MASCA Research Papers in Science and Archaeology, 12: 20–6.
——(1996). 'Production and Consumption in an Early Complex Society: Animal Use in Middle Saxon East Anglia', *World Archaeology*, 28: 58–75.
CREIGHTON, C. (1965). *A History of Epidemics in Britain*. 2nd edition. 7 vols. London: Cass.
CROSBY, J. H. (ed.) (1895–7). 'Manor of Tyd', *Fenland Notes and Queries*, 3: 63–8.
CROSSLEY-HOLLAND, N. (1996). *Living and Dining in Medieval Paris: The Household of a Fourteenth-Century Knight*. Cardiff: University of Wales Press.
CUMMINS, J. (1988). *The Hound and the Hawk: The Art of Medieval Hunting*. London: Weidenfeld and Nicolson.
CUNLIFFE, B. W. (ed.) (1976). *Excavations at Portchester Castle*, ii: *Saxon*. London: Thames and Hudson/Society of Antiquaries.
——(ed.) (1977). *Excavations at Portchester Castle*, iii: *Medieval: The Outer Bailey and its Defences*. London: Thames and Hudson/Society of Antiquaries.
——and MUNBY, J. (eds.) (1985). *Excavations at Portchester Castle*, iv: *Medieval: The Inner Bailey*. London: Thames and Hudson/Society of Antiquaries.

CURRIE, C. K. (1988). 'Medieval Fishponds in Hampshire', in Aston and Bond (1988*a*: ii. 267–89).

CUTTING, C. L. (1955). *Fish Saving*. London: Leonard Hill.

DALE, M. K. (ed.) (1950). *Court Roll of Chalgrove Manor, 1278–1313*. Streatley: Bedfordshire Historical Record Society, 28.

—— and REDSTONE, V. B. (eds.) (1931). *The Household Book of Dame Alice de Bryene of Acton Hall, Suffolk, Sept. 1412–Sept. 1413*. Ipswich: W. E. Harrison.

DANIELSSON, B. (ed.) (1977). *William Twiti: The Art of Hunting 1327*. Stockholm: Almquist and Wiksell International, Stockholm Studies in English, 37 (Cynegetica Anglica, 1).

DARBY, H. C. (1940). *The Medieval Fenland*. Cambridge: Cambridge University Press.

DARLING, F. F. (1937). *A Herd of Red Deer*. London: Oxford University Press.

DAVIDSON, A. (1999). *Oxford Companion to Food*. Oxford: Oxford University Press.

DAVIES, C. S. L. (1963). 'Les Rations alimentaires de l'armée et de la marine anglaise au xvie siècle', *Annales ESC*, 18: 139–41.

DAVIS, J. (2004). 'Baking for the Common Good: A Reassessment of the Assize of Bread in Medieval England', *EconHR*, 2nd series, 57: 465–502.

DAVIS, S. J. M. (1991). *Faunal Remains from the Late Saxon-Medieval Farmstead at Eckweek in Avon, 1988–1989 Excavations*. London: English Heritage, AML Report 35/91.

—— (1992). *Saxon and Medieval Animal Bones from Burystead and Langham Road, Northants: 1984–1987 Excavations*. London: English Heritage, AML Report 71/92.

—— (2002). 'British Agriculture: Texts for the Zoo-Archaeologist', *Environmental Archaeology*, 7: 47–60.

DAVRIL, A., and DONNAT, L. (eds.) (1984). *Consuetudines Floriacenses antiquiores saec. x ex*. Siegburg: F. Schmitt, Corpus Consuetudinum Monasticarum, 7/3: 3–60.

DAWES, M. C. B., and JOHNSON, H. C. (eds.) (1930–3). *Registers of Edward the Black Prince, 1348–1365*. 4 vols. London: HMSO.

DAWSON, J., JUSZCZAK, E., THOROGOOD, M., MARKS, S. A., DODD, C., and FITZPATRICK, R. (2003). 'An Investigation of Risk Factors for Symptomatic Osteoarthritis of the Knee', *Journal of Epidemiology and Community Health*, 57: 823–30.

De BOE, G., and VERHAEGHE, F. (eds.) (1997). *Environment and Subsistence in Medieval Europe*. Bruges: Institute for the Archaeological Heritage of Flanders.

DEMBIŃSKA, M., rev. WEAVER, W. W. (1999). *Food and Drink in Medieval Poland: Rediscovering a Cuisine of the Past*. Philadelphia: University of Pennsylvania Press.

De MOULINS, D. (forthcoming). *The Weeds from the Thatched Roofs of Medieval Cottages from the South of England*.

DENHOLM-YOUNG, N. (ed.) (1957). *Vita Edwardi Secundi*. London: Nelson.

DENKO, C. W., BOJA, B., and MOSKAWITZ, R. W. (1994). 'Growth Promoting Peptides in Osteoarthritis and Diffuse Idiopathic Skeletal Hyperostosis: Insulin, Insulin-Like Growth Factor-1, Growth Hormone', *Journal of Rheumatology*, 21: 1725–30.

DEWINDT, E. B. (ed.) (1976). *The Liber gersumarum of Ramsey Abbey*. Toronto: Pontifical Institute of Mediaeval Studies.

DICKSON, C. (1995). 'Macroscopic Fossils of Garden Plants from British Roman and Medieval Deposits', in D. Moe, J. H. Dickson, and P. M. Jørgensen (eds.), *Garden*

History: Garden Plants, Species Forms and Varieties from Pompeii to 1800. Rixensart: Pact, Pact 42 European Symposium/Ravello 1991, 47–71.

——(1996). 'Food, Medicinal and Other Plants from the 15th Century Drains of Paisley Abbey, Scotland', *Vegetation History and Archaeobotany*, 5: 25–31.

Dinter, P. (ed.) (1980). *Liber tramitis aevi Odilonis abbatis*. Siegburg: F. Schmitt, Corpus Consuetudinum Monasticarum, 10.

Dobney, K. (n.d.). *Review of Environmental Archaeology: Zooarchaeology in the North of England*. Unpublished report for English Heritage.

Dobney, K. M., and Jaques, S. D. (2002). 'Avian Signatures for Identity and Status in Anglo-Saxon England', *Acta zoologica Cracoviensia*, 45: 7–21.

————and Irving, B. G. (1995). *Of Butchers and Breeds: Report on Vertebrate Remains from Various Sites in the City of Lincoln*. Lincoln: City of Lincoln Archaeology Unit, Lincoln Archaeological Studies, 5.

Douie, D. L., and Farmer, D. H. (eds.) (1985). *Magna vita sancti Hugonis: The Life of St Hugh of Lincoln [by Adam of Eynsham]*. 2nd impression. 2 vols. Oxford: Clarendon Press.

Driver, J. C. (1990). 'Faunal Remains', in J. C. Driver, J. Rady, and M. Sparks (eds.), *Excavations in the Cathedral Precincts, 2. Linacre Garden, 'Meister Omers' and St Gabriel's Chapel*. Maidstone: Canterbury Archaeological Trust/Kent Archaeological Society, The Archaeology of Canterbury, 4: 228–57.

Drury, J. L. (1978). 'Durham Palatinate Forest Law and Administration, Specially in Weardale up to 1440', *Archaeologia Aeliana*, 5th series, 6: 87–105.

Dubois, J. (1974). 'Les Moines dans la société du Moyen Âge (950–1350)', *Revue d'Histoire de l'Église de France*, 60: 5–37.

Dufour, E., Bocherens, H., and Mariotti, A. (1999). 'Palaeodietary Implications of Isotopic Variability in Eurasian Lacustrine Fish', *Journal of Archaeological Science*, 26: 617–27.

Duke, P. G. (1979). 'The Animal Bones', in J. Hassall, 'St John's Street', in D. Baker, E. Baker, J. Hassall, and A. Simco (eds.), 'Excavations in Bedford 1967–77', *Bedfordshire Archaeological Journal*, 13: 97–126, at 125–6.

Dumond, H., Presle, N., Terlain, B., Mainard, D., Loeuille, D., Netter, P., and Pottie, P. (2003). 'Evidence for a Key Role of Leptin in Osteoarthritis', *Arthritis and Rheumatism*, 48: 3118–29.

Dyer, C. C. (1980). *Lords and Peasants in a Changing Society: The Estates of the Bishopric of Worcester, 680–1540*. Cambridge: Cambridge University Press.

——(1983). 'English Diet in the Later Middle Ages', in T. H. Aston, P. R. Coss, C. C. Dyer, and J. Thirsk (eds.), *Social Relations and Ideas: Essays in Honour of R. H. Hilton*. Cambridge: Cambridge University Press, 191–216.

——(1994*a*). *Everyday Life in Medieval England*. London: Hambledon Press.

——(1994*b*). 'Changes in Diet in the Late Middle Ages: The Case of Harvest Workers', in Dyer (1994*a*: 77–99).

——(1994*c*). 'The Consumption of Freshwater Fish in Medieval England', in Dyer (1994*a*: 101–11).

——(1994*d*). 'Gardens and Orchards in Medieval England', in Dyer (1994*a*: 113–31).

——(1998*a*). *Standards of Living in the Later Middle Ages: Social Change in England c.1200–1520*. 2nd edition. Cambridge: Cambridge University Press.

——(1998*b*). 'Did the Peasants Really Starve in Medieval England?', in Carlin and Rosenthal (1998: 53–71).

——(2003). *Making a Living in the Middle Ages: The People of Britain 850–1520*. London: Penguin Books.

——(2004). 'Alternative Agriculture: Goats in Medieval England', in R. W. Hoyle (ed.), *People, Landscape and Alternative Agriculture: Essays for Joan Thirsk*. Exeter: *Agricultural History Review*, Supplement Series, 3: 20–38.

Dymond, D. (ed.) (1995–6). *The Register of Thetford Priory*. 2 vols. London: British Academy, Records of Social and Economic History, new series, 24–5.

Eastham, A. (1976). 'The Bird Bones', in Cunliffe (1976: 287–96).

——(1977). 'Birds', in Cunliffe (1977: 233–9).

——(1985). 'Bird Bones', in Cunliffe and Munby (1985: 261–9).

Eisenstadt, S. N. (1969). 'Some Observation on the Dynamics of Traditions', *Comparative Studies in Society and History*, 11: 451–75.

Effros, B. (2002). *Creating Community with Food and Drink in Merovingian Gaul*. Basingstoke: Palgrave.

Ell, S. R. (1984). 'Immunity as a Factor in the Epidemiology of Medieval Plague', *Reviews of Infectious Diseases*, 6: 866–79.

Enghoff, I. B. (1996). 'A Medieval Herring Industry in Denmark and the Importance of Herring in Eastern Denmark', *Archaeofauna*, 5: 43–7.

——(1999). 'Fishing in the Baltic Region from the 5th Century BC to the 16th Century AD: Evidence from the Fish Bones', *Archaeofauna*, 8: 41–5.

——(2000). 'Fishing in the Southern North Sea Region from the 1st to the 16th Century AD: Evidence from the Fish Bones', *Archaeofauna*, 9: 59–132.

Enneking, D. (accessed May 2004). 'Common Vetch *Vicia sativa* ssp. *sativa* cv. Blanchefleur, a Mimic of Red Lentils'. Nedlands: Centre for Legumes in Mediterranean Agriculture, University of Western Australia, Nedlands WA 6907, at www.general.uwa.edu.au/u/enneking/VETCH.htm.

Erbe, T. (ed.) (1905). *Mirk's Festial*. London: Keegan Paul, Trench, Trübner, EETS, ES 96.

Ervynck, A. (1997*a*). 'Following the Rule? Fish and Meat Consumption in Monastic Communities in Flanders (Belgium)', in De Boe and Verhaeghe (1997: 67–81).

——(1997*b*). 'Detailed Recording of Tooth Wear (Grant, 1982) as an Evaluation of the Seasonal Slaughtering of Pigs? Examples from Medieval Sites in Belgium', *Archaeofauna*, 6: 67–79.

——(2004). '*Orant, pugnant, laborant*: The Diet of the Three Orders in the Feudal Society of Medieval North-Western Europe', in S. J. O'Day, W. Van Neer, and A. Ervynck (eds.), *Behaviour behind Bones: The Zooarchaeology of Ritual, Religion, Status and Identity*. Oxford: Oxbow, 215–33.

——Cooremans, B., and Neer, W. Van (1994). 'De voedselvoorziening in de Sint-Salvatorsabdij te Ename (Stad Oudenaarde, Prov. Oost-Vlanderen) 3. Een latrine bij de abstwoning (12de–begin 13de eeuw)', *Archeologie in Vlaanderen*, 4: 311–22.

——and Neer, W. Van (1994). 'A Preliminary Survey of Fish Remains in Medieval Castles, Abbeys and Towns of Flanders (Belgium)', *Offa*, 51: 303–8.

——Strydonck, V., and Boudin, M. (1999/2000). 'Dieetreconstructie en herkomstbepaling op basis van de analyse van de stabiele isotopen ^{13}C en ^{15}N uit dierlijk en menselijk skeletmateriaal: een eerste verkennend onderzoek op middeleurwse vondsten uit Vlaanderen', *Archeologie in Vlaanderen*, 7: 131–40.

Fain, O. (2005). 'Musculoskeletal Manifestations of Scurvy', *Joint, Bone, Spine: Revue du rheumatisme*, 72: 124–8.

Fair, J., and Moxom, D. (1993). *Abbotsbury and the Swannery*. Wimborne: Dovecote Press.

Farmer, D. L. (1988). 'Prices and Wages', in Hallam (1988*a*: 715–817).

——(1991*a*). 'Marketing the Produce of the Countryside 1200–1500', in Miller (1991: 324–430).

——(1991*b*). 'Prices and Wages, 1350–1500', in Miller (1991: 431–525).

Fegan, E. S. (ed.) (1914). *Journal of Prior William More*. Worcester: Worcestershire Historical Society.

Fenton, A. (1978). *The Northern Isles*. Edinburgh: John Donald.

Fenwick, C. C. (ed.) (1998–2001). *The Poll Taxes of 1377, 1379 and 1381*. 2 vols. London: British Academy, Records of Social and Economic History, new series, 27, 29.

Finberg, H. P. R. (1951). *Tavistock Abbey: A Study in the Social and Economic History of Devon*. Cambridge: Cambridge University Press.

Fiorato, V., Boylston, A., and Knüsel, C. (eds.) (2000). *Blood Red Roses: The Archaeology of a Mass Grave from the Battle of Towton AD 1461*. Oxford: Oxbow Books.

Fisher, J. (1966). *The Shell Bird Book*. London: Ebury Press.

Flandrin, J. L. (1983). 'Le Goût et la nécessité: sur l'usage des graisses dans les cuisines d'Europe occidentale (xive–xviiie siècles)', *Annales ESC*, 38: 369–401.

——(1992). 'Structure des menus français et anglais aux xive et xve siècles', in Lambert (1992: 173–92).

——and Montanari, M. (eds.) (1999). *Food: A Culinary History from Antiquity to the Present*, English edition, ed. A. Sonnenfeld. New York: Columbia University Press.

Fleming, R. (2000). 'The New Wealth, the New Rich and the New Political Style in Late Anglo-Saxon England', *Anglo-Norman Studies*, 23: 1–22.

Fletcher, J. M., and Upton, C. A. (eds.) (1996). *The Domestic Accounts of Merton College, Oxford, 1 August 1482–1 August 1494*. Oxford: Oxford Historical Society, new series, 34.

Fogel, R. W. (1992). 'Second Thoughts on the European Escape from Hunger: Famines, Chronic Malnutrition, and Mortality Rates', in Osmani (1992: 243–86).

Foreville, R., and Keir, G. (eds.) (1987). *The Book of St Gilbert*. Oxford: Clarendon Press.

Foster, B. (ed.) (1963). *The Local Port Book of Southampton for 1435–36*. Southampton: Southampton University Press, Southampton Record Series, 7.

Fox, H. S. A. (1991). 'Farming Practice and Techniques: Devon and Cornwall', in Miller (1991: 303–23).

——(2001). *The Evolution of the Fishing Village: Landscape and Society along the South Devon Coast, 1086–1550*. Oxford: Leopard's Head Press; Leicester Explorations in Local History, 1.

Frank, R. W., Jr. (1995). 'The "Hungry Gap", Crop Failure and Famine: The Fourteenth-Century Agricultural Crisis and *Piers Plowman*', in D. Sweeney (ed.), *Agriculture in the*

Middle Ages: Technology, Practice and Representation. Philadelphia: University of Pennsylvania Press, 227–43.

FRANKLIN, M. J. (ed.) (1988). *The Cartulary of Daventry Priory*. Northampton: Northamptonshire Record Society, 35.

FRANKLIN, P. (1982). 'Thornbury Woodlands and Deer Parks: The Earls of Gloucester's Deer Parks', *Transactions of the Bristol and Gloucestershire Archaeological Society*, 107: 149–69.

FRISK, G. (ed.) (1949). *A Middle English Translation of Macer Floridus De viribus herbarum*. Uppsala: A. B. Lundequist.

FRY, T. (ed.) (1980). *The Rule of St Benedict in Latin and English with Notes*. Minneapolis: Liturgical Press.

FRYDE, E. B. (1996). *Peasants and Landlords in Later Medieval England*. Stroud: Alan Sutton.

FULLER, B. T., RICHARDS, M. P., and MAYS, S. A. (2003). 'Stable Carbon and Nitrogen Isotope Variations in Tooth Dentine Serial Sections from Wharram Percy', *Journal of Archaeological Science*, 30: 1673–84.

FURNIVALL, F. J. (ed.) (1868). *Manners and Meals in Olden Time*. London: N. Trübner and Co.; EETS, OS 32.

——(ed.) (1870). Andrew Boorde, *The First Boke of the Introduction of Knowledge*. London: Kegan Paul, Trench, Trübner and Co.; EETS, ES 10.

GARMONSWAY, G. N. (ed.) (1967). *The Anglo-Saxon Chronicle*. London: Dent and Dutton.

——(ed.) (1978). *Aelfric's Colloquy*. Exeter: University of Exeter.

GERNAEY, A. M., MINNIKIN, D. E., COPLEY, M. S., DIXON, R. A., MIDDLETON, J. C., and ROBERTS, C. A. (2001). 'Mycolic Acids and Ancient DNA Confirm an Osteological Diagnosis of Tuberculosis', *Tuberculosis*, 81: 259–65.

GIDNEY, L. J. (1991*a*). *Leicester, The Shires, 1988 Excavations: The Animal Bones from the Medieval Deposits at Little Lane*. London: English Heritage, AML Report 57/91.

——(1991*b*). *Leicester, The Shires, 1988 Excavations: The Animal Bones from the Medieval Deposits at St Peter's Lane*. London: English Heritage, AML Report 116/91.

——(1999). 'The Animal Bones', in A. Connor and R. Buckley (eds.), *Roman and Medieval Occupation in Causeway Lane, Leicester*. Leicester Archaeology Monographs, 5: 310–29.

GILBERT, J. (1979). *Hunting and Hunting Reserves in Medieval Scotland*. Edinburgh: John Donald.

GIORGI, J. (1997). 'Diet in Late Medieval and Early Modern London: The Archaeobotanical Evidence', in D. Gaimster and P. Stamper (eds.), *The Age of Transition: The Archaeology of English Culture 1400–1600*. Oxford: Society for Medieval Archaeology, Monograph Series, 15: 197–213.

GODBOLD, S., and TURNER, R. C. (1994). 'Medieval Fishtraps in the Severn Estuary', *Medieval Archaeology*, 38: 19–54.

GOERING, J., and MANTELLO, F. A. C. (1986). 'The *Perambulavit Judas* ("*Speculum confessionis*") Attributed to Robert Grosseteste', *Revue Bénédictine*, 96: 125–68.

——and PAYER, P. J. (1993). 'The "Summa penitentie fratrum predicatorum": A Thirteenth-Century Confessional Formula', *Mediaeval Studies*, 55: 1–50.

GOLDBERG, P. J. P. (1986). 'Female Labour, Service, and Marriage in the Late Medieval Urban North', *Northern History*, 22: 18–38.

——(1988). 'Mortality and Economic Change in the Diocese of York, 1390–1514', *Northern History*, 24: 38–55.

——(1992*a*). *Women, Work and Life-Cycle*. Oxford: Oxford University Press.

——(1992*b*). '"For better, for worse": Marriage and Economic Opportunity for Women in Town and Country', in P. J. P. Goldberg (ed.), *Woman is a Worthy Wight: Women in English Society c.1200–1500*. Stroud: Alan Sutton, 108–25.

GOMME, G. L. (ed.) (1909). *Court Rolls of Tooting Beck Manor*. London: London County Council.

GOODE, H., WALDRON, T., and ROGERS, J. (1993). 'Bone Growth in Juveniles: A Methodological Note', *International Journal of Osteoarchaeology*, 3: 321–33.

GOODY, J. (1982). *Cooking, Cuisine and Class: A Study in Comparative Sociology*. Cambridge: Cambridge University Press.

GOPALAN, C. (1992). 'Undernutrition: Measurement and Implications', in Osmani (1992: 17–47).

GOTTFRIED, R. (1978). *Epidemic Disease in Fifteenth-Century England: The Medical Response and the Demographic Consequences*. New Brunswick, NJ: Rutgers University Press.

GRANSDEN, A. (ed.) (1963). *The Customary of the Benedictine Abbey of Eynsham in Oxfordshire*. Siegburg: F. Schmitt, Corpus Consuetudinum Monasticarum, 2.

——(ed.) (1973). *The Customary of the Benedictine Abbey of Bury St Edmund's in Suffolk*. London: Henry Bradshaw Society, 99.

GRANT, A. (1971). 'The Animal Bones', in G. F. Bryant and J. M. Steane (eds.), 'Excavations at the Deserted Medieval Settlement at Lyveden: A Third Interim Report', *Journal of the Northampton County Borough Museum and Art Gallery*, 9: 90–3.

——(1975). 'The Animal Bones', in J. M. Steane and G. F. Bryant (eds.), 'Excavations at the Deserted Medieval Settlement at Lyveden, Fourth Report', *Journal of the Northampton County Borough Museum and Art Gallery*, 12: 152–7.

——(1976). 'The Animal Bones', in Cunliffe (1976: 262–87).

——(1977). 'The Animal Bones, Mammals', in Cunliffe (1977: 213–33).

——(1979*a*). 'The Animal Bones', in D. Baker and E. Baker, 'The Excavations: Bedford Castle', in D. Baker, E. Baker, J. Hassall, and A. Simco (eds.), 'Excavations in Bedford 1967–77', *Bedfordshire Archaeological Journal*, 13: 58–62.

——(1979*b*). 'The Animal Bones', in J. Hassall, 'St John's Street', in D. Baker, E. Baker, J. Hassall, and A. Simco (eds.), 'Excavations in Bedford 1967–77', *Bedfordshire Archaeological Journal*, 13: 97–126, at 103–7.

——(1987). 'Some Observations on Butchery in England from the Iron Age to the Medieval Period', *Anthropozoologica*, special number, 1: 53–7.

——(1988). 'Animal Resources', in Astill and Grant (1988: 149–87).

——(2002). 'Food, Status and Social Hierarchy', in Miracle and Milner (2002: 17–23).

GRANT, E. (1988). 'Marine and River Fishing in Medieval Somerset: Fishbone Evidence from Langport', in Aston and Bond (1988*a*: ii. 409–16).

GRANT, J. (ed.) (1978). *La Passiun de seint Edmund*. London: ANTS, 36.

GRAUNT, J. (1662). *Natural and Political Observations Mentioned in a Following Index, and Made upon the Bills of Mortality*. London: Thomas Roycroft and Thomas Dicas.

GRAY, T. (ed.) (1995–6). *Devon Household Accounts, 1627–59*. 2 vols. Exeter: Devon and Cornwall Record Society, 38–9.

GREATREX, J. (ed.) (1984). *Account Rolls of the Obedientiaries of Peterborough*. Wellingborough: Northamptonshire Record Society, 33.

GREEN, J. (1979). 'Forests', in B. E. Harris (ed.), *VCH Chester II*. Oxford: Institute of Historical Research, University of London, 167–87.

GREIG, J. R. A. (1981). 'The Investigation of a Medieval Barrel-Latrine from Worcester', *Journal of Archaeological Science*, 8: 265–82.

——(1983). 'Plant Foods in the Past: A Review of the Evidence from Northern Europe', *Journal of Plant Foods*, 5: 179–214.

——(1988*a*). 'Plant Resources', in Astill and Grant (1988: 108–27).

——(1988*b*). 'Plant Remains', in S. Ward (ed.), 'Excavations at Chester: 12 Watergate Street in 1985: Roman Headquarters Building to Medieval', *Grosvenor Museum Archaeological Excavation and Survey Reports*, 5: 59–69.

——(1996). 'Archaeobotanical and Historical Records Compared: A New Look at the Taphonomy of Edible and Other Useful Plants from the 11th to the 18th Centuries A.D.', *Circaea*, 12: 211–47.

——(1997). 'Archaeobotany', in J. D. Hurst (ed.), *A Multi-Period Salt Production Site at Droitwich, Excavations at Upwich*. York: CBA, Research Report 107: 133–45, microfiche 92–4.

——(2002). 'The 13th–18th Century Plant Remains', in N. Baker (ed.), *Shrewsbury Abbey*. Shrewsbury: Shropshire Archaeological and Historical Society, Monograph Series, 2: 163–77.

GREWE, R. (1992). 'Hispano-Arabic Cuisine in the Twelfth Century', in Lambert (1992: 141–8).

GRIESSER, P. B. (1956). 'Die "Ecclesiastica officia Cisterciensis Ordinis" des Cod. 1711 von Trient', *Analecta sacri Ordinis Cisterciensis*, 12: 153–288.

GROSJEAN, P. (ed.) (1935). *Henrici VI Angliae regis miracula postuma ex codice Musei Britannici Regio 13. C. VIII*. Brussels: Société des Bollandistes, Subsidia Hagiographica, 22.

GUMERMAN, G. (1997). 'Food and Complex Societies', *Journal of Archaeological Method and Theory*, 4: 105–39.

GURNEY, J. H. (1921). *Annals of Ornithology*. London: Witherby.

HAGEN, A. (1998). *A Handbook of Anglo-Saxon Food: Processing and Consumption*. Frithgarth: Anglo-Saxon Books.

——(2002). *A Second Handbook of Anglo-Saxon Food and Drink: Processing and Distribution*. Frithgarth: Anglo-Saxon Books.

HALLAM, H. E. (ed.) (1988*a*). *The Agrarian History of England and Wales*, ii: *1042–1350*. Cambridge: Cambridge University Press.

——(1988*b*). 'Farming Techniques: Eastern England', in Hallam (1988*a*: 272–312).

——(1988*c*). 'Farming Techniques: Southern England', in Hallam (1988*a*: 341–68).

——(1988*d*). 'The Life of the People', in Hallam (1988*a*: 818–53).

HAMEROW, H., and MACGREGOR, A. (eds.) (2001). *Image and Power in the Archaeology of Early Medieval Britain: Essays in Honour of Rosemary Cramp*. Oxford: Oxbow.

HAMILAKIS, Y. (2003). 'The Sacred Geography of Hunting: Wild Animals, Social Power and Gender in Early Farming Societies', in E. Kotjabopdou, Y. Hamilakis, P. Halstead,

C. Gamble, and P. Elefanti (eds.), *Zooarchaeology in Greece: Recent Advances*. London: British School at Athens, British School at Athens Studies, 9: 239–47.

Hamilton, J. (1992). 'Animal Bone', in S. Cracknell and M. W. Bishop (eds.), 'Excavations at 23–33 Brook Street, Warwick, 1973', *Transactions of the Birmingham and Warwickshire Archaeological Society*, 97: 1–40, at 30–1 and microfiche.

Hamilton, N. E. S. A. (ed.) (1870). *William of Malmesbury: De gestis pontificum Anglorum*. London: Rolls Series.

Hamilton-Dyer, S. (1995). 'Fish in Tudor Naval Diet—with Reference to the *Mary Rose*', *Archaeofauna*, 4: 27–32.

Hanawalt, B. (ed.) (1976). *Crime in East Anglia in the Fourteenth Century*. London: Norfolk Record Society, 44.

Hanawalt, B. A. (1986). *The Ties that Bound: Peasant Families in Medieval England*. Oxford: Oxford University Press.

Hansell, P., and Hansell, J. (1988). *Doves and Dovecotes*. Bath: Millstream Press.

Hardy, A., Dodd, A., and Keevill, G. (eds.) (2003). *Aelfric's Abbey: Excavations at Eynsham Abbey, Oxfordshire 1989–92*. Oxford: Oxford University School of Archaeology, Thames Valley Landscape, 16.

Harman, M. (1979*a*). 'Mammalian Bone', in J. H. Williams (ed.), 'Excavations on Marefair, Northampton, 1977', *Northamptonshire Archaeology*, 14: 38–79.

——(1979*b*) 'The Mammalian Bones', in J. H. Williams (ed.), *St Peter's Street Northampton, Excavations 1973–1976*. Northampton: Northampton Development Corporation, Archaeological Monographs, 2: 328–32.

——(1985). 'The Other Mammalian Bones', in J. H. Williams, M. Shaw, and V. Denham (eds.), *Middle Saxon Palaces at Northampton*. Northampton: Northampton Development Corporation Archaeological Monograph, 4: 328–33.

——(1996). 'The Mammal Bones', in M. Shaw (ed.), 'The Excavation of a Late 15th- to 17th-Century Tanning Complex at The Green, Northampton', *Post-Medieval Archaeology*, 30: 63–127.

——and Baker, J. R. (1985). 'The Animal Bones', in M. Shaw (ed.), 'Excavations on a Saxon and Medieval Site at Black Lion Hill, Northampton', *Northamptonshire Archaeology*, 20: 113–38, at 135–6 and microfiche 86–93.

Harper-Bill, C., Rawcliffe, C., and Wilson, R. G. (eds.) (2002). *East Anglia's History: Studies in Honour of Norman Scarfe*. Woodbridge: Boydell Press.

Harris, M., and Thurgood, J. M. (eds.) (1984). 'The Account of the Great Household of Humphrey, First Duke of Buckingham, for the Year 1452–3', in *Camden Miscellany XXVIII*. London: Royal Historical Society, Camden, 4th series, 29: 1–57.

Harvey, B. F. (ed.) (1965). *Documents Illustrating the Rule of Walter de Wenlok, Abbot of Westminster, 1283–1307*. London: Royal Historical Society, Camden, 4th series, 2.

——(1993). *Living and Dying in England 1100–1540: The Monastic Experience*. Oxford: Clarendon Press.

——(1997). 'Monastic Diet, xiiith–xvith Centuries: Problems and Perspectives', in S. Cavaciocchi (ed.), *Alimentazione et nutrizione secc. xiii–xviii*. Florence: Le Monnier, Istituto Internazionale di Storia Economica 'F. Datini', Prato, Serie II—Atti delle 'Settimane di Studi' e altri Convegni, 28: 611–41.

——and Oeppen, J. (2001). 'Patterns of Morbidity in Late Medieval England: A Sample from Westminster Abbey', *EconHR*, 2nd series, 54: 215–39.

HARVEY, J. H. (1981). *Mediaeval Gardens*. London: Batsford.
——(1984). 'Vegetables in the Middle Ages', *Garden History*, 12: 89–99.
HARVEY, P. D. A. (1965). *A Medieval Oxfordshire Village: Cuxham, 1240 to 1400*. Oxford: Oxford University Press.
——(ed.) (1976). *Manorial Records of Cuxham, Oxfordshire circa 1200–1359*. London: HMSO, Historical Manuscripts Commission Joint Publications, 23.
——(1984). *The Peasant Land Market in Medieval England*. Oxford: Clarendon Press.
——(1991). 'Farming Practice and Techniques: The Home Counties', in Miller (1991: 254–68).
HARVEY, S. (1988). 'Domesday England', in Hallam (1988*a*: 45–138).
HATCHER, J. (1970). *Rural Economy and Society in the Duchy of Cornwall 1300–1500*. Cambridge: Cambridge University Press.
——(1977). *Plague, Population and the English Economy*. Basingstoke: Macmillan.
——(1986). 'Mortality in the Fifteenth Century: Some New Evidence', *EconHR*, 2nd series, 39: 19–38.
——(1994). 'England in the Aftermath of the Black Death', *Past and Present*, 144: 3–35.
——and BAILEY, M. (2001). *Modelling the Middle Ages*. Cambridge: Cambridge University Press.
HATTO, A. T. (trans.) (1967). Gottfried von Strassburg, *Tristan*. Harmondsworth: Penguin Classics.
HAYFIELD, C., and GREIG, J. R. A. (1989). 'The Excavation and Salvage Work on a Moated Site at Cowick, South Humberside, 1976', *Yorkshire Archaeological Journal*, 61: 41–70.
HEINRICH, D. (1994). 'Fish Remains of Two Medieval Castles and of an Urban Context: A Comparison', *Annales du Musée Royale de l'Afrique Centrale*, 274: 211–16.
HERRGOTT, M. (1726). *Ordo Cluniacensis per Bernardum saec. xi scriptorum*, in M. Herrgott, *Vetus disciplina monastica*. Paris: n. p. (reprinted Siegburg, 1999), 133–364.
HERRING, D. A., SAUNDERS, S. R., and KATZENBERG, M. A. (1998). 'Investigating the Weaning Process in Past Populations', *American Journal of Physical Anthropology*, 105: 425–39.
HERRSCHER, E., BOCHERENS, H., VALENTIN, F., and COLARDELLE, R. (2001). 'Comportements alimentaires au Moyen Âge à Grenoble: application de la biogéochimie isotopique à la nécropole Saint-Laurent (xiiie–xve siècles, Isère, France)', *Comptes rendus de l'Académie des Sciences, Series III—Sciences de la vie*, 324: 479–87.
HIEATT, C. B. (2002). 'Medieval Britain', in Weiss Adamson (2002: 19–45).
HIGGS, E., and GREENWOOD, W. (1979). 'Faunal Report' in P. Rahtz (ed.), *The Saxon and Medieval Palaces at Cheddar: Excavations 1960–2*. Oxford: BAR, British Series, 65: 354–62.
HIGHAM, M. C. (2002). 'Medieval Gardens in North West England, with Particular Reference to Lancashire', *Transactions of the Lancashire and Cheshire Antiquarian Society*, 98: 15–25.
HILLMAN, G. C. (1984). 'Interpretation of Archaeological Plant Remains: The Application of Ethnographic Models from Turkey', in Zeist and Casparie (1984: 1–41).
HILTON, R. H. (1976). 'Introduction', in R. H. Hilton (ed.), *The Transition from Feudalism to Capitalism*. London: Verso, 9–30.

Hilton, R. H. (1983). *A Medieval Society: The West Midlands at the End of the Thirteenth Century*. Cambridge: Cambridge University Press.

Hinton, D. A. (1990). *Archaeology, Economy and Society: England from the Fifth to Fifteenth Century*. London: Seaby.

——(2005). *Gold and Gilt, Pots and Pins: Possessions and People in Medieval Britain*. Oxford: Oxford University Press.

Hockey, S. F. (ed.) (1975). *The Account-Book of Beaulieu Abbey*. London: Royal Historical Society, Camden, 4th series, 16.

Hoffman, R. (1994). 'Medieval Cistercian Fisheries Natural and Artificial', in L. Pressouyre (ed.), *L'Espace cistercien*. Paris: Ministère de l'Enseignement Supérieur et de la Recherche, 404–14.

Hogg, J. (1980). 'Everyday Life in the Charterhouse in the Fourteenth and Fifteenth Centuries', in *Klösterliche Sachkultur des Spätmittelalters*. Vienna: Veröffentlichungen des Instituts für Mittelalterliche Realienkunde Österreichs, 3 = Sitzungsberichte der Österreichische Akademie der Wissenschaften, philosophisch-historische Klasse, 367: 113–46.

Holden, P., and Cleeves, T. (2002). *RSPB Handbook of British Birds*. London: Christopher Helm.

Holmes, J. (1992). 'The Animal Bones', in C. C. S. Woodfield (ed.), 'The Defences of Towcester, Northamptonshire', *Northamptonshire Archaeology*, 24: 53–5.

Homans, G. C. (1942). *English Villagers of the Thirteenth Century*. Cambridge, Mass.: Harvard University Press.

Hooke, D. (2001). *The Landscape of Anglo-Saxon England*. Leicester: Leicester University Press.

Horandner, E. (1986). 'Storing and Preserving Meat in Europe: Historical Survey', in A. Fenton and E. Kisban (eds.), *Food in Change: Eating Habits from the Middle Ages to the Present Day*. Edinburgh: Donald in association with National Museums of Scotland, 53–9.

Horrox, R. (ed.) (1994). *The Black Death*. Manchester: Manchester University Press.

Hudson, A. (1988). *The Premature Reformation: Wycliffite Texts and Lollard History*. Oxford: Clarendon Press.

——and Gradon, P. (eds.) (1983–96). *English Wycliffite Sermons*. 5 vols. Oxford: Clarendon Press.

Hudson, J. (ed.) (2002–). *Historia ecclesie Abbendonensis: The History of the Church of Abingdon*. 2 vols., in progress. Oxford: Oxford Medieval Texts.

Hull, P. L. (ed.) (1971). *The Caption of Seisin of the Duchy of Cornwall (1337)*. Torquay: Devon and Cornwall Record Society, new series, 17.

Hunnisett, R. F. (ed.) (1961). *Bedfordshire Coroners' Rolls*. Bedford: Bedfordshire Historical Record Society, 41.

Hunt, T. J., and Keil, I. (1959–60). 'Two Medieval Gardens', *Proceedings of the Somerset Archaeological and Natural History Society*, 104: 91–5.

Hunter, J. (ed.) (1833). *Magnum rotulum scaccarii, vel magnum rotulum pipae, anno tricesimo-primo regni Henrici primi*. London: Record Commission.

Huntley, J. (1995). 'The Plant Remains', in R. Fraser, C. Jamfrey, and J. Vaughan (eds.), 'Excavation on the Site of the Mansion House, Newcastle, 1990', *Archaeologia Aeliana*, 5th series, 23: 145–213.

Hybel, N. (1989). *Crisis or Change: The Concept of Crisis in the Light of Agrarian Structural Reorganization in Late Medieval England.* Aarhus: Aarhus University Press.

James, M. K. (ed.) (1971). *Studies in the Medieval Wine Trade.* Oxford: Clarendon Press.

James, M. R., rev., Brooke C. N. L., and Mynors, R. A. B. (eds.) (1983). *Walter Map: De nugis curialium: Courtiers' Trifles.* Oxford: Oxford Medieval Texts.

Janssen, H. A. M., and Maat, G. J. R. (1999). *Canons Buried in the Stiftskapel of the Saint Servaas Basilica at Maastricht AD 1070–1521: A Palaeopathological Study.* Leiden: Barge's Anthropologica, 5.

Jaritz, G. (1976). 'Die Reiner Rechnungsbücher (1399–1477) als Quelle zur klösterlichen Sachkultur des Spätmittelalters', in *Die Funktion der Schriftlichen Quelle in der Sachkulturforschung.* Vienna: Veröffentlichungen des Instituts für Mittelalterliche Realienkunde Österreichs, 1 = Sitzungsberichte der Österreichische Akademie der Wissenschaften, philosophisch-historische Klasse, 304: 145–249.

——(1980). 'Zur Sachkultur Österreichischer Klöster des Spätmittelalters', in *Klösterliche Sachkultur des Spätmittelalters.* Vienna: Veröffentlichungen des Instituts für Mittelalterliche Realienkunde Österreichs, 3 = Sitzungsberichte der Österreichische Akademie der Wissenschaften, philosophisch-historische Klasse, 367: 147–68.

——(1985). 'The Standard of Living in German and Austrian Cistercian Monasteries of the Late Middle Ages', in E. R. Elder (ed.), *Goad and Nail.* Kalamazoo: Cistercian Studies Series, 84: 56–70.

——(1990). 'Religiöse Stiftungen als Indikator der Entwicklung materieller Kultur im Mittelalter', in *Materielle Kultur und Religiöse Stiftung im Spätmittelalter.* Vienna: Veröffentlichungen des Instituts für Mittelalterliche Realienkunde Österreichs, 12 = Sitzungsberichte der Österreichische Akademie der Wissenschaften, philosophisch-historische Klasse, 554: 13–35.

Jessopp, A., and James, M. R. (eds.) (1896). *The Life and Miracles of St William of Norwich by Thomas of Monmouth.* Cambridge: Cambridge University Press.

Jones, A. (1977). 'Harvest Customs and Labourers' Perquisites in Southern England, 1150–1350: The Hay Harvest', *Agricultural History Review*, 25: 98–107.

Jones, A. K. G. (1988). 'Provisional Remarks on Fish Remains from Archaeological Deposits at York', in P. Murphy and C. French (eds.), *The Exploitation of Wetlands: Symposia of the Association for Environmental Archaeology No. 7.* Oxford: BAR, British Series, 186: 113–27.

Jones, G. (1984). 'Interpretation of Archaeological Plant Remains: Ethnographic Models from Greece', in Zeist and Casparie (1984: 43–61).

——and Halstead, P. (1995). 'Maslins, Mixtures and Monocrops: On the Interpretation of Archaeobotanical Crop Samples of Heterogeneous Composition', *Journal of Archaeological Science*, 22: 103–14.

——Straker, V., and Davis, A. (1991). 'Early Medieval Plant Use and Ecology', in A. Vince (ed.), *Aspects of Saxon and Norman London,* ii: *Finds and Environmental Evidence.* London and Middlesex Archaeological Society, Special Paper, 12: 347–88.

Jones, G. G. (1983). 'The Medieval Animal Bones', in D. Allen and C. H. Dalwood (eds.), 'Iron Age Occupation, a Middle Saxon cemetery, and 12th to 19th Century Urban Occupation: Excavations in George Street, Aylesbury, 1981', *Records of Buckinghamshire*, 25: 31–44.

Jones, G. G. (1984). 'Animal Bones', in A. Rogerson and C. Dallas (eds.), *Excavations at Thetford 1948–59 and 1973–80*. Norwich: East Anglian Archaeology, 22: 187–92.
——(1993). 'Animal Bone', in C. Dallas (ed.), *Excavations in Thetford by B. K. Davidson between 1964 and 1970*. Norwich: East Anglian Archaeology, 62: 176–91.
——(1994*a*). 'Animal Bones', in Allen (1994: 386–96).
——(1994*b*). 'Mammal and Bird Bone', in B. S. Ayers (ed.), *Excavations at Fishergate, Norwich 1985*. Norwich: East Anglian Archaeology, 68: 37–43.
——(2002). 'The Excavated Faunal Remains', in N. Baker (ed.), *Shrewsbury Abbey: Studies in the Archaeology and History of an Urban Abbey*. Shrewsbury: Shropshire Archaeological and Historical Society, Monograph Series, 145–58.
Jones, J. (2000). 'Plant Macrofossil Remains from Redcliffe Backs, Bristol'. Unpublished report for Bristol and Region Archaeological Services.
Jones, R. T., Levitan, B., Stevens, P. and Hocking, L. (1985). *Castle Lane, Brackley, Northamptonshire. The Vertebrate Remains*. London: English Heritage, AML Report 4812.
——and Ruben, I. (1987). 'Animal Bones, with Some Notes on the Effects of Differential Sampling', in G. Beresford (ed.), *Goltho: The Development of an Early Medieval Manor c.850–1150*. London: Historic Buildings and Monuments Commission for England, English Heritage Report 4: 197–206.
——and Serjeantson, D. (1983). *The Animal Bones from Five Sites at Ipswich*. London: English Heritage, AML Report 3951.
Jordan, W. C. (1996). *The Great Famine: Northern Europe in the Early Fourteenth Century*. Princeton: Princeton University Press.
Julkunen, H., Karava, R., and Viljanen, V. (1966). 'Hyperostosis of the Spine in Diabetes Mellitus and Acromegaly', *Diabetologia*, 28: 123–6.
Jurkowski, M., Smith, C. L., and Crook, D. (1998). *Lay Taxes in England and Wales 1188–1688*. Kew: PRO Publications, Public Record Office Handbook, 31.
Kane, G. (ed.) (1960). *Piers Plowman: The A Version*. London: Athlone Press.
Katzenberg, M. A. (2000). 'Stable Isotope Analysis: A Tool for Studying Past Diet, Demography and Life History', in M. A. Katzenberg and S. R. Saunders (eds.), *Biological Anthropology of the Human Skeleton*. New York: Wiley-Liss, 305–27.
Keene, D. (1985). *Survey of Medieval Winchester*. 2 vols. Oxford: Clarendon Press, Winchester Studies, 2.
Keevill, G. D. (1996). *In Harvey's House and in God's House: Excavations at Eynsham Abbey 1991–3*. Oxford: Oxford Archaeological Unit.
Kelly, F. (2000). *Early Irish Farming: A Study Based on the Law-Texts of the 7th and 8th Centuries AD*. Dublin: Dublin Institute for Advanced Studies.
Kent, S. (ed.) (1989). *Farmers as Hunters*. Cambridge: Cambridge University Press.
Kershaw, I. (1973*a*). *Bolton Priory: The Economy of a Northern Monastery 1286–1325*. Oxford: Oxford University Press.
——(1973*b*). 'The Great Famine and Agrarian Crisis in England 1315–1322', *Past and Present*, 59: 3–50.
——and Smith, D. M. (eds.) (2000). *The Bolton Priory Compotus 1286–1325 Together with a Priory Account Roll for 1377–1378*. Woodbridge: Boydell Press, Yorkshire Archaeological Society, Record Series, 154.

KIMBALL, G. (ed.) (1939). *Rolls of the Warwickshire and Coventry Sessions of the Peace, 1377–1397*. Stratford-upon-Avon: Dugdale Society, 16.

KING, E. (1991). 'Farming Practice and Techniques: The East Midlands', in Miller (1991: 210–22).

KIRK, R. E. G. (ed.) (1892). *Accounts of the Obedientiars of Abingdon Abbey*. London: Camden Society, new series, 51.

KITCHIN, G. W. (ed.) (1892). *Compotus Rolls of the Obedientiaries of St Swithun's Priory, Winchester*. London: Hampshire Record Society, 5.

KLEEREKOPER, M. (1996). 'Fluoride and the Skeleton', *Critical Reviews of Clinical and Laboratory Science*, 33: 139–61.

KNOWLES, D. (1963). *The Monastic Order in England: A History of its Development from the Times of St Dunstan to the Fourth Lateran Council 940–1216*. Cambridge: Cambridge University Press.

KNÜSEL, C., and BOYLSTON, A. (2000). 'How has the Towton Project Contributed to our Knowledge of Medieval and Later Warfare?', in Fiorato, Boylston, and Knüsel (2000: 169–88).

——CHUNDUN, Z., and CARDWELL, P. (1992). 'Slipped Proximal Femoral Epiphyses in a Medieval Priest', *International Journal of Osteoarchaeology*, 2: 109–19.

KONOMI, N., LEBWOHL, E., MOWBRAY, K., TATTERSALL, I., and ZHANG, D. (2002). 'Detection of Mycobacterial DNA in Andean Mummies', *Journal of Clinical Microbiology*, 40: 4738–40.

KOWALESKI, M. (ed.) (1993). *Local Customs Accounts of the Port of Exeter, 1266–1321*. Exeter: Devon and Cornwall Record Society, new series, 36.

——(1995). *Local Markets and Regional Trade in Medieval Exeter*. Cambridge: Cambridge University Press.

——(2000). 'The Expansion of the South-Western Fisheries in Late Medieval England', *EconHR*, 2nd series, 53: 429–54.

——and CHILDS, W. R. (2000). 'The Internal and International Fish Trades of Medieval England and Wales', in Starkey, Reid, and Ashcroft (2000: 29–35, 245–7).

KRISTOL, A. M. (ed.) (1995). *Manières de langage (1396, 1399, 1415)*. London: ANTS, 53.

LAMBERT, C. (ed.) (1992). *Du Manuscrit à la table: essais sur la cuisine au Moyen Âge et répertoire des manuscrits médiévaux contenant des recettes culinaires*. Montreal: Les Presses de l'Université de Montréal.

LANDERS, J. R. (2003). *The Field and the Forge: Population, Production and Power in the Pre-Industrial West*. Oxford: Oxford University Press.

LANDSBERG, S. (1996). *The Mediaeval Garden*. London: British Museum Press.

LANE, J. (1996). *John Hall and his Patients: The Medical Practice of Shakespeare's Son-in-Law*. Stratford-upon-Avon: Shakespeare Birthplace Trust.

LANGDON, J. (1986). *Horses, Oxen and Technological Innovation: The Use of Draught Animals in English Farming from 1066 to 1500*. Cambridge: Cambridge University Press.

LAUGHTON, J., and DYER, C. C. (2002). 'Seasonal Patterns of Trade in the Later Middle Ages: Buying and Selling at Melton Mowbray, Leicestershire, 1400–1520', *Nottingham Medieval Studies*, 46: 162–84.

Laurioux, B. (1988). 'Le Lièvre lubrique et la bête sanglante: réflexions sur quelques interdits alimentaires du haut Moyen Âge', in L. Bodson (ed.), *L'Animal dans l'alimentation humaine: les critères de choix. Anthropozoologica*, special number 2: 127–32.

Lauwerier, R. C. G. M. (1988). *Animals in Roman Times in the Dutch Eastern River Area*. Amersfoort: ROB.

Lawn, B. (ed.) (1979). *The Prose Salernitan Questions Edited from a Bodleian Manuscript (Auct. F.3.10)*. London: British Academy, Auctores Britannici Medii Aevi, 5.

Leccisotti, T., and Walker Bynum, C. (eds.) (1975). *Statuta Casinensia saeculi xiii*, in *Consuetudines Benedictinae variae (saec. xi–saec. xiv)*. Siegburg: F. Schmitt, Corpus Consuetudinum Monasticarum, 6: 211–27.

Lee, R. (1986). 'Population Homeostasis and English Demographic History', in R. I. Rothberg and T. K. Rabb (eds.), *Population and Economy*. Cambridge: Cambridge University Press, 75–100.

Lennard, R. (1959). *Rural England 1086–1135: A Study of Social and Agrarian Conditions*. Oxford: Clarendon Press.

Le Patourel, H. E. J. (1991). 'Rural Building in England and Wales', in Miller (1991: 820–919).

Letts, J. (1999). *Smoke Blackened Thatch: A Unique Source of Late Medieval Plant Remains from Southern England*. London: English Heritage and the University of Reading.

Levitan, B. (1982). 'The Faunal Remains', in P. Leach (ed.), *Ilchester 1: Excavations 1974–1975*. Bristol: Western Archaeological Trust, 269–85.

——(1984). 'The Vertebrate Remains', in S. Rahtz and T. Rowley (eds.), *Middleton Stoney: Excavation and Survey in a North Oxfordshire Parish, 1970–1982*. Oxford: University of Oxford Department for External Studies, 108–63.

Liddiard, R. (2003). 'The Deer Parks of Domesday Book', *Landscapes*, 4: 4–23.

Lilley, J. M., Stroud, G., Brothwell, D. R., and Williamson, M. H. (1994). *The Jewish Burial Ground at Jewbury*. York: York Archaeological Trust, The Archaeology of York, 12: The Medieval Cemeteries, fasc. 3.

Lilley, K. D. (2002). *Urban Life in the Middle Ages 1000–1450*. Basingstoke: Palgrave.

Littler, A. (1979). 'Fish in English Economy and Society down to the Reformation'. Unpublished Ph.D. thesis, University of Wales.

Livense, A. M., Bierma-Zeinstra, S. M., Verhager, A. P., Baar, M. E. van, Verhaar, J. H., and Koes, B. W. (2002). 'Influence of Obesity on the Development of Osteoarthritis of the Hip: A Systematic Review', *Rheumatology*, 41: 1155–62.

Livi-Bacci, M. (1991). *Population and Nutrition: An Essay on European Demographic History*. Cambridge: Cambridge University Press.

Liyanage, S. H., Gupta, C. M., and Cobb, A. R. (2003). 'Skeletal Tuberculosis', *Journal of the Royal Society of Medicine*, 96: 262–5.

Lloyd, T. H. (1977). *The English Wool Trade in the Middle Ages*. Cambridge: Cambridge University Press.

Lock, R. (1992). 'The Black Death in Walsham-le-Willows', *Proceedings of the Suffolk Institute of Archaeology and History*, 27: 316–37.

——(ed.) (1998). *The Court Rolls of Walsham le Willows, 1303–1350*. Woodbridge: Suffolk Records Society, 41.

——(ed.) (2002). *The Court Rolls of Walsham le Willows, 1351–1399*. Woodbridge: Suffolk Records Society, 45.

LOCKER, A. (1985). 'Animal and Plant Remains', in J. N. Hare (ed.) *Battle Abbey: The Eastern Range and the Excavations of 1978–1980*. London: Historic Buildings and Monuments Commission for England, 183–203.

——(1986). 'Trig Lane, City of London: The Fish Bones'. London: Museum of London, unpublished archive report.

——(1988*a*). 'The Animal Bone: 199 Borough High Street', in P. Hinton (ed.), *Excavations in Southwark 1973–76, Lambeth 1973–79*. London: Museum of London Department of Greater London Archaeology, London and Middlesex Archaeological Society, Surrey Archaeological Society, 427–33.

——(1988*b*). 'The Fish Bones', in R. Cowie, R. L. Whytehead, and L. Blackmore (eds.), 'Two Middle Saxon Occupation Sites: Excavations at Jubilee Hall and 21–22 Maiden Lane', *Transactions of the London and Middlesex Archaeological Society*, 39: 149–50.

——(1989). 'Fish Bones', in R. L. Whytehead, R. Cowie, and L. Blackmore (eds.), 'Excavations at the Peabody Site: Chandos Place and the National Gallery', *Transactions of the London and Middlesex Archaeological Society*, 40: 148–51.

——(1992*a*). 'Animal Bone', in S. Woodiwiss (ed.), *Iron Age and Roman Salt Production and the Medieval Town of Droitwich*. London: CBA, Research Report 81: 172–81.

——(1992*b*). 'The Fish Bones from St Mary Spital (from Excavations at 4 Spital Square, Norton Folgate and 4–12 Norton Place)'. London: Museum of London, unpublished archive report.

——(1996). 'The Fish Bones from St John's Priory, Jerusalem'. London: Museum of London, unpublished archive report.

——(1997*a*). *The Saxon, Medieval and Post Medieval Fish Bones from Excavations at Castle Mall, Norwich, Norfolk*. London: English Heritage, AML Report 85/97.

——(1997*b*). 'The Fish Bones', in P. Mills (ed.), 'Excavations at the Dorter Undercroft, Westminster Abbey', *Transactions of the London and Middlesex Archaeological Society*, 46: 111–13.

——(2001). *The Role of Stored Fish in England 900–1750 AD: The Evidence from Historical and Archaeological Data*. Sofia: Publishing Group Ltd.

——(2005). 'The Animal Bone', in M. Biddle (ed.), *Nonsuch Palace: The Material Culture of a Noble Household*. Oxford: Oxbow, 439–74.

LOSCO-BRADLEY, P. M., and SALISBURY, C. R. (1988). 'A Saxon and a Norman Fish Weir at Colwick, Nottinghamshire', in Aston and Bond (1988*a*: ii. 329–51).

LOVELUCK, C. (2001). 'Wealth, Waste and Conspicuous Consumption: Flixborough and its Importance for Mid- and Late Saxon Settlement Studies', in Hamerow and MacGregor (2001: 78–130).

LOVETT, J. (1990). 'Animal Bone', in J. Hughes, 'Survey and Excavation at Evesham Abbey', *Transactions of the Worcestershire Archaeological Society*, 3rd series, 12: 184–9.

LOYN, H. R. (1970). *Anglo-Saxon England and the Norman Conquest*. London: Longman.

Lucas, G. (1998). 'A Medieval Fishery on Whittlesea Mere, Cambridgeshire', *Medieval Archaeology*, 42: 19–44.

Lucy, S. (2000). *The Anglo-Saxon Way of Death: Burial Rites in Early England*. Gloucester: Sutton.

Luders, A., Tomlins, T. E., Raithby, J., et al. (eds.) (1810–28). *Statutes of the Realm*. 11 vols. London: Record Commission.

Luff, R. (1993). *Animal Bones from Excavations in Colchester, 1971–85*. Colchester: Colchester Archaeological Trust, Colchester Archaeological Report, 12.

Lunardon, P., and Spinelli, G. (eds.) (1977). *Pontida, 1076–1976: documenti per la storia del monastero di S. Giacomo*. Bergamo: Monumenta Bergomensia, 45.

Lupton, A. (2000). 'Friar's Gate Warrington: Site Narrative. Lancaster University Archaeology Unit Excavation, January–April 2000'. Lancaster: Lancaster University Archaeology Unit, unpublished excavation report.

Maat, G. J. R., Mastwijk, R. W., and Sarfatij, H. (1998). *Een fyisch anthropologisch onderzoek van begravenen bij het Minderbroederklooster aate Dordricht, circa 1275–1572 AD*. Amersfoort: ROB.

McCance, R. A., and Widdowson, E. M. (1998) *The Composition of Foods*. 5th edition. London: Royal Society of Chemistry and the Ministry of Agriculture, Fisheries and Food.

McClean, T. (1981). *Medieval English Gardens*. London: Collins.

McCormick, F. (1991). 'The Effect of the Anglo-Norman Settlement on Ireland's Wild and Domesticated Fauna', in P. Crabtree and K. Ryan (eds.), *Animal Use and Cultural Change*. Philadelphia: University of Pennsylvania, MASCA Research Papers in Science and Archaeology, Supplement to Volume 8: 41–52.

——(2002). 'The Distribution of Meat in a Hierarchical Society: The Irish Evidence', in Miracle and Milner (2002: 25–31).

MacDonald, K. C. (1992). 'The Domestic Chicken (*Gallus gallus*) in Sub-Saharan Africa: A Background to its Introduction and its Osteological Differentiation from Indigenous Fowls (*Numindinae* and *Francolinus* sp.)', *Journal of Archeological Science*, 19: 303–18.

Macdougall, E. (ed.) (1986). *Mediaeval Gardens*. Washington: Trustees for Harvard University, Dumbarton Oaks Colloquium on the History of Landscape Architecture, 9.

McIntosh, M. K. (1986). *Autonomy and Community: The Royal Manor of Havering 1200–1500*. Cambridge: Cambridge University Press.

McKenna, W. J. B. (1992). 'The Environmental Evidence', in D. H. Evans and D. G. Tomlinson (eds.), *Excavations at 33–35 Eastgate, Beverley, 1983–86*. Sheffield: Sheffield Archaeological Unit/Humberside County Council, Sheffield Excavation Reports, 3: 227–33.

——(1993). 'The Plant Macrofossils from Queen Street', in D. H. Evans (ed.), 'Excavations in Hull 1975–76', *East Riding Archaeologist*, 4 (Hull Old Town Report Series, 2): 198–201.

McKeown, T. (1976). *The Modern Rise of Population*. London: Edward Arnold.

——(1983). 'Food, Infection and Population', *Journal of Interdisciplinary History*, 14: 227–47.

Maddicott, J. R. (1975). *The English Peasantry and the Demands of the Crown, 1294–1341*. Oxford: *Past and Present*, Supplement 1.

MADGE, S. J. (ed.) (1903). *Abstracts of Inquisitiones post mortem for Gloucestershire.* London: British Record Society, 30.

MAGENNIS, H. (1999). *Anglo-Saxon Appetities: Food and Drink and their Consumption in Old English and Related Literature.* Dublin: Four Courts Press.

MAITLAND, F. W. (ed.) (1889). *Select Pleas in Manorial and Other Seignorial Courts.* London: Selden Society, 2.

——and BAILDON, W. P. (eds.) (1891). *The Court Baron.* London: Selden Society, 4.

MALTBY, M. (1979). *The Animal Bones from Exeter 1971–1975.* Exeter: Exeter Archaeological Reports, 2.

——(1993). 'Animal Bones', in P. J. Woodward, S. M. Davies, and A. H. Graham (eds.), *Excavations at the Old Methodist Chapel and Greyhound Yard, Dorchester.* Dorchester: Dorset Natural History and Archaeological Society, Monograph 12: 315–40.

MALTHUS, T. R. (1966). *First Essay on Population: A Reprint in Facsimile of An Essay on the Principle of Population, 1798.* London: Macmillan.

MARESH, M. M. (1955). 'Linear Growth in Long Bones of Extremities from Infancy through Adolescence', *American Journal of the Diseases of Children*, 89: 725–42.

MARKHAM, G. (1668). *Farewel to Husbandry.* London: George Sawbridge.

MARTIN, D. J. (2002). 'Fish Consumption in a Medieval English Bishop's Household, 1406–7', *aqua*, 5: 81–8.

MATE, M. (1991). 'The Agrarian Economy of South-East England before the Black Death: Depressed or Buoyant?', in Campbell (1991: 90–103).

MAUNDE THOMPSON, E. (ed.) (1902–4). *Customary of the Benedictine Monasteries of Saint Augustine, Canterbury, and Saint Peter, Westminster.* 2 vols. London: Henry Bradshaw Society, 23, 28.

MAYS, S. A. (1991). *The Medieval Burials from the Blackfriars Friary, School Street, Ipswich, Suffolk.* London: English Heritage, AML Report 16/91.

——(1997). 'Carbon Stable Isotope Ratios in Mediaeval and Later Human Skeletons from Northern England', *Journal of Archaeological Science*, 24: 561–7.

——RICHARDS, M. P., and FULLER, B. T. (2002). 'Bone Stable Isotope Evidence for Infant Feeding in Mediaeval England', *Antiquity*, 76: 654–6.

MEAD, W. (1967). *The English Medieval Feast.* Woking: Unwin.

MEECH, S. B., and ALLEN, H. E. (eds.) (1940). *The Book of Margery Kempe: The Text from the Unique MS. Owned by Colonel W. Butler Bowdon.* London: Oxford University Press, EETS, os 212.

MENNELL, S. (1985). *All Manners of Food: Eating and Taste in England and France from the Middle Ages to the Present.* Oxford: Blackwell.

MERTES, K. (1988). *The English Noble Household, 1250–1600: Good Governance and Politic Rule.* Oxford: Blackwell.

MIGNE, J. P. (ed.) (1853). *Antiquiores consuetudines Cluniacensis monasterii collectore Udalrico monacho Benedictino.* Paris: Patrologia Latina, 149, cols. 635–778.

MIKKERS, E. (1946). 'Un "Speculum novitii" inédit d'Étienne de Salley', *Collectanea Cisterciensis Ordinis reformatorum*, 8: 17–68.

MILLER, D., DARCH, E. and PEARSON, E. (2002). 'Archaeological Watching Brief at 15/19 Fish Street, Worcester'. Worcestershire County Council Archaeological Service, internal report 964.

Miller, E. (ed.) (1991). *The Agrarian History of England and Wales,* iii: *1348–1500.* Cambridge: Cambridge University Press.

Miller, N., Williams, D., and Kenward, H. K. (1993). 'Plant Macrofossils and Insect Remains from Mytongate', in D. H. Evans (ed.) 'Excavations in Hull 1975–76', *East Riding Archaeologist,* 4 (Hull Old Town Report Series, 2): 195–8.

Miracle, P., and Milner, N. (eds.) (2002). *Consuming Passions and Patterns of Consumption.* Cambridge: McDonald Institute for Archaeological Research.

Miyoshi, Y. (1981). *A Peasant Society in Economic Change: A Suffolk Manor, 1279–1437.* Tokyo: University of Tokyo Press.

Moffett, L. (1988). 'The Archaeobotanical Evidence for Saxon and Medieval Agriculture in Central England Circa 500 AD to 1500 AD'. Unpublished M.Phil. thesis, University of Birmingham.

——(1991*a*). 'The Archaeobotanical Evidence for Free-Threshing Tetraploid Wheat in Britain', in *Palaeoethnobotany and Archaeology (International Workgroup for Palaeoethnobotany, 8th Symposium at Nitra-Nové Vozokany 1989).* Nitra: Slovak Academy of Sciences (*Acta Interdisciplinaria Archaeologica,* 7), 233–43.

——(1991*b*). *Plant Economy at Burton Dassett, a Deserted Medieval Village in South Warwickshire.* London: English Heritage, AML Report 111/91.

——(1994). 'Charred Plant Remains', in Allen (1994: 398–406).

——(1997). 'Plant Remains', in C. Jones, G. Eyre-Morgan, S. Palmer, and N. Palmer (eds.), 'Excavations in the Outer Enclosure of Boteler's Castle, Oversley, Alcester, 1992–3', *Transactions of the Birmingham and Warwickshire Archaeological Society,* 101: 74–85.

——(in preparation). 'Report on the Plant Remains from Higham Ferrers, Northamptonshire'.

Monckton, A. (1999). 'The Environmental Evidence', in A. Connor and R. Buckley (eds.), *Roman and Medieval Occupation in Causeway Lane, Leicester, Excavations 1980 and 1991.* Leicester: University of Leicester Archaeological Services, Leicester Archaeology Monographs, 5: 309–65.

——(2003). 'Charred Plant Remains from a Medieval Suburb at St Mary's Gate, Derby: The Magistrates' Courts Redevelopment (2001.59)'. Leicester: University of Leicester Archaeological Services, Report 2003-048 (unpublished).

Montanari, M. (1988). *Alimentazione e cultura nel Medioevo.* Rome: Laterza.

——(1999*a*). 'Production Structures and Food Systems in the Early Middle Ages', in Flandrin and Montanari (1999: 168–77).

——(1999*b*). 'Towards a New Dietary Balance', in Flandrin and Montanari (1999: 247–50).

Morgan, M. (1968). *The English Lands of the Abbey of Bec.* Oxford: Oxford University Press.

Morris, C. (1989). *The Papal Monarchy: The Western Church from 1050 to 1250.* Oxford: Oxford University Press.

Morris, G. (1990). 'Animal Bone and Shell', in S. Ward (ed.), *Excavations at Chester: The Lesser Medieval Religious Houses: Sites Investigated 1964–1983.* Chester: Chester City Council, 178–90.

Müldner, G. (2001). 'Fast or Feast: Social Differences in Later Medieval Diet from Stable Isotope Analyses'. Unpublished M.Sc. thesis, University of Bradford.

——(2003). 'Hereford Cathedral Stable Isotope Data', in A. Boylston and D. Weston, 'Proposal for Hereford Cathedral Human Remains Research Project'. Bradford: Calvin Wells Laboratory, University of Bradford, unpublished report for Worcestershire County Council Historic Environment and Archaeology Service.

——and Richards, M. P. (2005). 'Fast or Feast: Reconstructing Diet in Later Medieval England by Stable Isotope Analysis', *Journal of Archaeological Science*, 32: 39–48.

————Albarella, U., Dobney, K., Fuller, B., Jay, M., Pearson, J., Rowley-Conwy, P., and Schibler, J. (in preparation). 'Stable Isotope Evidence of *Sus* Diets from European and Near Eastern Archaeological Sites'.

Mulkeen, S., and O'Connor, T. P. (1997). 'Raptors in Towns: Towards an Ecological Model', *International Journal of Osteoarchaeology*, 7: 440–9.

Mulville, J. (2003). 'Phases 2a–2e: Anglo-Saxon Occupation', in Hardy, Dodd, and Keevill (2003: 343–60).

———— and Ayres, K. (2004). 'Animal Bone', in G. Hey (ed.), *Yarnton: Saxon and Medieval Settlement and Landscape: Results of Excavations 1990–96*. Oxford: Oxford University School of Archaeology, Thames Valley Landscape, 20: 325–50.

Munby, L. M. (ed.) (1986). *Early Stuart Household Accounts*. Hertford: Hertfordshire Record Society, 2.

Murphy, M. (1998). 'Feeding Medieval Cities: Some Historical Approaches', in Carlin and Rosenthal (1998: 117–31).

Muus, B. J., and Dahlstom, P. (1974). *Sea Fishes of Britain and North-Western Europe*. London: Collins.

Myers, A. R. (ed.) (1959). *The Household of Edward IV: The Black Book and the Ordinance of 1478*. Manchester: Manchester University Press.

Neer, W. Van, and Ervynck, A. (1996). 'Food Rules and Status: Patterns of Fish Consumption in a Monastic Community (Ename, Belgium)', *Archaeofauna*, 5: 155–64.

Newman, L. F., Boegehold, A., Herlihy, D., Kates, R. W., and Raaflaub, K. (1990). 'Agricultural Intensification, Urbanization, and Hierarchy', in L. F. Newman (ed.), *Hunger in History: Food Shortage, Poverty and Deprivation*. Oxford: Blackwell, 101–25.

Nicholson, R. A. (1996). 'Fish Bone Diagenesis in Different Soils', *Archaeofauna*, 5: 79–91.

Nicol, A. (ed.) (1991). *Curia Regis Rolls, 1242–3*. London: HMSO.

Nicolas, N. H. (ed.) (1830). *Privy Purse Expenses of Elizabeth of York: Wardrobe Accounts of Edward the Fourth, with a Memoir of Elizabeth of York, and Notes*. London: Pickering.

Nightingale, P. (1995). *A Medieval Mercantile Community: The Grocers' Company and the Politics and Trade of London 1000–1485*. New Haven: Yale University Press.

Noble, C., Moreton, C. E., and Rutledge, P. (eds.) (1997). *Farming and Gardening in Late Medieval Norfolk*. Norwich: Norfolk Record Society, 61.

Noddle, B. (1976). 'Report on the Animal Bones from Walton, Aylesbury', in M. Farley (ed.), 'Saxon and Medieval Walton, Aylesbury: Excavations 1973–4', *Records of Buckinghamshire*, 20: 269–87.

——(1977). 'Mammal Bone', in Clarke and Carter (1977: 378–99).

——(1980). 'Identification and Interpretation of Animal Bones', in P. Wade-Martins (ed.), *North Elmham Park, vol. ii*. Gressenhall: Norfolk Archaeological Unit, East Anglian Archaeology, 9: 375–408.

NODDLE, B. (1985*a*). 'The Animal Bones: Berrington St. Sites 1, 2 and 3', in Shoesmith (1985: microfiche M8.D9).

——(1985*b*). 'The Animal Bones: Bewell House', in Shoesmith (1985: microfiche M8.F1).

——(1986). 'Mammalian Bone Report', in J. Bateman and M. Redknap (eds.), *Coventry: Excavations on the Town Wall 1976–78*. Coventry: Coventry Museums Monograph Series, 2: 100–19.

NORTHEAST, P. (ed.) (2001). *Wills of the Archdeaconry of Sudbury, 1439–1474*, Volume i: *1439–1461*. Woodbridge: Suffolk Records Society, 44.

O'CONNOR, T. P. (1982). *Animal Bones from Flaxengate, Lincoln, c.870–1500*. Lincoln: Lincolnshire Archaeological Trust, The Archaeology of Lincoln XVIII–1.

——(1989). 'Deciding Priorities with Urban Bones: York as a Case Study', in Serjeantson and Waldron (1989: 189–200).

——(1991). *Bones from 46–54 Fishergate*. London: CBA, The Archaeology of York, 15: The Animal Bones, fasc. 4.

——(1992). 'Provisioning Urban Communities: A Topic in Search of a Model', *Anthropozoologica*, 16: 101–6.

——(1994). '8th–11th Century Economy and Environment in York', in Rackham (1994*a*: 136–47).

——(2000). 'Bones as Evidence of Meat Production and Distribution in York', in E. White (ed.), *Feeding a City: York*. Totnes: Prospect Books, 43–60.

——(2001). 'On the Interpretation of Animal Bone Assemblages from *Wics*', in D. Hill and R. Cowie (eds.), *Wics: The Early Medieval Trading Centres of Northern Europe*. Sheffield: Sheffield Academic Press, 54–60.

OSCHINSKY, D. (ed.) (1971). *Walter of Henley and Other Treatises on Estate Management and Accounting*. Oxford: Clarendon Press.

OSMANI, S. R. (ed.) (1992). *Nutrition and Poverty*. Oxford: Clarendon Press.

O'SULLIVAN, D. (2001). 'Space, Silence and Shortages on Lindisfarne: The Archaeology of Asceticism', in Hamerow and MacGregor (2001: 33–52).

OVERTON, M. (1996). *Agricultural Revolution in England: The Transformation of the Agrarian Economy 1500–1850*. Cambridge: Cambridge University Press.

——and CAMPBELL, B. (1992). 'Norfolk Livestock Farming 1250–1740: A Comparative Study of Manorial Accounts and Probate Inventories', *Journal of Historical Geography*, 18: 377–96.

OWEN, D. M. (1972). 'Bacon and Eggs: Bishop Buckingham and Superstition in Lincolnshire', in G. J. Cuming and D. Baker (eds.), *Popular Belief and Practice*. Cambridge: Cambridge University Press, Ecclesiastical History Society, Studies in Church History, 8: 139–42.

——(ed.) (1984). *The Making of King's Lynn: A Documentary Survey*. London: British Academy, Records of Social and Economic History, new series, 9.

OWST, G. R. (1933). *Literature and Pulpit in Medieval England: A Neglected Chapter in the History of English Letters and of the English People*. Cambridge: Cambridge University Press.

PAAP, N. A. (1984). 'Palaeobotanical Investigations in Amsterdam', in Zeist and Casparie (1984: 339–44).

PAGE, F. M. (ed.) (1936). *Wellingborough Manorial Accounts, 1258–1323*. Kettering: Northamptonshire Record Society, 8.

PAGE, M. (ed.) (1996). *The Pipe Roll of the Bishopric of Winchester, 1301–2*. Winchester: Hampshire Record Series, 14.
—— (ed.) (1999). *The Pipe Roll of the Bishopric of Winchester, 1409–10*. Winchester: Hampshire Record Series, 16.
PÁLFI, G., DUTOUR, O., DEÁK, G., and HUTÁS, I. (1999). *Tuberculosis: Past and Present*. Budapest: Golden Book Publisher.
PALMER, W. M. (1927). 'Argentine's Manor, Melbourn, Cambridgeshire, 1317–18', *Cambridge Antiquarian Society Communications*, 28: 16–79.
PANNETT, D. J. (1988). 'Fish Weirs of the River Severn with Particular Reference to Shropshire', in Aston and Bond (1988*a*: ii. 371–89).
PATRICK, P. (2004). 'Greed, Gluttony and Intemperance? Testing the Stereotype of the Obese Medieval Monk'. Unpublished Ph.D. thesis, University of London.
PATRICK, S. J. (1975). 'Animal Bone', in R. Hall, 'Excavations at Full Street, Derby, 1972', *Derbyshire Archaeological Journal*, 92: 29–77, at 41–2.
PEARSALL, D. (ed.) (1979). *Piers Plowman by William Langland: An Edition of the C-Text*. London: Edward Arnold.
PEGG, P. G. (2001). *The Corruption of Angels: The Great Inquisition of 1245–1246*. Princeton: Princeton University Press.
PENN, S. A. C. (1987). 'Female Wage-Earners in Late Fourteenth-Century England', *Agricultural History Review*, 35: 1–14.
—— and DYER, C. C. (1994). 'Wages and Earnings in Late Medieval England: Evidence from the Enforcement of the Labour Laws', in Dyer (1994*a*: 167–89).
PERCY, T. (ed.) (1905). *The Regulations and Establishment of the Household of Henry Algernon Percy, the Fifth Earl of Northumberland, at his Castles of Wressle and Leckonfield, in Yorkshire. Begun Anno Domini MDXII*. New edition. London: A. Brown.
PERDIKARIS, S. (1997). 'The Transition to a Commercial Economy: Lofoten Fishing in the Middle Ages, a Preliminary Report', *Anthropozoologica*, 25–6: 505–10.
—— (1999). 'From Chiefly Provisioning to Commercial Fishery: Long-Term Economic Change in Arctic Norway', *World Archaeology*, 30: 388–402.
PETIT-DUTAILLIS, C. (1908–29). *Studies and Notes Supplementary to Stubbs' Constitutional History*. trans. W. E. Rhodes, W. T. Waugh, M. I. E. Robertson, and R. F. Treharne. 3 vols. Manchester: Manchester University Press.
PHELPS-BROWN, E., and HOPKINS, S. V. (1962). 'Seven Centuries of the Price of Consumables, Compared with Builders' Wage Rates', in E. Carus-Wilson (ed.), *Essays in Economic History*. 3 vols. London: Edward Arnold, ii. 168–96.
PINTER-BELLOWS, S. (2000). 'The Animal Remains', in P. A. Stamper and R. A. Croft (eds.), *Wharram: A Study of Settlement on the Yorkshire Wolds*, viii: *The South Manor Area*. York: University of York, 167–84.
PLOT, R. (1686). *The Natural History of Stafford-shire*. Oxford: The Theater.
—— (1705). *The Natural History of Oxford-shire*. 2nd edition. Oxford: L. Lichfield.
POLET, C., and KATZENBERG, M. A. (2003). 'Reconstruction of the Diet in a Mediaeval Monastic Community from the Coast of Belgium', *Journal of Archaeological Science*, 30: 525–33.
POLLARD, A. J. (1989). 'The North-Eastern Economy and the Agrarian Crisis of 1438–40', *Northern History*, 25: 88–105.

POLLARD, A. J. (1990). *North-Eastern England during the Wars of the Roses: Lay Society, War and Politics, 1450–1500.* Oxford: Clarendon Press.

POLLINGTON, S. (2003). *The Mead-Hall: Feasting in Anglo-Saxon England.* Frithgarth: Anglo-Saxon Books.

POOS, L. R. (1985). 'The Rural Population of Essex in the Later Middle Ages', *EconHR*, 2nd series, 38: 515–30.

——(1991). *A Rural Society after the Black Death: Essex 1350–1525.* Cambridge: Cambridge University Press.

——(ed.) (2001). *Lower Ecclesiastical Jurisdiction in Late-Medieval England: The Courts of the Dean and Chapter of Lincoln, 1336–1349, and the Deanery of Wisbech, 1458–1484.* London: British Academy, Records of Social and Economic History, new series, 32.

POSTAN, M. M. (1966). *The Cambridge Economic History of Europe*, i: *The Agrarian Life of the Middle Ages.* Cambridge: Cambridge University Press.

——(1972). *The Medieval Economy and Society.* Harmondsworth: Penguin.

——(1973*a*). *Essays on Medieval Agriculture and General Problems of the Medieval Economy.* Cambridge: Cambridge University Press.

——(1973*b*). 'The Economic Foundations of Medieval Society', in Postan (1973*a*: 3–27).

——(1973*c*). 'The Chronology of Labour Services', in Postan (1973*a*: 89–106).

——(1973*d*). 'Some Agrarian Evidence of Declining Population in the Later Middle Ages', in Postan (1973*a*: 186–213).

——(1973*e*). 'Village Livestock in the Thirteenth Century', in Postan (1973*a*: 214–48).

——(1973*f*). 'Glastonbury Estates in the Twelfth Century', in Postan (1973*a*: 249–77).

——and TITOW, J. (1973). 'Heriots and Prices on Winchester Manors', in Postan (1973*a*: 150–85).

POSTLES, D. (2003). 'Pittances and Pittancers', *Thirteenth Century England*, 9: 175–86.

POULTON, R. (ed.) (1998). *The Lost Manor of Hextalls, Little Pickle, Bletchingley.* Kingston: Surrey County Archaeological Unit.

POWELL, A. (2002). 'Animal Bone', in S. Foreman, J. Hiller, and D. Petts (eds.), *Gathering the People, Settling the Lands: The Archaeology of a Middle Thames Landscape,* iii: *Anglo-Saxon to Post-Medieval.* Oxford: Oxford Archaeology, Thames Valley Landscape Monograph 14: 44–9.

——and SERJEANTSON, D. (in preparation). 'The Animal Bones from the Saxon Shore Fort and the Medieval Occupation', in M. Fulford (ed.), *Excavations at Pevensey Castle.*

————and SMITH, P. (2001). 'Food Consumption and Disposal: The Animal Remains', in M. Hicks and A. Hicks (eds.), *St Gregory's Priory, Northgate, Canterbury: Excavations 1988–1991.* Canterbury: Canterbury Archaeological Trust, 289–333.

POWELL, E. (ed.) (1910). *A Suffolk Hundred in 1283.* Cambridge: Cambridge University Press.

POWICKE, F. M., and CHENEY, C. R. (eds.) (1964). *Councils and Synods with Other Documents Relating to the English Church. II AD 1205–1313.* 2 vols. Oxford: Clarendon Press.

PREECE, M. A. (1996). 'The Genetic Contribution to Stature', *Hormone Research*, Supplement 2: 56–8.

PRESTWICH, M. (1967). 'Victualling Estimates for English Garrisons in Scotland during the Early Fourteenth Century', *English Historical Review*, 82: 536–43.

PRIVAT, K. L., O'CONNELL, T., and RICHARDS, M. P. (2002). 'Stable Isotope Analysis of Human and Faunal Remains from the Anglo-Saxon Cemetery at Berinsfield, Oxfordshire: Dietary and Social Implications', *Journal of Archaeological Science*, 29: 779–90.

PROBERT, W. (1823). *The Ancient Laws of Cambria*. London: E. Williams.

PRUMMEL, W. (1997). 'Evidence of Hawking (Falconry) from Bird and Mammal Bones', *International Journal of Osteoarchaeology*, 7: 333–8.

QUERRIEN, A. (2003). 'Pêche et consommation du poisson en Berry au Moyen Âge', *Bibliothèque de l'École des Chartes*, 161: 409–35.

RACKHAM, J. (ed.) (1994*a*). *Environment and Economy in Anglo-Saxon England*. York: CBA, Research Report 89.

——(1994*b*). 'Economy and Environment in Saxon London', in Rackham (1994*a*: 126–36).

——(2003). 'Animal Bones', in G. Taylor (ed.), 'An Early to Middle Saxon Settlement at Quarrington, Lincolnshire', *Antiquaries Journal*, 83: 231–80, at 258–73.

RACKHAM, O. (1980). *Ancient Woodland*. London: Edward Arnold.

——(1997). *The History of the Countryside: The Classic History of Britain's Landscape, Flora and Fauna*. London: Phoenix Press.

RATCLIFF, S. C. (ed.) (1946). *Elton Manorial Records, 1279–1351*. Cambridge: Roxburghe Club.

RAVENSDALE, J. (1974). *Liable to Floods: Village Landscape on the Edge of the Fens, 450–1850*. Cambridge: Cambridge University Press.

RAWCLIFFE, C. (1999). *Medicine for the Soul: The Life, Death and Resurrection of an English Medieval Hospital: St Giles's, Norwich, c.1249–1550*. Thrupp: Sutton Publishing.

——(2002). '"On the Threshold of Eternity": Care for the Sick in East Anglian Monasteries', in Harper-Bill, Rawcliffe, and Wilson (2002: 41–72).

RAZI, Z. (1980). *Life, Marriage and Death in a Medieval Parish: Economy, Society and Demography in Halesowen (1270–1400)*. Cambridge: Cambridge University Press.

REDSTONE, L. J. (ed.) (1944). *The Cellarer's Account for Bromholm Priory, Norfolk, 1415–1416*. London: Norfolk Record Society, 17.

REES, U. (ed.) (1985). *The Cartulary of Haughmond Abbey*. Cardiff: University of Wales Press.

REICHSTEIN, H., and PIEPER, H. (1986). *Untersuchungen an Skelettresten von Vogeln aus Haithabu: Ausgrabung 1966–1969*. Neumünster: Wachholz.

RHODES, J. (ed.) (2002). *A Calendar of the Registers of the Priory of Llanthony by Gloucester 1457–1466, 1501–1525*. Bristol: Bristol and Gloucestershire Archaeological Society, Gloucestershire Record Series, 15.

RICHARDS, M. P. (2000). 'Human Consumption of Plant Foods in the British Neolithic: Direct Evidence from Bone Stable Isotopes', in A. S. Fairbairn (ed.), *Plants in Neolithic Britain and Beyond*. Oxford: Oxbow Books, 123–35.

Richards, M. P. (2002). 'Isotope Analysis of Hair and Bone Samples from St Martin's, Birmingham for BUFAU'. Bradford: Stable Isotope Laboratory, University of Bradford, unpublished report for Birmingham University Field Archaeology Unit.

——Fuller, B. T., Sponheimer, M., Robinson, T., and Ayliffe, L. (2003). 'Sulphur Isotopes in Palaeodietary Studies: A Review and Results from a Controlled Feeding Experiment', *International Journal of Osteoarchaeology*, 13: 37–45.

——Hedges, R. E. M., Molleson, T. I., and Vogel, J. C. (1998). 'Stable Isotope Analysis Reveals Variations in Human Diet at the Poundbury Camp Cemetery Site', *Journal of Archaeological Science*, 25: 1247–52.

——Mays, S., and Fuller, B. T. (2002). 'Stable Carbon and Nitrogen Isotope Values of Bone and Teeth Reflect Weaning Age at the Medieval Wharram Percy Site, Yorkshire, UK', *American Journal of Physical Anthropology*, 119: 205–10.

Richardson, H. G., and Sayles, G. O. (eds.) (1955–83). *Fleta*. 3 vols. London: Selden Society, 72, 89, 99.

Richmond, C. (1981). *John Hopton: A Fifteenth Century Suffolk Gentleman*. Cambridge: Cambridge University Press.

Ridgard, J. (ed.) (1985). *Medieval Framlingham: Select Documents, 1270–1524*. Woodbridge: Suffolk Records Society, 27.

Rierai Melis, A. (2000). ' "Transmarina vel orientalis especies magno labore quaesita, multo precio empta": especias y sociedad en el Mediterráneo noroccidental en el siglo xii', *Anuario de estudios medievales*, 30: 1015–87.

Riley, H. T. (ed.) (1866). *Johannis de Trokelowe et Henrici de Blaneforde chronica et annales*. London: Rolls Series.

——(ed.) (1867–9). *Gesta abbatum monasterii sancti Albani*. 3 vols. London: Rolls Series.

——(ed.) (1868). *Memorials of London and London Life, 1276–1419*. London: Longmans.

Rippon, S. (2004). 'Making the Most of a Bad Situation? Glastonbury Abbey, Meare, and the Medieval Exploitation of Wetland Resources in the Somerset Levels', *Medieval Archaeology*, 48: 91–130.

Rivers, J. P. W. (1988). 'The Nutritional Biology of Famine', in G. A. Harrison (ed.), *Famine*. Oxford: Oxford University Press, 57–106.

Rixson, D. (2000). *The History of Meat Trading*. Nottingham: Nottingham University Press.

Roberts, E. (1986). 'The Bishop of Winchester's Fishponds in Hampshire, 1150–1400: Their Development, Function and Management', *Proceedings of the Hampshire Field Club and Archaeological Society*, 42: 125–38.

——(1988). 'The Bishop of Winchester's Deer Parks in Hampshire, 1200–1400', *Proceedings of the Hampshire Field Club and Archaeological Society*, 44: 67–86.

Robertson, J. C. (ed.) (1875–85). *Materials for the History of Thomas Becket, Archbishop of Canterbury*. 7 vols. London: Rolls Series.

Rogers, J., and Waldron, T. (2001). 'DISH and the Monastic Way of Life', *International Journal of Osteoarchaeology*, 11: 357–65.

Rogers, J. E. T. (1866–1902). *A History of Agriculture and Prices in England 1259–1793*. 7 vols. Oxford: Clarendon Press.

Rooney, A. (1993). *Hunting in Middle English Literature*. Cambridge: D. S. Brewer.

Ross, B. (ed.) (2003). *Accounts of the Stewards of the Talbot Household at Blakemere 1392–1425*. Keele: University of Keele, Shropshire Record Series, 7.

Rowley, T. (1999). *The Normans*. Stroud: Tempus.

Ruston, D., Hoare, J., Henderson, L., Gregory, J., Bates, C. J., Prentice, A., Birch, M., Swas, G., and Farron, M. (2004). *The National Diet and Nutrition Survey: Adults Aged 19 to 64 Years*. London: The Stationery Office.

Ryder, M. L. (1974). 'Animal Remains from Wharram Percy', *Yorkshire Archaeological Journal*, 46: 42–52.

——(1983). *Sheep and Man*. London: Duckworth.

Sadler, P. (1990). 'The Faunal Remains', in J. R. Fairbrother (ed.), *Faccombe Netherton: Excavations of a Saxon and Medieval Manorial Complex II*. London: British Museum, British Museum Occasional Paper, 74: 462–508.

——(1991). 'The Animal Bone', in A. Hawkins, 'Medieval and Post-Medieval Occupation at the Teachers' Centre, Walton Road', *Records of Buckinghamshire*, 31: 209–17.

Salisbury, J. E. (1994). *The Beast Within: Animals in the Middle Ages*. London: Routledge.

Saltman, A. (ed.) (1962). *Cartulary of Tutbury Priory*. London: HMSO, Historical Manuscripts Commission Joint Publications, 2.

Salzman, L. F. (1931). *English Trade in the Middle Ages*. Oxford: Clarendon Press.

——(ed.) (1955). *Ministers' Accounts of the Manor of Petworth*. Lewes: Sussex Record Society, 55.

Santich, B. (1992). 'Les Éléments distinctifs de la cuisine médiévale méditerranéenne', in Lambert (1992: 133–9).

Sass, L. J. (ed.) (1975). *To the King's Taste: Richard II's Book of Feasts and Recipes*. New York: Metropolitan Museum of Art.

Sayer, A. A., Poole, J., Cox, V., Kuh, D., Hardy, R., Wadsworth, M., and Cooper, C. (2003). 'Weight from Birth to 53 Years: A Longitudinal Study of the Influence on Clinical Hand Osteoarthritis', *Arthritis and Rheumatism*, 48: 1030–3.

Scargill-Bird, S. R. (1887). *Custumals of Battle Abbey*. London: Camden Society, new series, 41.

Schmidt, A. V. C. (ed.) (1992). W. Langland, *Piers Plowman*. Oxford: Oxford University Press.

Schreiner, K. (1982). 'Zisterzienisches Mönchtum und soziale Umwelt: Wirtschaftlicher und sozialer Strukturwandel in hoch- und spätmittelalterlichen Zisterzienserkonventen', in K. Elm (ed.), *Die Zisterzienser: Ordensleben zwischen Ideal und Wirklichkeit, Ergänzungsband*. Cologne: Rheinland-Verlag, 79–135.

Schwarcz, H. P., and Schoeninger, M. J. (1991). 'Stable Isotope Analyses in Human Nutritional Ecology', *Yearbook of Physical Anthropology*, 34: 283–321.

Scott, J. (ed.) (1981). *The Early History of Glastonbury: An Edition, Translation and Study of William of Malmesbury's De antiquitate Glastonie ecclesie*. Woodbridge: Boydell Press.

Scott, S. (1986). *Animal Bones from West Parade, Lincoln*. London: English Heritage, AML Report 4820.

——(1991). 'The Animal Bones', in D. H. Evans and D. G. Tomlinson (eds.), *Excavations at 33–35 Eastgate, Beverley, 1983–86*. Sheffield: Collis, 236–50.

SCOTT, S. and DUNCAN, C. J. (2001). *Biology of Plagues: Evidence from Historical Populations*. Cambridge: Cambridge University Press.

SCULLY, T. (1995). *The Art of Cookery in the Middle Ages*. Woodbridge: Boydell Press.

SEARLE, E., and ROSS, B. (eds.) (1967). *Accounts of the Cellarers of Battle Abbey, 1275–1513*. Sydney: Sydney University Press.

SEQUIRA, W., CO, H., and BLOCK, J. A. (2000). 'Osteoarticular Tuberculosis: Current Diagnosis and Treatment', *American Journal of Therapy*, 7: 393–8.

SERJEANTSON, D. (1993). 'Harry Stoke, Avon: Assessment of the Animal Bones', Southampton: University of Southampton, Faunal Remains Unit, unpublished report.

——(2000). 'Bird Bones', in C. Young (ed.), *Excavations at Carisbrooke Castle, Isle of Wight, 1921–1996*. Salisbury: Wessex Archaeology Report 18: 182–5.

——(2001*a*). 'A Dainty Dish: Consumption of Small Birds in Late Medieval England', in H. Buitenhuis and W. Prummel (eds.), *Animals and Man in the Past: Essays in Honor of Dr A. T. Clason*. Groningen: ARC-Publicatie, 41: 263–74.

——(2001*b*). 'The Great Auk and the Gannet: A Prehistoric Perspective on the Extinction of the Great Auk', *International Journal of Osteoarchaeology*, 11: 43–55.

——(2002). 'Goose Husbandry in Medieval England and the Problem of Ageing Goose Bones', *Acta zoologica Cracoviensia*, 45: 39–54.

——and REES, H. (in press). *Food, Craft and Status in Medieval Winchester: The Plant and Animal Remains from the Suburbs and City Defences*. Winchester: Winchester City Council.

——and SMITH, P. L. (in press). 'Medieval and Post-Medieval Animal Bone from the Northern and Eastern Suburbs and the City Defences', in Serjeantson and Rees (in press).

——and WALDRON, T. (eds.) (1989). *Diet and Crafts in Towns: The Evidence of Animal Remains, the Roman to the Post-Medieval periods*. Oxford: BAR, British Series, 199.

SHARP, M. (ed.) (1982). *Accounts of the Constables of Bristol Castle in the Thirteenth and Early Fourteenth Centuries*. Bristol: Bristol Record Society, 34.

SHARPE, R. R. (ed.) (1913). *Calendar of Coroners' Rolls of the City of London AD 1300–1378*. London: Richard Clay.

SHOESMITH, R. (ed.) (1985). *Hereford City Excavations*, iii: *The Finds*. London: CBA, Research Report 56.

SIMON, A. (1944). *A Concise Encyclopedia of Gastronomy*. London: Curwen.

SIMOONS, F. J. (1994). *Eat Not This Flesh: Food Avoidances from Prehistory to the Present*. Madison: University of Wisconsin Press.

SKEAT, W. W. (ed.) (1882). *The Book of Husbandry by Master Fitzherbert*. London: Trübner and Co., English Dialect Society, 37.

SKINNER, P. (1997). *Health and Medicine in Early Medieval Southern Italy*. Leiden: Brill.

SLADE, C. F, and LAMBRICK, G. (eds.) (1990–1). *Two Cartularies of Abingdon Abbey*. 2 vols. Oxford: Oxford Historical Society, new series, 32–3.

SMITH, C. (2000). 'A Grumphie in the Sty: An Archaeological View of Pigs in Scotland, from the Earliest Domestication to the Agricultural Revolution', *Proceedings of the Society of Antiquaries of Scotland*, 130: 705–24.

SMITH, P. L. (1994). *The Early Norman Animal Bone from Carisbrooke Castle, the Isle of Wight*. London: English Heritage, AML Report 49/94.

——(1995). *The Fish Bone from Launceston Castle, Cornwall*. London: English Heritage, AML Report 56/95.

SMITH, R. M. (1991). 'Demographic Developments in Rural England, 1300–1348: A Survey', in Campbell (1991: 25–77).

——(1993). 'Demography and Medicine', in W. F. Bynum and R. Porter (eds.), *Companion Encyclopedia of the History of Medicine*. 2 vols. London: Routledge, ii. 1665–7.

SNEYD, C. A. (ed.) (1847). *A Relation, or Rather a True Account, of the Island of England; with Sundry Particulars of the Customs of These People, and of the Royal Revenues under King Henry the Seventh, about the Year 1500*. London: Camden Society, 1st series, 37.

SNOOKS, G. D. (1995). 'The Dynamic Role of the Market in the Anglo-Norman Economy and Beyond, 1086–1300', in Britnell and Campbell (1995: 27–54).

SOUTHERN, R. W (1982). 'Between Heaven and Hell', *Times Literary Supplement*, 18 June: 651–2.

SOWERS, M. (2001). 'Epidemiology of Risk Factors for Osteoarthritis: Systemic Factors', *Current Opinions in Rheumatology*, 13: 447–51.

SPANNAGEL, A., and ENGELBERT, P. (eds.) (1974). *Smaragdi abbatis expositio in regulam sancti Benedicti*. Siegburg: F. Schmitt, Corpus Consuetudinum Monasticarum, 8.

SPÄTLING, L. G., and DINTER, P. (eds.) (1985). *Consuetudines Fructuarienses-Sanblasianae*. Siegburg: F. Schmitt, Corpus Consuetudinum Monasticarum, 12/1.

STALLIBRASS, S. (1993). *Animal Bones from Excavations at a Medieval Hospital and a Post-Medieval Farmstead at St Giles by Brompton Bridge, North Yorkshire, 1989–1990*. London: English Heritage, AML Report 95/93.

STAMPER, P. (1988). 'Woods and Parks', in Astill and Grant (1988: 128–48).

STARKEY, D. J., REID, C., and ASHCROFT, N. (eds.) (2000). *England's Sea Fisheries: The Commercial Sea Fisheries of England and Wales since 1300*. London: Chatham.

STEANE, J. M. (1988). 'The Royal Fishponds of Medieval England', in Aston and Bond (1988*a*: i. 39–68).

STEPHENSON, M. (1977–8). 'The Role of Poultry Husbandry in the Medieval Agrarian Economy, 1200–1450', *Veterinary History*, 10: 17–24.

——(1986). 'The Productivity of Medieval Sheep on the Great Estates, 1100–1500'. Unpublished Ph.D. thesis, University of Cambridge.

STEVENS, D. (1985). 'A Somerset Coroner's Roll, 1315–1321', *Somerset and Dorset Notes and Queries*, 31: 451–72.

STEVENS, M., and CAWLEY, A. C. (eds.) (1994). *The Towneley Plays*. 2 vols. Oxford: Oxford University Press; EETS, ss 13–14.

STEVENS, P. (1992). 'The Vertebrate Remains from Site 39', in G. Milne and J. D. Richards (eds.), *Two Anglo-Saxon Buildings and Associated Finds*. York: York University Archaeology Publication 9: 74–9.

STEVENSON, J. (ed.) (1847). Reginald of Durham, *Libellus de vita et miraculis sancti Godrici, heremitae de Finchale*. London: J. B. Nichols, Surtees Society, 20.

——(ed.) (1858). *De abbatibus Abbendoniae*, in J. Stevenson (ed.), *Chronicon monasterii de Abingdon*. 2 vols. London: Rolls Series, ii. 267–95.

STOKES, P., and ROWLEY-CONWY, P. (2002). 'Iron Age Cultigen? Experimental Return Rates for Fat Hen (*Chenopodium album* L.)', *Environmental Archaeology*, 7: 95–9.

STONE, D. J. (2005). *Decision-Making in Medieval Agriculture*. Oxford: Oxford University Press.

STONE, R., and APPLETON-FOX, N. (1996). *A View from Hereford's Past: A Report on the Archaeological Excavation in Hereford Cathedral Close in 1993*. Little Logaston: Logaston Press.

STOUFF, L. (1970). *Ravitaillement et alimentation en Provence aux xive et xve siècles*. Paris: Mouton.

STYLES, D. (ed.) (1969). *Ministers' Accounts of the Collegiate Church of St Mary, Warwick 1432–85*. Stratford-upon-Avon: Dugdale Society, 26.

SULLIVAN, R. J. (ed.) (1975). *Consuetudines Affligenienses (saec. xiii)*, in *Consuetudines Benedictinae variae (saec. xi–saec. iv)*. Siegburg: F. Schmitt, Corpus Consuetudinum Monasticarum, 6: 107–99.

SURTEES, R. (1816–40). *The History and Antiquities of the County Palatine of Durham*. 4 vols. London: J. B. Nichols and Son.

SYKES, N. J. (2001). 'The Norman Conquest: A Zooarchaeological Perspective'. Unpublished Ph.D. thesis, University of Southampton.

——(2004). 'The Introduction of Fallow Deer (*Dama dama*): A Zooarchaeological Perspective', *Environmental Archaeology*, 9: 75–83.

——(n.d.). 'The Animal Remains from Reading Oracle Sites 12 and 29'. Unpublished report to Oxford Archaeology.

——(in press, *a*). *The Norman Conquest: A Zooarchaeological Perspective*. Oxford: BAR, International Series.

——(in press, *b*). 'The Animal Bones', in R. Poulton (ed.), *Guildford Castle and Royal Palace*. Guildford: Surrey Archaeological Society.

——(in preparation). 'Hunting for the Normans: Zooarchaeological Evidence for Medieval Identity', in A. Pluskowski (ed.), *Just Skin and Bones? New Perspectives on Human–Animal Relations in the Historic Past*. Oxford: BAR.

TANAKA, I. I., ANNOP, I. S., LEITE, S. R., COOKSEY, R. C., and LEITE, C. Q. (2003). 'Comparison of a Multiplex-PCR Assay with Mycolic Acids Analysis and Conventional Methods for the Identification of Mycobacteria', *Microbiology and Immunology*, 47: 307–12.

TANNER, J. M. (1981). *A History of the Study of Human Growth*. Cambridge: Cambridge University Press.

TANNER, N. P. (ed.) (1977). *Heresy Trials in the Diocese of Norwich, 1428–31*. London: Royal Historical Society, Camden, 4th series, 20.

TAYLOR, G. M., CROSSEY, M., SALDANHA, J., and WALDRON, T. (1996). 'DNA from *Mycobacterium tuberculosis* Identified in Medieval Human Skeletal Remains Using Polymerase Chain Reaction', *Journal of Archaeological Science*, 23: 789–98.

——GOYAL, M., LEGGE, A. J., SHAW, R. J., and YOUNG, D. (1999). 'Genotypic Analysis of *Mycobacterium tuberculosis* from Medieval Human Remains', *Microbiology*, 145: 899–904.

TERROINE, A., and FOSSIER, L. (eds.) (1966–98). *Chartes et documents de l'abbaye de Saint-Magloire*. 3 vols. Paris: Éditions du CNRS.

TEVERSHAM, T. F. (ed.) (1959). *Sawston Estate Accounts of the Fourteenth Century*. Sawston: duplicated typescript, in CUL.

Thawley, C. R. (1981). 'The Mammal, Bird and Fish Bones', in J. E. Mellor and T. Pearce (eds.), *The Austin Friars, Leicester*. London: CBA, Research Report 35: 173–5.

Thirsk, J. (1997). *Alternative Agriculture: A History from the Black Death to the Present Day*. Oxford: Oxford University Press.

——(ed.) (2000). *The English Rural Landscape*. Oxford: Oxford University Press.

Thomas, K. (1984). *Man and the Natural World: Changing Attitudes in England, 1500–1800*. Harmondsworth: Penguin.

Thomas, R. M. (2002). 'Animals, Economy and Status: The Integration of Historical and Zooarchaeological Evidence in the Study of a Medieval Castle'. Unpublished Ph.D. thesis, University of Birmingham.

Thompson, A. H. (ed.) (1914–29). *Visitations of Religious Houses in the Diocese of Lincoln*. 3 vols. Lincoln: Lincoln Record Society, 7, 14, 21.

——(ed.) (1924). 'Household Roll of Bishop Ralph of Shrewsbury (1337–8)', in T. F. Palmer (ed.), *Collectanea I*. Frome: Somerset Record Society, 39: 72–174.

Thompson, F. (1939). *Lark Rise*. Oxford: Oxford University Press.

Thompson, J. S. (ed.) (1990). *Hundreds, Manors, Parishes and the Church: A Selection of Early Documents for Bedfordshire*. Bedford: Bedfordshire Historical Record Society, 69.

Thomson, J. A. F. (1965). *The Later Lollards 1414–1520*. Oxford: Oxford University Press.

Thornton, C. (1992). 'Efficiency in Medieval Livestock Farming: The Fertility and Mortality of Herds and Flocks at Rimpton, Somerset, 1208–1349', *Thirteenth Century England*, 4: 25–46.

Threlfall-Holmes, M. (2005). *Monks and Markets: Durham Cathedral Priory 1460–1520*. Oxford: Oxford University Press.

Thrupp, S. (1965). 'The Problem of Replacement Rates in Medieval England', *EconHR*, 2nd series, 18: 101–19.

Ticehurst, N. F. (1957). *The Mute Swan in England*. London: Cleaver-Hulme Press.

Tillotson, J. H. (1989). *Marrick Priory: A Nunnery in Late Medieval Yorkshire*. York: University of York, Borthwick Paper, 75.

Titow, J. (1960). 'Evidence of Weather in the Account Rolls of the Bishopric of Winchester 1209–1350', *EconHR*, 2nd series, 12: 360–407.

——(1972). *Winchester Yields: A Study in Medieval Agricultural Productivity*. Cambridge: Cambridge University Press.

Tomlinson, P. (1991). 'Vegetative Plant Remains from Waterlogged Deposits Identified at York', in J. M. Renfrew (ed.), *New Light on Early Farming*. Edinburgh: Edinburgh University Press, 109–19.

Tout, T. F. (1936). *The Place of the Reign of Edward II in History*, rev. H. Johnstone. 2nd edition. Manchester: Manchester University Press.

Trigg, S. (ed.) (1990). *Wynnere and Wastoure*. Oxford: Oxford University Press; EETS, os 297.

Trotter, M. (1970). 'Estimation of Stature from Intact Limb Bones', in T. D. Stewart (ed.), *Personal Identification in Mass Disasters*. Washington: Smithsonian Institution Press, 71–97.

Trow-Smith, R. (1957). *A History of British Livestock Husbandry to 1700*. London: Routledge and Kegan Paul.

TUCK, J. A. (1991). 'Farming Practice and Techniques: The Northern Borders', in Miller (1991: 175–82).

TURTON, R. (ed.) (1894–7). *The Honor and Forest of Pickering*. 4 vols. London: North Riding Record Society, new series, 1–4.

TWIGG, G. (1984). *The Black Death: A Biological Reappraisal*. London: Batsford.

——(2003). 'The Black Death: A Problem of Population-Wide Infection', *Local Population Studies*, 71: 40–52.

VALE, M. (2001). *The Princely Court: Medieval Courts and Culture in North-West Europe 1270–1380*. Oxford: Oxford University Press.

VANDERZEE, G. (1807). *Nonarum inquisitiones in curia scaccarii temp. regis Edwardi III*. London: Record Commission.

VAQUIER, A. (1923). 'Une réforme de Cluny en 1428', *Revue bénédictine*, 35: 157–98.

VEEN, M. Van Der (1995). 'The Identification of Maslin Crops', in H. Kroll and B. Pasternak (eds.), *Res archaeobotanicae: 9th Symposium of the IWGP*. Kiel: Institut für Ur- und Frühgeschichte der Christian-Albrecht-Universität, 335–43.

WALDRON, T. (1985). 'DISH at Merton Priory: Evidence for a "New" Occupational Disease?', *British Medical Journal*, 291: 1762–3.

——(1998). 'A Note on the Estimation of Height from Long-Bone Measurements', *International Journal of Osteoarchaeology*, 8: 75–7.

——(2001). *Shadows in the Soil: Human Bones and Archaeology*. Stroud: Tempus.

WALKER BYNUM, C. (1999). *Holy Feast and Holy Fast: The Religious Significance of Food to Medieval Women*. Berkeley and Los Angeles: University of California Press.

WALL, S. M. (1980). 'The Animal Bones from the Excavation of the Hospital of St Mary of Ospringe', *Archaeologia Cantiana*, 96: 227–66.

WALTER, J., and SCHOFIELD, R. (eds.) (1989*a*). *Famine, Disease and the Social Order in Early Modern Society*. Cambridge: Cambridge University Press.

————(1989*b*). 'Famine, Disease and Crisis Mortality in Early Modern Society', in Walter and Schofield (1989*a*: 1–73).

WARD, J., and MAINLAND, I. (1999). 'Microwear in Modern Rooting and Stall-Fed Pigs: The Potential of Dental Microwear Analysis for Exploring Pig Diet and Management in the Past', *Environmental Archaeology*, 4: 25–32.

WATKINS, A. (1993). 'The Woodland Economy of the Forest of Arden in the Later Middle Ages', *Midland History*, 18: 19–36.

——(1997). 'Landowners and their Estates in the Forest of Arden in the Fifteenth Century' *Agricultural History Review*, 45: 18–33.

WATZL, H. (1978). 'Über Pitanzen und Reichnisse für den Konvent des Klosters Heiligenkreuz, 1431', *Analecta Cisterciensia*, 34: 40–147.

WEBB, J. (ed.) (1854–5). *A Book of Household Expenses of Richard de Swinfield, Bishop of Hereford during Part of the Years 1289 and 1290*. 2 vols. London: Camden Society, 1st series, 59, 62.

WEDGWOOD, J. C. (ed.) (1911). 'Inquisitions "post mortem", "ad quod damnum" etc., Staffordshire 1223–1327', *Collections for a History of Staffordshire*, 3rd series, 2.

WEISS, R. (2001). *The Yellow Cross: The Story of the Last Cathars 1290–1329*. London: Penguin Books.

WEISS ADAMSON, M. (ed.) (2002). *Regional Cuisines of Medieval Europe: A Book of Essays*. London: Routledge.

WEST, J. R. (ed.) (1932). *St Benet of Holme, 1020–1210: The Eleventh and Twelfth Century Sections of Cott. MS Galba E. ii, the Register of the Abbey of St Benet of Holme*. Fakenham: Norfolk Record Society, 2.

WHEELER, A. (1978). *Key to the Fishes of Northern Europe*. London: Warne.

——(1979*a*). 'Fish Remains from Priory Barn and Benham's Garage', in P. Leach (ed.), *The Archaeology of Taunton: Excavations and Fieldwork*. Bristol: Western Archaeological Trust, 193–4.

——(1979*b*). *The Tidal Thames*. London: Routledge.

——(1979*c*). 'Fish Remains', in M. Parrington (ed.), 'Excavations at Stert Street, Abingdon, Oxon.' *Oxoniensia*, 44: 21–3.

——(1982). 'Medieval Fish Bones', in P. Leach (ed.), *Ilchester*, i: *Excavations 1974–1975*. Gloucester: Western Archaeological Trust: 284–5.

——and JONES, A. K. G. (1989). *Fishes*. Cambridge: Cambridge University Press (Cambridge Manuals in Archaeology).

WHITEHEAD, K. G. (1972). *Deer of the World*. London: Constable.

WHITELOCK, D. M., BRETT, M., and BROOKE, C. N. L. (eds.) (1981). *Councils and Synods with Other Documents Relating to the English Church. IAD 871–1204*. 2 vols. Oxford: Clarendon Press.

WILCOX, R. (2002). 'Excavation of a Monastic Grain-Processing Complex at Castle Acre Priory, Norfolk, 1977–82', *Norfolk Archaeology*, 44: 15–58.

WILKIE, T., MAINLAND, I., ALBARELLA, U., DOBNEY, K. and ROWLEY-CONWY, P. (in preparation). 'A Microwear Study of Pig Husbandry Practices in Urban and Rural Contexts in Romano-British, Anglo-Scandinavian and Medieval England', in U. Albarella, K. Dobney, and P. Rowley-Conwy (eds.), *Pigs and Humans: Proceedings of the Conference Held at Walworth Castle (UK) 26th–28th September 2002*.

WILKINS, D. (ed.) (1737). *Concilia Magnae Britanniae et Hiberniae AD 446–1717*. 4 vols. London: n.p.

WILKINSON, M. (1979). 'The Fish Remains', in J. M. Maltby (ed.), *Faunal Studies on Urban Sites: The Animal Bones from Exeter 1971–1975*. Huddersfield: H. Charlesworth and Co., 74–81.

——(1982). 'Fishbones', in R. A. Higham, J. P. Allan, and S. R. Blaylock (eds.), 'Excavations at Okehampton Castle, Devon', *Proceedings of the Devon Archaeological Society*, 40: 135–8.

WILLIAMS, A., and MARTIN, G. H. (eds.) (2002). *Domesday Book: A Complete Translation*. London: Penguin Books, 2002.

WILLIAMS, R. J. (1993). *Pennyland and Hartigans: Two Iron Age and Saxon Sites in Milton Keynes*. Aylesbury: Buckinghamshire Archaeological Society, Monograph Series, 4.

WILSON, B. (1986). 'Faunal Remains: Animal Bones and Marine Shells', in D. Miles (ed.), *Archaeology at Barton Court Farm, Abingdon, Oxon*. Oxford: CBA, Research Report 50: microfiche 8A1–8G14.

WILSON, J. M., and GORDON, C. (eds.) (1908). *Early Compotus Rolls of the Priory of Worcester*. Oxford: Worcestershire Historical Society, 21.

WILSON, R. (1979). 'The Mammal Bones and Other Environmental Records', in M. Parrington (ed.), 'Excavations at Stert Street, Abingdon, Oxon.', *Oxoniensia*, 44: 16–20.

Wiseman, J. (2000). *The Pig: A British History*. London: Duckbacks.

Wood-Legh, K. L. (ed.) (1956). *A Small Household of the XVth Century being the Account Book of Munden's Chantry, Bridport*. Manchester: Manchester University Press.

Woolf, D. A., Riach, I. C., Derweesh, A., and Vyas, H. (1990). 'Lead Lines in Young Infants with Acute Lead Encephalopathy: A Reliable Diagnostic Test', *Journal of Tropical Pediatrics*, 36: 90–3.

Woolgar, C. M. (ed.) (1992–3). *Household Accounts from Medieval England*. 2 vols. London: British Academy, Records of Social and Economic History, new series, 17–18.

——(1995). 'Diet and Consumption in Gentry and Noble Households: A Case Study from around the Wash', in R. E. Archer and S. Walker (eds.), *Rulers and Ruled in Late Medieval England: Essays Presented to Gerald Harriss*. London: Hambledon Press, 17–31.

——(1999). *The Great Household in Late Medieval England*. New Haven: Yale University Press.

——(2000). '"Take this Penance now, and Afterwards the Fare will Improve": Seafood and Late Medieval Diet', in Starkey, Reid, and Ashcroft (2000: 36–44, 247–8).

——(2001). 'Fast and Feast: Conspicuous Consumption and the Diet of the Nobility in the Fifteenth Century', in M. Hicks (ed.), *Revolution and Consumption in Late Medieval England*. Woodbridge: Boydell Press, 7–25.

——(in press). 'Medieval Europe', in P. Freedman (ed.), *Food: The History of Taste*. London: Thames and Hudson.

Wormald, F. (ed.) (1939–46). *English Benedictine Kalendars after AD 1100*. 2 vols. London: Henry Bradshaw Society, 77, 81.

Wright, L. (1996). *Sources of London English: Medieval Thames Vocabulary*. Oxford: Clarendon Press.

Wrigley, E. A. (1989). 'Some Reflections on Corn Yields and Prices in Pre-Industrial Economies', in Walter and Schofield (1989*a*: 235–78).

——and Schofield, R. S. (1981). *The Population History of England and Wales, 1541–1871: A Reconstruction*. Cambridge: Cambridge University Press.

Wrottesley, G. (ed.) (1891). 'Extracts from the Plea Rolls of the Reign of Edward III', *Collections for a History of Staffordshire*, 12, part 1: 3–173.

——(ed.) (1896). 'Extracts from the Plea Rolls of the Reigns of Henry V and Henry VI', *Collections for a History of Staffordshire*, 17: 3–153.

Young, C. R. (1979). *The Royal Forests of Medieval England*. Leicester: Leicester University Press.

Yvinec, J.-H. (1993). 'La Part du gibier dans l'alimentation du haut Moyen Âge', in *Exploitation des animaux sauvage à travers le temps*. Juan-les-Pins: Association pour la Promotion et la Diffusion des Connaissances Archéologiques (Société de Recherche Interdisciplinaire, XIIIe Rencontres internationales d'archéologie et d'histoire d'Antibes, IVe Colloque international de l'homme et l'animal), 491–504.

Zeder, M. A. (1991). *Feeding Cities: Specialized Animal Economy in the Ancient Near East*. Washington: Smithsonian Institution Press.

Zeist, W. van, and Casparie, W. A. (eds.) (1984). *Plants and Ancient Man*. Rotterdam: A. A. Balkema.

ZIMMERMANN, G. (1971). *Ordensleben und Lebensstandard: Die Cura Corporis in den Ordensvorschriften des abendländischen Hochmittelalters*. Münster: Aschendorffsche Verlagsbuchhandlung, Beiträge zur Geschichte des alten Mönchtums und des Benediktinerordens, 32.

ZINK, A. R., GRABNER, W., REISCHL, U., WOLF, H., and NEHRLICH, A. G. (2003). 'Molecular Study of Human Tuberculosis in Three Geographically Distinct and Time Delineated Populations from Ancient Egypt', *Epidemiology and Infection*, 130: 239–49.

Index

Lightning Source UK Ltd.
Milton Keynes UK
UKHW02f1007150718
325717UK00012B/576/P